EXPERIMENTAL PRACTICE

EXPERIMENTAL FUTURES

TECHNOLOGICAL LIVES,
SCIENTIFIC ARTS,
ANTHROPOLOGICAL VOICES

A SERIES EDITED BY
Michael M. J. Fischer and Joseph Dumit

EXPERIMENTAL PRACTICE

TECHNOSCIENCE,
ALTERONTOLOGIES,
AND
MORE-THAN-SOCIAL
MOVEMENTS

Dimitris Papadopoulos

DUKE UNIVERSITY PRESS

Durham and London 2018

Designed by Matthew Tauch
Typeset in Warnock Pro and Trade Gothic LT Std by
Westchester Publishing Services

Library of Congress Cataloging-in-Publication Data
Names: Papadopoulos, Dimitris, author.
Title: Experimental practice : technoscience, alteron-
tologies, and more-than-social movements / Dimitris
Papadopoulos.
Description: Durham : Duke University Press, 2018. |
Series: Experimental futures | Includes bibliographical
references.
Identifiers: LCCN 2017060623 (print)
LCCN 2018000305 (ebook)
ISBN 9781478002321 (ebook)
ISBN 9781478000655 (hardcover : alk. paper)
ISBN 9781478000846 (pbk. : alk. paper)
Subjects: LCSH: Social movements. | Technology—Social
aspects. | Discoveries in science—Social aspects. |
Philosophical anthropology. | Financialization—Social
aspects.
Classification: LCC HM881 (ebook) | LCC HM881 .P354
2018 (print) | DDC 303.48/4—dc23
LC record available at https://lccn.loc.gov/2017060623

Cover art: Thomas Jackson, *Straws no. 4*, Mono Lake,
California, 2015.

CONTENTS

INTRO-
DUCTION

THE TWO BEGINNINGS

This book has two beginnings. The first one is the decentering of the human in its relations to other species, machines, and the material world. What matters for this project is that the early twenty-first century saw the vision of "a more-than-human world," of "humans no longer being in control," and of a human-nonhuman continuum becoming slowly part of how we (some humans, that is) imagine ourselves (the human species) in the future. Let's hope that our species-being will be forever marked by this realization—although it may be too late.[1] I use the term *posthuman culture* to refer to this decentering of the human (and the humanist subject and its politics) into its relations to other living beings and the material world as well as the wider cultural realization of this decentering.[2] Technoscience has been the main force shaping posthuman culture: the continuous folding of science, technology, and the everyday into each other. Rather than the focus of this book, the histories and current formations of technoscience and posthuman culture constitute its first beginning, the stage on which its arguments are played out.

The second beginning is an affect rather than a phenomenon. A commitment rather than a thought. An obligation rather than an interest. A feeling of urgency to grasp the incapacity of the extraordinary social mobilizations that took place in countries across the North Atlantic and beyond since 2006 to instill social change.[3] Seasoned social movements analysts tell us that social movements have a *longue durée*—their effects are not clearly visible immediately, and what they achieve is often transposed in time and in different, often remote, fields of life. However true this may be, it is difficult not to feel that the mobilizations, struggles, and uprisings of the last ten years changed so many things and yet the transformational potential of these movements toward a materialization of socioeconomic and ecological justice was not accomplished. A map of this cycle of struggles would have many action points, sites of conflict, squares and plazas, linkages, transnational exchanges, alliances, and virtual meeting spaces,[4] but I barely can count any broad effects in the direction of what these movements hoped[5]—knowing that it may be too early to look for these or to even have the conceptual tools and perceptual skills to see and grasp them. However, neither these mobilizations nor the eclipse of their claims for justice are the target of my analysis in this book. These social movements, their efforts, and their achievements shape the intellectual background and affective tonality of this book. I am gambling on a feeling with this second beginning.

This feeling tells me that there is a connection between the limited range of transformations that these movements have achieved and the displacement of the human and of human politics in posthuman culture. *Experimental Practice* attempts to investigate this connection. Consider, for example, the 2011 riots in Britain. They came unexpectedly. Speaking shortly after they happened, Paul Gilroy (2011) concluded by highlighting that many black communities drawn in the vortex of privatization and the intense neoliberal disintegration of British society are fragmented and often unable to defend and organize themselves.[6] Later Gilroy (2013) expanded his analysis of the 2011 riots and attempted a comparison with the 1981 riots, emphasizing the very limited effects that the 2011 events had. The cry for change vanished soon after the riots stopped. They did not transform British society in the way this happened in the 1980s.[7]

I take Gilroy's diagnosis seriously and translate it to my words: these recent struggles show that there was no infrastructure that could hold together and protect the communities and perpetuate and multiply the effects of their actions. How can an ontology of community and infrastructures of communal connectivity be created?[8] I am aware that the term *ontology* may be unexpected here. A notion of ontology will unfold across this book, but for now I mean the shared, durable, open material spaces—tangible and virtual—that can be inhabited autonomously by these communities.

Ontology and infrastructure are about something much greater than social relations. There are many social relations in our lives, probably more than enough, but there are not many material spaces where social and political autonomy can be performed. My bet in this book is that creating spaces of political autonomy and self-organization is not just a social affair. It is a practical and ontological affair that goes as far as to change the materiality of the lived spaces and the bodies, human and nonhuman, of communities. In fact, we may need to disconnect[9] from the ubiquitous networks of social relations in order to create these autonomous material spaces of existence. If this is what needs to be done, then what is social movement politics today? What if we approach social movement action not as targeting existing political power but as experimenting with worlds? What if we see social movement action not as addressing existing institutions for redistributing justice but as the creation of alternative forms of existence that reclaim material justice from below? And, what if this becomes possible not when social movements engage in resistance to power but when they experiment with the materiality of life?

Experimental practice in this sense is about modes of intuition, knowledges, and politics that trigger intensive material changes and mobilize energies in ways that generate alternative and autonomous spaces of existence.[10] Autonomy is meant as autonomous politics here, not as the modernist humanist value of individual independence and the seclusion of the personal and the private. The opposite is the case: autonomous politics requires material interconnectedness, practical organizing, everyday coexistence and the fostering of ontological alliances. And these are always more than human, more than social. They entail interactions, ways of knowing, forms of practice that involve the material world, plants and the soil, chemical compounds and energies, other groups of humans and their surroundings, and other species and machines.

The notion of experimental practice emerged as I was exploring if/where the two beginnings mentioned earlier meet. Bringing technoscience into the picture by retrieving the posthuman experimentations that are an undisclosed part of social movement action shows their politics under a new light: as *more-than-social* movements. I mean, as movements that do much more than just targeting visible and recognized social institutions; as movements that immerse into the human-nonhuman continuum and change society practically by engaging with both the human and the nonhuman world. And they can do that only to the extent that they involve some part of technoscience. This is the ideal formula that this book seeks to advance: the metamorphosis of social movements to *movements*— that is, movements of matter and the social simultaneously. Movements of ontology. The book investigates this alternative perspective by slowly weaving the posthuman and technoscience into social movement politics and, the other way around, by weaving social movement politics into the practices of technoscience.

These arguments prolong work developed with Niamh Stephenson and Vassilis Tsianos in two previous books that attempted to show that creative social transformation since the 1960s and 1970s has been rarely the outcome of pure resistance or of opposition to power but of the remaking of everyday existence below the radar of control in mundane and yet unexpected ways.[11] Following the autonomous social movements at that time, we described how social conflict and social mobilizations drive social transformation instead of just being a mere response to (economic and social) power. In *Analysing Everyday Experience: Social Research and Political Change* (2006) we tried to show how this is possible at the level of individual subjectivity and experience. In *Escape Routes: Control and Subversion in the 21st Century* (2008), we tried to reconstruct this type of politics at a collective and community level. *Experimental Practice* closes the trilogy and addresses the same question on the level of matter. Experience, community, matter. The book seeks to put forward a form of politics that addresses these three aspects of our lives simultaneously: experimental practice.

BAROQUE FIELDWORKING

Making the case for an experimental practice of more-than-social movements is both empirically grounded in today's realities of social movement politics and simultaneously a deeply speculative undertaking. It is empirically grounded because it follows developments in the actions and politics of social movements that are already happening now. However, my aim is not to just follow and describe current social movements but to magnify specific aspects of their actions that can foster an experimental view of politics and point toward its transformative potential. Although all chapters attempt to contribute to this objective, they are all located in different fields and debates, have their own internal logic and argumentation, and engage diverse methodologies and data. Introducing such a varying set of materials and methodologies allowed me to find a path for navigating the terrain that the mingling of the two beginnings mentioned earlier has created. As I look back to the years of work since I started doing research for this book in 2010, I feel that the methodologies I worked with and the materials I gathered and analyzed were almost imposed by the nature of the terrains I was delving into and the questions I was confronted with as I was moving through them.

Two of the chapters (3 and 8) present fieldwork studies: one is based on empirical materials from my long-standing engagement with migration and precarious work politics up to the end of the 2010s as well as fieldwork materials that were collected between 2009 and 2012 with my collaborator on this project, Vassilis Tsianos. The other one is based in my involvement in the maker and hacker communities across the East Midlands in the UK, primarily in Leicester and Derby and with different hacker and maker groups or activist groups that engaged with technological and ecological issues across the world. Chapters 6 and 7 present case studies that expand on the ontological and experimental implications of two cases of embodied technoscience that I have been investigating for several years now; the first one is on neuroplasticity, epigenetics, and the embodied brain, and the second discusses AIDS activism in the 1980s. Finally, the remaining four chapters collect and mix different materials: films, magazine and book covers, advertisements, historical accounts of specific events from secondary sources and different types of images, theories, Internet sites, scholarly texts, concepts, and science fiction literature.[12]

This approach allowed me to put together a speculative vocabulary of different experimental practices. Compositional politics, decolonial politics of matter, and alterontologies are some of the concepts I use to describe aspects of experimental practice from different angles. I would call the underlying methodological approach baroque fieldworking, a mix of politically engaged research, speculative historiography, and social science fiction. Although much of the work presented here is the outcome of different combinations of research and activism, this book is not an ethnographic account of this political participation. Rather, it is a theoretically motivated project grounded in sustained and lengthy political involvements as a committed practitioner and activist. Through these activities, the problems that I discuss in the book were presented to me and then took shape and form. One could even say that the whole book is an attempt to negotiate problematics that arose through these practical and political involvements.[13]

One of the aims of this book is to explore different narratives of emergence of experimental practice. This necessitates engaging speculatively with the historicity of the topics and events that are central for my argument: I bring together different, often discrepant, historical events, artifacts or chronotopes in order to re-create a possible historical trajectory that unearths a present phenomenon that has not fully emerged yet. I try to explore the otherwise, the not yet fully materialized, the unknown from significant proximity. I am less interested in exploring the unknown from distance or even in exploring the known by creating (critical) distance through a predetermined methodology. I am drawn to getting as close as possible to something that is already there and yet has not fully emerged. And this involves getting as close as possible to its past by unlocking its speculative potential in order to reconstruct promising alternative histories and virtual futures.

The combination of empirics and theory and the development of a speculative historiography are part of my attempt to emphasize the fictional side of social science writing. I discuss this in length in chapter 2 but for now it may be important to say that rather than trying to bring fictional tropes, contents, and genres into social science, the social science fiction I am pursuing here attempts to write social science *itself* as fiction. That is, the objective is to write social science in a scholarly social science and social theory fashion (by following standard citational and stylistic conventions and by docking onto existing debates) and to incorporate in it, almost imperceptibly, the fictional and speculative dimensions that emerged through my political and practical engagements.

Social science fiction helped me to elevate something that is happening already but is still not a defining moment of social movements—that is, experimental practice—to a form of politics that is forcefully present in our realities. I see social science fiction as 100 percent empirically grounded and 100 percent fictional. It is both at the same time. It is almost as if there is a spiritual dimension in experimental practice, a "material spirituality" (Puig de la Bellacasa, 2014a), that this process tries to reveal in our realities. The implicit philosophy of this project is therefore neither realist nor social constructivist, neither critical realist nor a constructivism without adjectives either. It is an approach as experimental as the worldings it engages with require it to be.[14] Thus, despite the intensive engagement within the fields I was involved in, I did not have a predetermined methodology and I did not *do* fieldwork. Rather, I was, as are many other humans, politically active *in* these fields for reasons that are far greater than this study. I was (and still am in some cases) fieldworking.

Instead of trying to create some unity and permanence while putting together the materials, concepts, and ideas that I gathered while fieldworking, I have developed a baroque piece containing a transitory, often expressively dissonant and extensively ornamental main text folding into an infinitely cavernous, continuously curvilinear and often disjunctive set of relations to other texts and thoughts presented in the notes of the book.[15] In this sense, I constructed a transfigured objective baroque world where speculative occasions, spiritual meanings, empirical materials, and social research have equal place.[16] The aim of baroque fieldworking and social science fiction is to present a world of abundance—and I see this not only as a stylistic attempt but as a political obligation of the text.[17] Instead of dearth, which is the outcome of traditional management and control, I looked for abundance as the outcome of material self-organization and autonomy and, of course, of experimental practice.

ESCAPING HUMANITY (OVERVIEW)

I trace experimental practice in different fields of life and through different cases and occasions. In all of them I try to retain those aspects that allow me to understand the material practices of social movements in their quest for justice. These practices that fuse justice and ontology will allow me to unearth the more-than-human and more-than-social aspects of social movements and their slow path to become *movements*. I start this journey in chapter 1, "Decolonial Politics of Matter," which serves as the theoretical introduction to the book and provides a conceptual diagram of its key questions and arguments. The chapter argues that in order to be able to address questions of justice, social movements engage in a decolonial politics of matter: they learn and experiment with changing—literally—the material composition of life in ways that delink from the Western epistemic appropriation of matter as an open frontier for exploration and enclosure.

Chapter 2, "Biofinancialization as Terraformation," and chapter 3, "Ontological Organizing," make up the first part of the book and attempt to grasp the role of these experimental material practices in contemporary social movement politics. Chapter 2 aims to establish why the experimental moment in social movement politics is necessary and, indeed, unavoidable. The chapter describes the ascent of financialization since the 1980s, which brought with it a culture of valuation that spread well beyond financial markets and came to pervade everyday activities, subjectivity, ecology, and materiality—biofinancialization. I argue that the assetiza-tion of life and the financialization of bios has made the current regime of production and accumulation untouchable in political terms. This, of course, poses significant challenges to traditional politics, including social movement politics, because it interrupts established channels for social change through existing political institutions. The question then is how to imagine the autonomy of politics when biofinance becomes molecularized in code and in matter.

Chapter 3, "Ontological Organizing," approaches social movement ac-tion from the reverse perspective: from the practices of movements on the ground. In this chapter I work with materials from a long fieldwork study of migration activism and migration movements developed and analyzed with my collaborator Vassilis Tsianos. Here we argue that many social movements today, and certainly the migration movement, increas-ingly change the ways they perform politics. They avoid targeting directly institutions of power and organize outside existing political channels by setting up alternative ways of being that support their aims. In the case of migration, this means the creation of imperceptible but durable infra-structures and ontologies of existence that facilitate the freedom of move-ment of migrants. The sovereign regime of mobility control is displaced on the level on which it attempts to take hold: the everyday movements of transmigrants. The securitization of borders and spaces is challenged by organizing common ontologies of existence below the radar of pervasive political control and by creating alternative everyday worlds: the mundane ontologies of transmigration, the *mobile commons* of migration.

Chapter 4, "Activist Materialism," and chapter 5, "Insurgent Posthumanism"—the second part of the book—mix and remix different historical incidences, conceptual resources, and perceptual strategies in ways that allow me to trace the experimental dimension in historical forms of political mobilization and social movement action long before anything like posthuman culture and technoscience existed. Chapter 4 specifically discusses the adventures of the concept of materialism and its uneasy relation to political activism. Here I explore possibilities for en-acting activist interventions in conditions where materialist politics is not primarily performed as a politics of institutions but as the fundamental capacity to remake and transform processes of matter and life. What is activism when materialism is not about a politics of history but a politics of matter? What is materialism when it comes to an activist engagement with matter itself?

Chapter 5, "Insurgent Posthumanism," traces experimental practice as the posthuman making and remaking of alternative material conditions of existence in different historical cases of social movements and theoretical accounts of politics. But coupling politics and experimental practice and fusing social movements and the posthuman defies much of posthumanism's current assumptions as well as many of the theoretical presuppositions of the politics of social movements. The chapter gradually weaves together politics and posthumanism along three distinct experimental practices: the making of common material worlds (not just a common humanity); the embodiment—literally—of radical politics; and, finally, the enactment of justice through a materialist, nonanthropocentric history.

Chapters 6, 7, and 8 constitute the third part of the book. They bring back the discussion of experimental practice to the present by performing the rapprochement between social movements and the more-than-human worlds of technoscience. How do we perform experimental practice today? If the experimental cannot be thought independently of technoscience, the aim of this third part of the book is then to explore different ways to conceive politics in technoscience. As I argue throughout this part, the experimental practice of more-than-social movements is enacted with, within, and occasionally against, but never outside, technoscience.

Chapter 6, "Brain Matter," is a case study of the notion of neuroplasticity and the emergent epigenetic nature of the brain. At the heart of the vision of plasticity circulating equally in popular culture and in the sciences of the brain lies the possibility of recombining brain matter and understanding the making of ecologically dependent morphologies in a nondeterminist manner. But plasticity as recombination is not only a radical challenge to predominant determinist assumptions of the brain, it becomes also one of the major avenues through which politics becomes articulated within neuroscience. The chapter explores different prevalent versions of such politics related to neoliberal markets, processes of governance, and traditional visions of social liberation. Engaging with these different forms of politics allows me to explore their specificities and, indeed, limitations in establishing questions of justice in the relation of humans to their brain. Here I ask what an experimental practice of the brain is or would look like and introduce the notion of composition as a key feature of experimentation with matter.

Chapter 7, "Compositional Technoscience," picks up the idea of composition and proposes to reexamine how politics has been conceived within science and technology studies: the politics of credibility and expertise, institutional participation, the governance of human-nonhuman relations, and the inclusion of marginalized experiences. I debate these different approaches to politics through a case study of AIDS activism in the 1980s and show that more than an organized response of the gay community targeting inclusion in existing institutions and expert committees, AIDS activism was the product of the community's efforts to survive the epidemic and the prevalent social hostility. In fact, the community did not preexist these efforts, it became a community only by engaging in an extremely wide spectrum of everyday material experimental practices that made its very existence possible. I refer to these experimental practices as compositional: the creation of alternative forms of life that primarily allow a certain actor to be able to exist, to articulate with other actors and forms of life, and to address questions of justice by changing its everyday material conditions.

The last chapter of the book, chapter 8, "Crafting Ontologies," presents a larger fieldwork study of the maker and hacker culture in the British East Midlands and beyond. The main thesis of the chapter is that the achievement of invention in technoscience takes the form of dispersed experimentation within more-than-human and more-than-scientific worlds: distributed invention power. In maker and hacker spaces, the focus of this chapter, we can trace how distributed invention power is organized: the topological stacking of materials and processes; ecological transversality and the emergence of new compositions of code and matter; the proliferation of commensal relations between the participating actors; the complex traffic between instituted and community technoscience; the involution of human, animal, and inorganic actors; the centrality of craft and experimental labor in makers' ethopoiesis; the precarization of work and multiplication of free labor; and, finally, the continuous folding of the commons and the private and public spheres into each other. What is constitutive of these diverse practices of making and other movements of community technoscience is that they change the conditions of knowledge production by recomposing the fabric of everyday life: the stacking and forking of worlds into alternative ontologies. What is at stake here is not just technoscience itself but the ontological constitution of life and the attempt to defend it. In posthuman conditions, traditional politics and the corresponding social movements can support us in this endeavor only to a limited extent. The alterontologies of more-than-social movements do not just represent a new form of political organizing. Something else, something existential is at stake here: alterontological politics is a possible way to survive a world that is disintegrating through human action. Alterontologies may be a way to escape humanity.

01

DECOLONIAL POLITICS OF MATTER

ONTOLOGICAL POLITICS

The rise of technoscience in its contemporary configuration since the 1960s and 1970s resonates with a cultural imaginary marked by the idea that social transformation is primarily driven by material transformation.[1] Technoscience creates new ontologies; it is world-making and history-making.[2] The simultaneous production of society and ontology is often described as ontological politics: there exist multiple ontologies rather than just one, and these different ontologies are enacted by the actors involved in them.[3] As I discuss later in the book (especially in chapters 7 and 8), this is not an epistemological question, it is a practical one. Depending on the specific actors involved, ontologies are practiced differently and thus are materially different as such. This is a "multinatural" world.[4] And it is ultimately a political question which ontologies a certain actor participates in, and how. Politics here means that by performing ontology in a single concrete way rather than any other, we change the very constitution of being and its material organization in a specific direction. Ontology is not a description of the state of things, but of ways of being that include alternative possibilities of world-making.

What absences and what silences are produced when we conceive our engagement with and within technoscience as ontological politics? And what ways of being and acting, what voices does this understanding of technoscientific politics privilege? Along with Boaventura de Sousa Santos (2001) I am interested in the study of absences. Rather than exploring what ontological politics is or how it is theorized, I aim to discuss what ontological politics actively produces as absent and nonexistent.[5] Is the politics of (more-than-social) social movements that I mentioned in the introduction a form of ontological politics? If not, what are the alternatives? As de Sousa Santos (2004, p. 178) writes, "There is no single, univocal way of not existing. The logic and process through which hegemonic criteria of rationality and efficiency produce the non-existence of what does not fit them are various. Non-existence is produced whenever a certain entity is disqualified or rendered invisible, unintelligible, or irreversibly discardable. What unites the different logics of production of non-existence is that they are all manifestations of the same rational monoculture." With Susan Leigh Star (1991) we can think of the production of the nonexistent as an accumulation of residuals created in the process of making the world: residuals produced though work that has been invisibilized. What are the absences, the residues, the invisibilized labors that cannot be considered in ontological politics? What is rendered silent in the process of performing ontological politics?

THE FRONTIER OF MATTER

In ontological politics, matter is not just raw material for other social or political ends; rather, matter itself is opened up as a space of expansion.[6] Matter in all its expressions and formations is not a substance with fixed qualities and given potentials; matter is active and creative, complex and enlivened. The shift to the ontological denotes an interest in immersing into this process of self-ordering and in co-acting within it—to let other possible ontologies emerge.

Consider, for example, the nascent field of epigenetics. Over the past decade, a new set of theories and experiments exploring the unpredictable dynamics of gene expression has begun to take center stage in genetics research. The *epigenome* describes the overall state of a cell in flux, each point in time yielding multiple cascading possibilities for divergence of individual phenotypes. The inert genome is thereby supplemented by a softer, more adaptable, epigenome, incorporating mechanisms capable of responding to the environment, sensing "time" and retaining "memories" that regulate subsequent development. Within myriad microscopic epigenomes, the effects of the wider social and physical environment are translated via biochemical interactions to become an integral part of a fluctuating landscape of gene expression.

I discuss epigenetics in chapter 6 (for an extensive discussion, see the work with my collaborators on this project, Emma Chung, John Cromby, Chris Talbot, and Cristina Tufarelli)[7] but for now I want to highlight that the epigenome as an emerging object of study as well as a theoretical framework (epigenetics) could be described as a set of alternate ontologies that are made by a multiplicity of actors who contribute to the field from many different and often conflicting perspectives and positions. Their specific ways to do research are not primarily shaped by their commitment to some external politics to the field (such as left or right, liberal or emancipatory, or conservative or progressive), not least because many of the actors involved are nonhuman others and do not have such politics. The politics that one can find are intrinsic to epigenetic research, yet their connections go far beyond the field of epigenetics and address broader questions of politics and justice.

I wonder what the ontologies of the epigenome will look like in the next decade when this nascent field will start to take a more definite shape. Which ontologies of the epigenome will develop further, and what will they look like? Which will disappear? How and why? Ontological politics captures this process. Performing ontological politics means to open up matter as a field of exploration, experimentation, and, ultimately, appropriation. Ontological politics conceives matter as a *frontier*. In every frontier, expansion takes place as inclusion of new territories and entities in a process of continuous creation of new worlds.

COLONIALITY AND PRODUCTIONISM

Every frontier has a promise: to liberate the one who moves into the open space from the limitations that preoccupy life before the frontier opens. The promise of the new frontier of matter is to liberate material action from being dominated solely by social imperatives—class, sexuality, race, power, religion, culture, inequality—and to develop a radical commitment to matter (and to its multiple expressions). But simultaneously every frontier has a secret: in order to expand and include new spaces, entities, and actors, it needs to exercise power over them and silence or oppress some of them in order to make them fit. Ontological politics on the one hand expresses this liberating move—that is, the openness of ontology to co-action of all actors involved and to the multiple possibilities within it—and on the other hand administers its secrets: How many actors can be included? How? What is going to happen with the rest of them?

Every moving frontier is contemporaneous with a form of liberation, and simultaneously it enacts oppression. In the American frontier, the laborers who escape the wage labor market of British America to become independent peasants and artisans from the seventeenth century onward move into the frontier to the West only to bring savagery, destruction, and the dispossession of land.[8] In the Eastern Cape frontier of South Africa, the opposition of the Afrikaners to British colonial power and their subsequent migration came only to consolidate militarized white supremacy and extreme nationalism.[9]

The frontier opens as moving people (most often violently) appropriate new territories and then include and integrate them into their realities. And as people move, the new territories absorb them into their workings and transform them. The logic of every frontier is the logic of inclusion, the colonial inclusion of hinterland and the outside into some form of control that spreads well beyond the actual frontier itself. The term *coloniality of power* describes an order of power that emerged as a result of colonialism and colonial administrations but survived their later demise to structure culture, knowledge, economy, and our ways of being.[10] Colonialism shaped modernity to such extent that modernity itself would be impossible without it: modernity/coloniality. And with it also our epistemic practices and systems of knowledge are constituted by the modernity/coloniality nexus. The frontier of matter, this newly opened frontier that operates as an epistemic appropriation of matter, exists within the practical logic of the coloniality of power. The frontier of matter operates within "modernity's epistemic territory," inescapably.[11]

In the movement of the frontier, territory is considered to be a space on which nobody has property rights and therefore it is open for appropriation, *terra nullius*.[12] It is only because of this rendering of the territory as unclaimed that the frontier legitimizes its moves and the erasure of already existing systems of knowledge, local epistemic traditions, and the communities that sustain them. In modernity's epistemic territory, mapping space, representing its entities, and shaping new political institutions in order to include these entities (and exclude and, often, erase others) is the way the frontier expands. Through these processes the frontier is turned *productive*, in a double sense: the space of the frontier is turned into a space of production, that is, a space that is produced through relations of forces and power as well as a space that is gradually docked to existing systems of the production of goods, knowledge, and commodities.[13]

The appropriation of the frontier is a double act of organization: it is organized as a political space through processes of representation, and it is organized as a productive space by turning it to an accumulation system. Inclusion is about enclosing the new spaces into a regime of representation and a regime of accumulation that sustains the power of coloniality.[14] So also in our case: when matter becomes a frontier, the attempt is to make it productive—politically productive—as well as render it compatible with the existing mode of production. Ontological politics are these specific practices that perform the inclusion of new formations of matter into the accumulation regime of current economies; I discuss specific ways that this happens in chapter 2 and again in chapter 7. As the frontier of matter moves, its political institutions also move. This necessitates the creation of new political forums to accommodate these emerging formations. Including nonhuman actors into polity is about rendering them amenable to being represented;[15] as they become representable, that is, as they become identifiable within the coordinates of an existing mode of political organization, they pressure existing political institutions to change and to accommodate these new actors.

Ontological politics is a description of practices that portray how the frontier of matter advances and that unfold within spaces marked by colonialism. I am not saying that ontological politics is a colonial enterprise (although this may be true in many cases). What I am saying is that ontological politics exists within the modern epistemic territory that is constituted by its coloniality: ontological politics makes the frontier of matter, modernity's ultimate frontier, move. Thinking politics in technoscience with ontological politics allows us, as I discuss in chapter 5, to challenge widespread anthropocentric understandings of what matter is and to introduce the idea of the coexistence of multiple contingent possibilities that emerge in the process of movements of matter. Ontological politics describes the forces exercising pressure on the outer limits of constituted political institutions to differentially include more human and nonhuman actors, relations, and ontologies in a given political configuration in order to allow for these ontologies to emerge and become embedded in our social and material worlds.

The epigenome, a neuron, an MRI scanner, or a new subsea fiber-optic cable are not just actors that are differentially enacted by existing social institutions and people to create new forms of existence and new ecological configurations; they are also embedded in actual epistemic and value production processes that maintain the existence of the frontier.[16] In other words, ontological politics is politics performed in order to keep the frontier of matter moving by activating new material processes, allowing multiple possibilities, including new material actors, representing them, assigning them specific rights and positions, inserting them into the political sphere, extracting value from them, and speculating on their value to multiply financial yields. Ontological politics is not just a theoretical tool depicting politics within technoscience, but also the modus operandi of rendering matter productive, in the double sense of productionism: politically representable as well as creating value and innovation.[17]

MATTER AND JUSTICE

What would it mean to question the implicit coloniality of the frontier of matter? To what extent can material actors object to their enlisting in the productive machine of the frontier? Is it possible to think of actors escaping modernity's epistemic territory? How can the epigenome object to specific meanings and functions that it is expected to perform in the emerging research on epigenetics? Can a transatlantic fiber-optic cable challenge its specific use by a certain consortium of companies that maintain it and exploit its labors? What labors do these actors do that remain nonexistent, absent, invisible while many other of their labors are fully absorbed in the productive regime of the frontier? As I argue throughout this book, in order to be able to start looking for answers to these questions we will need to challenge the assumption that the *inclusion* of these actors in constituted political institutions will securely bring their absent voices and their invisibilized work to light.

With Star (1983, 1991) I am interested in the organization of the invisible and erased work that remains hidden when ontological politics is performed, rather than the organization of visible work as such (and this pertains to human and nonhuman actors alike, as I discuss later in the book). Erasure and inclusion coexist in the modern/colonial epistemic territory. Rather than inclusion, then, in the chapters that follow I aim to explore alternative political practices that attempt to disconnect and redraw the conditions and terms of this epistemic territory, "a delinking that leads to de-colonial epistemic shift and brings to the foreground other epistemologies, other principles of knowledge and understanding and, consequently, other economy, other politics, other ethics" (Mignolo, 2007, p. 453).

Following Jacques Rancière (1998), my starting point is not inclusion but the emergence of the invisibilized and the imperceptible, of those who have no place within existing normalizing political institutions. I understand politics here as a deep dispute over the existence of those who have no part in constituted institutions or even of those who refuse to participate in them.[18] And this form of politics within the frontier of matter happens when those who have no part change the material conditions of existence in a way that cannot be overheard or simply included in existing political institutions.[19] In chapter 3, for example, I discuss this in the case of people's migration and in chapter 7 in relation to AIDS activism. I am thinking with Starhawk (2002) of a form of politics that is not exercised as the *power over* a territory or simply a power that appropriates what is within it; rather it is politics as *power with*, the power of creating alternative common forms of life that reorder the language and practices of existing political arrangements of constituted power.

Instead of asking questions such as "How can we include different humans and nonhumans in our institutions?" "What is matter in itself?" or "Who is not included?" I focus on how actors create alternative ecologies of existence that become inhabited by these silenced and absent others, by those who have been rendered residual and invisible.[20] Rather than exploring with ontological politics the making of different ontologies, I am interested in a *decolonial politics of matter*: politics that, by instituting direct changes on the material level of existence, challenge existing conditions of inclusion *and* the idea of inclusion that rests on epistemic coloniality. Deleuze in his homage to François Châtelet points out: "No science, but rather a politics of matter, since man is entrusted with matter itself" (Deleuze, 2005, p. 717). This is a politics of matter not because humans are in charge of matter but because certain groups of humans and nonhumans can continue to exist only to the extent that they develop other alternative entanglements with matter. Matter is hope. Ontology *is* desire (see chapter 4 for further discussion).[21]

Beyond ontological politics as a general description of politics within the frontier of matter, I want to think of a decolonial politics of matter as a committed politics to make alternative worlds of existence with and against and outside the productionist moves of the frontier. But unlike traditional decolonial approaches that I discussed earlier in this chapter and throughout this book, my approach here is not primarily the epistemological unsettling of modern/colonial knowledge production. The unsettling of the frontier of matter can only take place on the material level, and it can only be a practical ontological question. I am less interested in how many ontologies exist and whether it is possible to create a pluriversal world (something that I discuss in length in chapter 8); I aim to explore practices that destabilize, delegitimize, and, ultimately, decolonize the frontier of matter through the alternative production of materiality: alternative ontologies. This decolonial politics of matter attempts to restore justice step by step through everyday practice (something that I discuss in chapter 5). In politics of matter, justice becomes ingrained in the materiality of being: in the soil, in the water, in our bodily tissues, limbs, organs, cells, genes, and molecules. Decolonizing settler land is one thing; decolonizing matter is another (even if they are tightly connected).

Material justice is a form of justice that happens even before its epistemological representation and political inclusion into constituted governance has taken place. This creation of alternative ontologies characterizes the actions of what I refer throughout this book as *more-than-social* movements: those that, rather than aiming at social and political power, primarily change the immediate ontological conditions of life. This book provides different materials—historical (chapter 4 and 5), empirical (chapter 3 and 8), and conceptual (chapter 2, 6, and 7)—that allow us to grasp how classic social movements can be conceived as more-than-social movements.

We are somewhat trained to believe that inclusion comes first: that inclusion in structures of social power is what politics is about. It is probably the other way around: politics is when certain actors, imperceptible actors, emerge in the political scene and change the very constitution of being by—literally—materializing ordinary relations of justice; inclusion can only follow this move. Instead of prescriptive justice, I am searching for material, processual, and generative justice,[22] which, rather than being focused on normative issues, is concerned with fusing justice and matter: practical justice, thick justice.

ALTERNATIVE FORMS OF LIFE

How does thick justice materialize? How can justice be installed through bodies, with other animals, things, and matter? How can we think matter and justice in a nondualistic manner? Let's turn upside down Clifford Geertz's "thick description" (1993): thickness for Geertz is semantic, let's seek material thickness; practice for Geertz is text, let's read practice as the material worlding of existence. Thick justice is about reclaiming materialist politics in the age of technoscience by making alternative ontologies: alterontologies (a concept that I discuss throughout this book and in particular in chapter 7). When ontological politics goes to formal institutions, politics of matter goes to the everyday: the space where alterontologies are crafted. Alterontologies = alternative forms of life.

I borrow the term *forms of life* from Langdon Winner (1986, especially chap. 1), who traces it back to Ludwig Wittgenstein as well as to Karl Marx. In forms of life we encounter a reweaving of the social and the material through the development of new practices, knowledges, and technologies. A practice, a set of practices, a device, a new form of connection becomes part of a form of life *by changing it*. There are no users, no tools, no symmetrical representations, no assemblages, no networks. There are just forms of life that set up the material constraints to what we are, what we can become, and how we co-construct each other. Wittgenstein (1958, p. 226) writes: "What has to be accepted, the given, is—so one could say—*forms of life*." I understand this as the making and sustaining of forms of life that have to be accepted because they transform the material order in ways that cannot be bypassed or neglected. Every social context, every sociotechnical environment, every ecosystem has enough space for alterontologies: conflicting alternative forms of life.

In previous work we called the politics of these social movements *imperceptible*.[23] Imperceptible politics are not invisible, yet they neither aim for their inclusion in constituted institutions nor claim visibility in the existing regimes of polity; rather, they transform the immediate conditions of existence without having as their central target their own representation in main political institutions. Imperceptible politics is the creation of new speculative figurations, new deliberate actual constructions, which put us right in the heart of new material and experiential forms of life. Günther Anders's call to train our capacity for "moral fantasy" and Walter Benjamin's "speculative experience" are behind these ideas. Anders discusses the inadequacy (what he later develops as a philosophy of discrepancy) between our feelings and the unforeseeable effects of things and demands that we train the elasticity and capacity of our imagination (Anders, 2002, p. 271ff).[24] Benjamin (1996b) discusses also the magical and spiritual language of things and develops the idea of speculative experience as a means of recognizing the wholeness of life beyond a naïve utopian idealism or blunt versions of materialist dialectics (I discuss this in chapter 4).

The quest, then, is how speculative experience becomes materialized in the making of alternative forms of life and simultaneously escapes the pressures of productionism—in its double sense as discussed earlier: inclusion into the political representation and into the current regime of accumulation. Instead of seeing technoscience as the instance against which claims need to be addressed and instead of seeing that claims of justice need to be imported from outside into technoscience, the aim is to, literally, *craft* alternative words. Crafting—that is, the everyday making of alterontologies—is one of the main themes that this book tries to explore. In chapters 4 and 5 I discuss different historical instances of making alternative ontologies of existence, and in chapters 6, 7, and, in particular, chapter 8 I collect different aspects that delineate crafting within contemporary (more-than-)social movements.

DECOLONIZING CRAFT

Craft is the result of a skill developed to such a degree that what matters is not the skill itself but the very moment of making. Craft is an excess of practice, "the desire to do a job well for its own sake," as Richard Sennett (2008, p. 9) writes. Craft and artisanal production were always in the heart of making things; indeed, they defined the birth of modern science until its gradual absorption into big science.[25] Craft in this picture becomes a necessary prerequisite and the absent mediator of what Barad (2007) calls an intra-active relation with matter. However, this tight interconnection between craft and the modern epistemic territory reveals also its limits, the need to approach craft from a decolonial perspective. Craft and artisanal skills are the invisibilized labors, the erased residuals that sustain any situated relation to matter. Simultaneously, craft is inescapably located within the frontier of matter. This specificity of craft defines the approach to decolonization that I use throughout this book. It is not primarily about the decolonization of seized lands, ways of being, and Eurocentric epistemologies or the creation of a pluriversal world. All of these are surely important parts of the decolonial project. But my attention is on something less central but crucial for my specific project here: the decolonization of our relation to matter and materiality from a position situated in northern Europe, where I live and I am politically active.

This relation to matter and materiality within northern Europe and more broadly the North Atlantic is one that is mediated in multiple ways through technoscience, the contemporary inheritor of scientific practice of the modern epistemic territory. Decolonizing matter means, then, decolonizing technoscience and its absorption into the frontier of matter. However, I approach this decolonizing project not by deconstructing technoscience from a perspective that would, for example, reveal how technoscience helps to perpetuate the coloniality of power—how it is used to sustain coloniality in geopolitical relations, how infrastructures maintain social power over indigenous communities, or how technoscientific knowledge reproduces colonial relations by other means, to name just few examples. Instead of approaching technoscience from the outside as an object of study—and eventually as an object of critique—in this book I rethink the practices of technoscience *itself*, hoping that this investigation will contribute to such broad decolonial projects. To what extent is it possible for social (or more-than-social) movements to abandon the split between politics, justice, and technoscience that pervades the ontological politics of the frontier of matter? I interrogate craft as the moment where the bifurcation between technoscience and politics, matter and justice can be potentially abolished. Inasmuch as craft is an inheritor of the colonial architectures of the modern epistemic territory, it is also a practice that can contribute to its destabilization.

Consider, for example, how grassroots ecological activism has contested the externalization of the costs of production to the environment by crafting a multiplicity of alternative forms of life (even if the relations between these different forms of life may conflict at times): repatriations of indigenous land, revegetation, biodynamic principles of farming, water conservation, inner-city food gardens, recuperation of traditional and indigenous systems of land use and land care, cooperative production, organizing against extractivism and the agroindustrial frontier, creation of alternative seed banks, permaculture activism, soil regeneration, whole-farm organization, urban gardening, ecofeminist advocacy, bioremediation projects, disruption of agribusiness, open-source agriculture, experimentation with biofuels, reclaiming of dispossessed spaces, production of alternative research, making of alternative collectives, setting up local systems of exchange and transactions, the early years of Seeds of Change, Earth Activism—all examples of crafting alternative material-ecological justice on the ground, multiplying livable worlds, or making alterontologies. The political organization of social movements here does not pre-exist the making of alternative forms of life; rather, political organizing *is* the crafting of alterontologies. Ontological organizing matters within social movements (I discuss this in detail in chapter 3). Ecological activism of the kind I describe here has addressed ecological destruction and the ontological politics of technoscience not primarily by opposing it or by exposing its connections to big agribusiness but also by occupying its very activity, by practically democratizing knowledge, and by organizing with technoscience for alternative worlds of ecological life.

But one could object here that craft as effective political organizing seems to be untenable when scale is at stake. How can alterontologies work with regard to the vast magnitude and impact of mainstream research in technoscience? How can alterontologies contribute to a decolonial politics of matter that develops alternative practices of justice within and against the frontier of matter? Following the earlier example, can grassroots ecological activism ever develop a viable alternative to big agribusiness and the technologies that it deploys? Infrastructural changes or large techno-scientific projects require extensive mobilization in order to be open to democratic politics, and craft seems to be unable to achieve that; big science is too big to change under the pressure of alterontologies, as the argument goes. But is not the question about *scale-making*[26] the real issue here, rather than the entrancing size of technoscience? Craft is about rescaling the geographies of technoscience in ways that matter.

Consider the free software movement.[27] What mattered in the birth of the free software movement was not primarily the question of scale in terms of its size (that is, if free software could become so widespread as to potentially challenge the domination of proprietary software) but the attempt to change the conditions in which software was made. The primary goal of the movement was to allow for alternative forms of software creation to emerge: How is software owned (or not owned)? How can software become freely distributed? How can it efficiently engage a large number of actors in the global virtual space? How can peer-to-peer nets be set up? How can the gift economy of cyberspace be defended and maintained?—all questions that indicate that crafting new code and new forms of cooperation was not a strategy to attack proprietary software but to create livable digital worlds. The creation of alterontologies is not primarily about size but about remaking the scale, not only crafting objects and processes but crafting scales too—that is, changing the values that a scale measures and not just its ratio, or else changing the scale itself. Scale-making in alterontologies is a *minoritarian* move: it focuses on the intensity of the actual moment of crafting rather than the extensity of the critique of mainstream technoscience. And as I discuss later in this book (in chapter 4 and in particular chapter 8), the relation between mainstream technoscience and alterontological practice is not clear-cut and oppositional, but characterized by multidirectional traffic, arduous exchanges, and endless collaborations and conflicts.

If a decolonial politics of matter is possible, then craft and artisanal skill lies in the ability to recognize the constraints of a situation, where to stop, and how to stop. Craft is about caring for the worlds we live in by acting in accordance with the intensities and the limits that matter imposes in each concrete situation. Maria Puig de la Bellacasa (2017) approaches care not as a moralistic stance but as a practical everyday engagement with the mundane worlds we inhabit that seeks to create more just and livable worlds. This is a generic notion of care: it can shape different organizational ontologies, but it is not universal. It needs to be materialized anew in each situation. In fact, the one who crafts alterontologies knows what the constraints are, where to stop, and how to leave himself or herself behind. Craft is not about DIY, but about DIWY: do it without yourself. Craft at its core is not about making things or producing relations but about leaving yourself aside for the sake of viably coexisting with other things and beings.

I. MOVEMENTS

BIOFINANCIAL- IZATION AS TERRA- FORMATION

GEOCIDE AND GEOENGINEERING

Politics is a human affair. It sits uneasily with other forms of life and movements of matter. But there are reasons for extending the reach of politics; the most compelling one is to challenge the stubborn persistence of humanist universalism in more-than-human worlds. Consider the Anthropocene narrative:[1] *We* have changed the planet irreversibly! *We* have been terraforming Earth by changing its ontological constitution. Terraforming is extraterrestrial geoengineering, but the extent of human impact on Earth is so extensive that we could say that we are now terraforming our own planet.

But who is this "we"? Is it human beings, is it the human species, is it groups of humans? We know that human action is as unified as the actions of any other species; which means that it is not. Would someone claim seriously that cyanobacteria acted in a conscious and concerted way when 2.3 billion years ago they oxygenated the atmosphere, caused a vast mass extinction, and changed organic life irrevocably? A particular species never acts as a unified subject by intention. The individuals of a species act according to relations each one of them establishes with other members, the environment, and other individuals from other species. They form populations and interspecies communities not by design but by chance, contingency, and involvement.

And as if this "we" is not problematic enough, it suddenly takes even more dangerous proportions: the fantasy that we can act rationally and become the guarantor of Earth's future. Cyanobacteria did not know that they are changing the planet, but we do. What follows is that humans have damaged Earth so badly that we need to terraform it. Humans the destroyers, humans the saviors; both combined elevate humans to the makers of one unified single world, a grand universalism of a terraformed world. Terraformation appears to be the source of destruction and the remedy simultaneously. Geocide and geoengineering.

This combination—terraformation as changing the planet to such an extent that it can be codified as a new geological epoch, and terraformation as a futuristic enterprise that can supposedly save the planet—captures a salient moment of contemporary technoscience. How is it possible at all to think of our past and current actions as well as future solutions as being on one unified linear trajectory? How it is possible to combine the destruction of Earth and the imaginary of catastrophe—geocide—and the promise of redemption in a single matrix?

THE UNIVERSALIZING MATRIX OF FINANCIALIZATION

What underlies the universalizing matrix of terraformation is, as discussed in the previous chapter, the making of matter as a frontier: the belief that we are able to make matter, to the extent that it becomes part of our existing systems of political representation and of the production process of current economies in the Global North. As discussed in the previous chapter, production here has a double sense: the construction of new ontologies and the insertion of these ontologies into scales of value. How is it possible to conceive the production of matter and its value in one conceptual framework? This chapter explores how the financialization of everyday life, subjectivity, ecology, and materiality—biofinancialization—provides this framework for unifying the double meaning of productionism in engagements with matter. However, I do not discuss financialization as such but the implications of this universalizing approach to life and matter for politics.

The current financialized regime of production has become so embodied in the ontology of our everyday lives that even social groups that can challenge its legitimacy cannot do so without challenging their very existence. This is more than a mere intellectual problem, or an issue of (lack of) parrhesia, courage, and responsibility; biofinancialization is an issue that poses questions about how can we engage with questions of political power and justice and what are alternative political scenarios.[2] In other words, it raises questions about social movements, the nature of their politics, how they are constituted, and what they try to achieve.

Biofinancialization has propelled the engulfment of large segments of Global North societies into a mode of existence that has changed the conditions for the articulation of possible political alternatives. Despite the predicament of politics that we are experiencing in the Global North and despite the powerful mobilizations[3] that have shaken many countries since the economic crisis of 2008, a turning point seems almost impossible. "We have a situation here"[4] that is defined not only by economic and political exigencies but also by the fact that alternative forms of politics that were put in motion in many Global North societies by broad autonomous social campaigns and social movements of the past decades seem unable to create fertile conditions for social change.

I focus my discussion on these mobilizations and briefly discuss what came to be called autonomist politics. However, I do not introduce autonomist politics as such[5] but describe the starting point of the argument that I will develop in the rest of this book: The *more-than-social* becoming of social movements is a response to the closure of current forms of autonomous politics. Biofinancialization—the entanglement of everyday existence, socioeconomic life, and the frontier of matter—has made this closure tangible and has created conditions that can no longer be addressed by current social movement politics. In the following sections I reconstruct this argument by discussing the emergence of biofinancialization and how it has neutralized traditional autonomist political practice in the Global North. Traditional autonomist politics gravitate around questions of class and labor, and I retain this line of thought in the next sections, but I also open up these debates to an alternative understanding of autonomism that could contribute to the decolonial politics of matter in the Global North that I discussed in the previous chapter: the capacity of social movements to become movements in the material sense and to promote alternative ontological configurations of life. This is the magic formula of this book:[6] terraformation from below.

Biofinancialization has its roots in the ways the unruly social and political movements after the 1960s were gradually reinserted into the current regime of accumulation. The contentious mobilizations of the working classes[7] and subaltern populations that started in the 1960s and 1970s put an end to the Fordist-Keynesian accumulation regime by challenging the cultural, racial, and sexual organization of labor on a local and global scale. These struggles transformed gradually and diversified through the 1980s and 1990s to a multiplicity of social mobilizations and conflicts across the Global North that forced social regulation to reorganize itself in order to capture the new exiting subjectivities: antiracist and migrant mobilizations, feminist struggles and social rights mobilizations, the anticolonial and alter-globalization movement, and ecological movements.[8]

These contentious politics intensified with the relocation of production outside the Global North and with the elevation of financial markets to one of the primary engines for economic recovery after the economic crisis of the 1970s.[9] This crisis never ended in the destruction of value or in the creation of a new system for the invigoration of demand by changing the living conditions of the populations through a new "New Deal."[10] Rather, Global North societies entered into a perpetual crisis[11] that paradoxically became the cause as well as the consequence of a new regime of social production dominated by stagnant wages, underemployment, and the flexibilization of labor markets[12] as well as of finance-led accumulation[13] with the introduction of securitization and increased consumer[14] and corporate lending.[15] Financialization on the one hand and the deregulation of labor markets on the other became the key features of the neoliberal turn after the 1970s.[16]

A crucial point in this chapter is that these transformations are not mere instruments for social regulation or economic tools for counteracting the socioeconomic troubles that took place after the 1970s. Their effects are far more important than their economic performativity. In other words, the economic quandaries were not primarily resolved through financial means but through social, cultural, and technomaterial transformations (which were enhanced by new economic devices and accounting techniques). The virtualization of the economy that dominated the post-1970s crises was not just an economic strategy and a new regime of accumulation. It is culture. Financialization is culture, not only because it came to pervade society and the everyday, as Randy Martin (2002) has described,[17] but also because it contributed to the consolidation of an ever-expanding culture of financial valuation of goods and services.[18] This is the tendency to translate disparate judgments about value to financial measurements that introduced a *culture of valuation* into everyday life. Any and all aspects of sociomaterial life and the environment enter into this indeterminate and unstable process of evaluation that feeds the movements of financial markets and financialized societies.

The underlying logic of the culture of (financial) valuation is that the worth of goods, things, activities, spaces, and other species can be *essentially* translated into financial evaluations.[19] Although different scales of evaluation are by definition incommensurable,[20] the culture of valuation in Global North societies presupposes that the worth of almost everything— including the present and *future* appreciation of assets, goods, services, intangibles, the health and subjective capacities of individuals, the physical environment, human artifacts, animals and plants, and urban space—is transferable into one logic of financial value that is potentially tradable in the market; this is biofinancialization.[21] Neither valuation as such nor the cultures of valuation are novel and distinctive features of the current world; what *is* distinctive, however, is that the different forms of value— and, indeed, radically divergent values—are imagined as convertible into the universalizing matrix of financial value. *Financial value* is used here to express the primacy of investment value over other values (aesthetic, use, moral, ecological, material, cultural) that predominantly assess the future monetary profit to be gained from potentially any field of life or the environment. The principle of investment value hinges on the belief that the future is universal and exploitable.

BIOFINANCIALIZATION

The imaginary of measurable future value lies at the heart of the current culture of valuation that traverses many fields of life. For example, we have studied how young precarious workers "invest" in themselves by shaping their activities according to possible *future* gains in unstable labor markets;[22] consider also how biomatter is evaluated according to future monetary gains from its potentially scientific or commercial exploitations,[23] or Joe Dumit's (2012) remarkable study of treating possible future biomedical risks instead of diseases. These are just few examples; I discuss more later in this chapter and in the notes. What is crucial here is that value becomes an intrinsically indeterminate magnitude that has to be calculated by creating appropriate measuring tools[24] and then defended and negotiated between experts in designated public spaces or in the secluded spaces of the markets.[25] Future value is by definition unpredictable, and in order for it to be realized, the actors involved need to experiment, to manage conflicting information, and to create knowledge in action.[26] Future value and investment value are recombining other forms of value into a process of uncertainty. These valuation clashes and the frictions between different systems and scales of values constitute the emerging culture of valuation in biofinancial societies.

The ascendance of biofinancialization and the concomitant culture of valuation goes hand in hand with another major response to the social conflicts and the crises of the 1970s and 1980s: the deregulation of labor markets and the changes in the process of value production. Despite the general slowdown of the economy[27] and the turbulences of the profit rate, the levels of labor productivity remained high after the crisis of the 1970s.[28] In fact, productivity of labor per hour has risen steadily at an average of about 2 percent every year for more than one hundred years.[29] Deregulation of labor markets, the retreat of organized labor, the collapse of Fordism, deindustrialization, the steep increase of the service and retail sector over the steady decline of manufacturing,[30] and the proliferation of atypical and precarious labor[31] all become core features of the decline and transformation of industrial production in Global North societies into what many authors have described as a third stage in the development of the system of production.[32]

Industrialism in the Global North became increasingly hybridized by the expansion of the service sector (professional, high-skilled work as well as low-paid, nonstandard, and insecure work),[33] the rise of the knowledge and culture economy, the expropriation of goods and resources from countries outside the Global North, the accumulation of wealth through the production of intangible goods, and the extraction of surplus value from consumption, communication, and social reproduction. The biofinancialized regime of accumulation relies on a double architecture of production. On the one hand, it mobilizes the existing system of value production in its industrial formation in order to regulate the immediate labor process (through the traditional system of wage labor and remuneration); on the other hand, it relies on the appropriation of broader aspects of social and material life, everyday activities, resources of cooperation, working people's general skills, and subjective capacities that are not strictly involved in the immediate labor process. There is no comprehensive research about how these two dimensions of the accumulation regime contribute to value production in Global North societies.[34] But many studies[35] have highlighted that the architecture of value production in biofinancial accumulation extends beyond the workplace and relies on the expropriation of *res communes*: the commons, common pool resources, and common forms of sociality.[36]

The commons here not only refers to a possible social force that could resist its own expropriation, as broadly used in social movements, but also constitutes the underlying system of production of biofinancial accumulation. *Biofinancialization*—the financialization of life and matter—is here specifically used to describe how the commons becomes the ground and the material substratum on which biofinancial accumulation thrives. Openly and commonly used infrastructures (information and communication technologies, collaboratively produced knowledge, and cultural networks); the material and ecological commons (within the Global North but most importantly the often violent extraction of value from outside); and structures of cooperation, everyday sociality, and exchange between producers and consumers make up some of the main sites of value production. Biofinancialization flourishes because it extracts value from reproduction, distribution, and consumption as well as other activities that do not directly belong to the immediate sphere of production; this is possible because exploitation in the workplace is organized through the specificities of the lives of working people beyond the workplace itself.[37]

EMBODIED VALUE PRODUCTION

The externalization of production from the workplace to the social and the material does not mean that the site of value production is transferred "outside" living labor. Duration of the working day and intensity of work are the main dimensions that define the degree of exploitation of living labor.[38] But this intensification of exploitation affects—and to a large extent entails and necessitates—a wider set of activities that extend beyond immediate work-related activities. One could say that the supposed intensification of exploitation is complemented by its extensification. Value production expands across the existential conditions of living labor in the Global North; the lines of fight and control multiply and traverse different domains of life.

Activities that people perform as part of their nonwork life or secondary activities of their work life become directly productive.[39] Beyond that it also means that working people mobilize multiple social and personal investments in order to remain in the labor market (such as social relations, general skills, informal networks, ideas, their subjectivity, their mobility, their health, their self-organized structures for cooperation, their potential for development)[40]—some of this is entailed in the "final product" of their labor, but much remains outside it. The epicenter of value production is the workplace, but it is only the epicenter. If we focused on the workplace only, we would miss the important and sometimes defining broader conditions in which work and employment take place.[41] However, the extensified mode of value production does not mean that work becomes simply dispersed and socialized, that it moves "outside" the singular worker. Rather, it means that value production becomes embodied: it becomes an indissoluble characteristic of the whole situated existence of each worker. The situated and embodied quality of work includes all things and artifacts that constitute the worlds in which we exist, our social relations as well as the broader networks of the commons—material commons, ecological commons, social commons, informational commons—that we rely on to maintain everyday life.

The separation of labor power and living labor, which was possible in the system of production of industrialism, becomes the major source of conflict in biofinancial societies, simply because living labor in its embodied and situated wholeness is the source and vehicle of labor power. Control over embodied production is taking place along several lines that attempt to cut across and appropriate the existential continuum of people: first, the attempt to measure labor power and to quantify it despite the fact that it mobilizes the whole embodied conditions of life;[42] second, the expropriation of the infrastructures of cooperation through property rights, patents, rent, extractivism, and the reprivatization of access to and circulation of information and material objects;[43] third, the individualization of the costs of social reproduction and privatization of forms of social reproduction that cannot be taken up by the individual;[44] and fourth, the transformation of citizenship to a tool for creating various tiers of working people whose degree of exploitation depends on their varied access to citizenship rights.[45] One could say that all these lines break the horizontal and continuous lived experience of working people and create some form of separate vertical segments that are the productive motor of the biofinancial regime of accumulation.[46] Vertical life. Walls of value.

The attempt to impose a separation between product of work and process of work, of cooperative ownership and proprietary ownership, of production and social reproduction that is, in other words, an attempt to impose the law of value and the system of wage labor constitute the fault lines of embodied value production. The tensions on these lines is multiplied as financialization enters every aspect of work: through the financialization of everyday life (debt) as workers substitute falling or stagnant wages and the dismantling of welfare provision with lending; through the extensive valuation of work outputs; through the exploitation of one's own future in postcontractual employment;[47] and finally, as discussed earlier, through the transformation of the nonwork spheres of life into value-producing activities.

The main conflict is between an extended process of value production that is experienced as indissoluble from everyday existence and the attempt to control, organize, and remunerate this system according to the immediate labor needed for the production of value. In other words, the conflict lies in the fact that value production is embodied but is treated as if it is external to the worker. This conflict—which threatens to erupt anytime and to throw the balance of power into disorder[48]—is *not* the result of the uncontrollable economic forces that emerged with the turn to financialization; rather, it is the result of the deep social conflict that traverses embodied value production as a whole.

The instability of biofinancial societies "has nothing to do with any presumed instability per se of the mechanisms of the financial system; quite the contrary, the ambition of those mechanisms is precisely to absorb shocks and to smooth out discontinuities in the economic cycle" (Moulier Boutang, 2011, p. 152). Here is where the culture of valuation described earlier in this chapter meets the production of value in today's conditions. The culture of valuation not only underpins the system of finance-led accumulation but also absorbs the shocks of the social conflicts that traverse embodied value production. Thus the instability and the conflicts that lie in the heart of value production are not intensified by the indeterminacy of the culture of valuation, but rather the opposite is the case: the culture of valuation that comes to dominate the sociomaterial regime of accumulation is the main tool through which conflict in value production is regulated (and potentially also contested, as I discuss later).

The pervasiveness of the culture of valuation with the crucial indeterminacy of value that lies at the core of this culture is the main way to control the outputs of work in the extensified mode of production and is the means that working people themselves use to modify and change their position in the social nexus. Financialization turns "bio" not only because it is actively embedded in people's lives, bodies, and environments but also because this embeddedness, this becoming fleshly of financialization, comes to constitute anew how social conflict unfolds and social struggles are performed. As financialization becomes an integral part of value production, it becomes also the vehicle for articulating social demands and political claims; it becomes a tool for creating and maintaining social hierarchy. Financialization is culture because it has come to dominate our imaginary to such an extent that even social justice can be fought for with financial means. Financialization is not ideology; it is as real as something can be, ingrained in the everyday ontology of life.

Biofinancialization not only created the ground for a new phase of expansion (and crisis) but also, with the culture of valuation, created a tool for managing the conflicts that traverse embodied value production. Through biofinancialization and more broadly the culture of valuation, specific segments of the elites and the middle classes maintained and strengthened their position in the social order over the past forty years. The ascent of managerial and professional classes and their privileged access to education and sociocultural capital have contributed to the expansion and consolidation of the middle class in the postwar period.[49] While the middle class became a global phenomenon,[50] it has also been under increased pressure in the societies of the Global North, and definitely since the 2008 financial crisis.[51] And this pressure, the fear of falling that Barbara Ehrenreich (1989) diagnosed almost thirty years ago, reinforces today's increasing reliance of the middle class on the culture of valuation to maintain their social mobility and sustain their class position.

The reluctance to challenge the architecture of the financial system after the 2008 crisis is not only imposed but also desired by a broad social coalition that includes the elites and middle classes (and certain segments of the emergent working classes and service workers). But at the same time, this acceptance of the predominance of the financial system challenges the Global North's liberal democratic principles. Paradoxically, biofinance is the outcome of extreme liberalism and simultaneously signals its demise. Elsewhere we have called this condition *postliberalism*: the condensation of segments of the state together with specific private interests, segments of social classes, groups, or subjectivities into large formations that coalesce along an imagined commonality of social domination.[52] In postliberalism we have formations of vertical aggregates of power reassembled from parts of the fragmented society that was the outcome of forty years of neoliberal policies. And the devices that are deployed for erecting and maintaining these postliberal aggregates vary; we can see the reemergence of strong nationalisms, territorial aggression, escalation of geopolitical conflicts, and the resurgence of traditional conservative ideologies and values.

THE END OF THE REFUSAL OF WORK

This political situation is defined by a deep immanent conflict: the vertical-ization and appropriation of the commons and everyday life (as discussed earlier, through measure, proprietary regimes, the individualization of so-cial reproduction, and the transformation of citizenship into a tool for the regulation of labor markets) undermines working people's everyday lives and the flow of embodied value production; at the same time, the verti-calization of the commons is the condition for maintaining the current sociomaterial regime of accumulation as well as the balance and stability of political power in biofinancial societies. The main political responses to this situation are guided by some form of revival of autonomous po-litical practices that played a role in the 1970s and 1980s (and offered the framework for the analysis that I presented in the previous sections), in particular the politics of the refusal of work and the self-organization of social reproduction: an exit from work toward activities that lie outside capitalist valorization and the organization of immediate social life outside formal public services or private provision.[53] But when value production becomes embodied in the existence of working people, as argued earlier, these political alternatives seem almost impossible. Production no longer operates through an externality between the subject and his or her work but through accumulation of the embodied totality of one's own biofinan-cialized existence. Equally, large-scale self-organized social reproduction in the sense that has been described in places outside the metropolises of the Global North[54] seems an untenable political scenario simply because it is impossible to give up work in its embodied configuration in order to free space for self-organizing social reproduction.[55]

Bifo Berardi delivers an intriguing description of the mixture of every-day life and the biofinancial regime, but his vision that "autonomy is the independence of social time from the temporality of capitalism" (Berardi,

2009, p. 75) does not seem to hold against the carnal orgy of contemporary biofinancialization's feasting on the commons and everyday life. One cannot say as an expression of autonomy today, "I don't *want* to go to work because I *prefer* to sleep." The refusal of work is impossible not only de facto—that is, because work is indissoluble from the body of working people, animals, and things—but also because it is not desired: verticalized value production has become the condition for maintaining everyday existence within the social order.

We can exist and make a life only through the biofinancialized bodies we have. One can only say "I no longer *can* work" and be punished, as Berardi so aptly describes, with stigma, panic, depression, and the deactivation of one's own capacity for empathy (Berardi, 2011). But neither can empathy be infused to the social body nor panic and stigma just simply extracted from it, as if they are external to it; neither Prozac nor poetry, neither Ritalin nor mindfulness are enough to do this—they inhabit the social body and when they move from one singular body to the next, they leave their traces on them; they mark life forever.

Jackie Orr's (2006) work shows how panic became institutionalized through its systematic use for "preparing" citizens for national emergencies after the 1950s and 1960s. And along with its institutionalization, panic also became individualized by entering the psychiatric diagnostic manuals as "panic disorder," which later came to be also a medical disorder as blockbuster pharmaceuticals entered into the lucrative battle for commercializing its medical treatment. "Psychopower"—that is, "technologies of power and techniques of knowledge developed by a normalizing society to regulate the psychological life, health and disorders of individual and entire populations" (Orr, 2006, p. 11)—transforms materially the very being of societies and bodies. It becomes incorporated in us; it flows in people's blood and across the social tissue in an era of extreme medical "treatment maximization" (Dumit, 2012). Biofinancialization becomes an embodied "psychopolitics"; it shapes perception, affects, desires, and our self-crafting (Orr, 2012). Biofinancialization is in us and in our ecologies. We live from it and it lives from us: a carnal feast.

Thus, cultures of valuation are inextricably linked to everyday life and the creation of value, while many of the alternative political analyses and responses mentioned earlier are trapped in the increasingly exceeded logic that value is primarily created in a system of wage labor that is supposed to be external to the broader conditions of people's everyday existence. Even approaches that try to resuscitate some form of subtraction from labor toward self-organized alternative productive activities[56] neglect the fact that these activities sustain the broader system of embodied value production *even if* they are not directly implicated in market activities. In biofinancialized societies value is not only created in the production process and even less only through labor. The culture of valuation shifts the site of value creation to a multiplicity of activities that overdetermine the practice of labor. Labor does not cease to be one of the major sites of value production; rather, labor and value production cannot exist without all these practices that contribute in complex ways to sustain a form of existence that is governed by an intense culture of valuation connecting different aspects and activities of one's life. In the same way that political economy (whether traditional, critical, or autonomist) as we know it does not offer adequate concepts to grasp this political impasse, the refusal of work does not offer a political alternative to biofinancialized existences. In order to establish how alternative political responses to biofinancialization have been rendered obsolete, it is important to explore how the culture of valuation came to permeate the ontological constitution of life.

PERFORMING BIOFINANCIALIZATION

A possible way to start thinking about the political impasse of biofinancialized societies is offered by the analyses of social studies of finance that investigate clashes over the valuation of circulating financial objects and reveal that these objects are active agents that shape the institutions, markets, and social spaces in which they operate.[57] Their valuation is not a straightforward process. Social studies of finance allow us to understand that these phenomena cannot be approached with the means of political economy because their technoscientific ontology,[58] their very semiotic-material existence, creates worlds in which pricing cannot be simply imposed from an "outside" (be it political power or an instance of capital); rather, valuation is the outcome of a complex set of intra-actions inside the worlds in which the clashes of valuation unfold. In *The Laws of the Markets*, Michel Callon (1998, p. 10) describes this process: in situations of extreme uncertainty, actors do not perform their calculations (and subsequently recognize opportunities) by getting help from outside the networks in which they operate but use their connections and knowledge to arrive at the best possible calculations.[59] Economics is just one of these calculative tools. In other words, economics does not describe markets but performs, modifies, and revises ways to judge in the markets. In the culture of valuation, different calculative agencies operate alongside other social institutions to economize the world and make it tradable.

Central to this argument is that every tangible or intangible object or activity can be potentially valued. What it takes to realize this potential is the design of an appropriate technology of counting and measuring that ensures the comparability of values. Measuring technologies are essentially *technologies of temporality*. Cultures of valuation are sustained by technologies for appropriating the future. Social studies of finance weave temporalities, technologies, social interactions, and market uncertainties[60] into stories about managing ontological contingency that defy both neoclassical economics and the received stories of critical political economy. Calculative agencies are not an instrument of control "in the hands of capital," as critical political economy would assert, nor are they mere instruments for achieving and maintaining perfect markets, as neoclassical economists would assume. Calculative agencies and the resulting contingency are "ontologically real" (C. W. Smith, 2011, p. 278); they are embedded in the everydayness, the materiality of current societies, and the systems of reproduction and production of the Global North. The fact that contingency is ontological means that it is *before* the actors that engage with it.[61] What one can take from social studies of finance before moving ahead, then, is that our semiotic-ontological access to the world is organized through cultures of valuation to such extent that one cannot simply withdraw from these cultures without dismantling one's own existence.

Although social studies of finance have provided vivid accounts of the inner life of financial markets, their self-proclaimed view that they provide a better insight[62] into how financial markets and the involved actors work, think, and act, or that they provide even better explanations of the recent crises and the uncontrollability of those markets, is misleading.[63] Instead of explaining biofinancialization, a social approach to finance constitutes one of the components that perform it. Social studies of finance are not just a response to the proliferation of complex technologies of valuation and their inherent instability but a continuation of them. The attempt to investigate and understand finance as a social activity also performs and reproduces biofinancialization, the mode of existence that made social finance possible. But because of that we find ourselves in an impasse: the ontological contingency and indeterminacy of valuation does not allow us to fall back on Marxist, autonomist, or post-Marxist political economic analyses of value production. Simultaneously, in political terms, we can also not remain within the realist preoccupation of social studies of finance that is transfixed on delivering neat accounts of the intricacies of pricing and valuation without providing a political analysis of the social and material conflicts traversing biofinancial societies. So, what is the meaning of politics when it comes to grasp a situation in which a specific mode of existence, financialization, has been ingrained into the ontologies of existence? Is there a possibility to develop an autonomous politics in these conditions?

ASSETIZATION AND RENT

Autonomy refers to the idea that social conflicts and social movements drive social transformation instead of just being a mere response to (economic and social) power.[64] Böhm and colleagues (2010) discuss how autonomous politics in various configurations—for example, autonomy as a self-valorizing process of one's own labor from capital, autonomy as a negative relation to state power, or autonomy as independence from global hegemonic development policies—are often implicated in reproducing the conditions that they seek to challenge. On the other hand, though, autonomy produces an excess[65] of practices and social spaces that "opens up frontiers of resistance and change toward radical practices, an equal society and self-organization" (Böhm et al., 2010, p. 28). How can this excess be conceived when biofinancialization becomes both ever present and untouchable? How can autonomy be practiced when financialization changes the ontological tissue of our everyday lives?

As argued earlier, the core characteristic of biofinancial accumulation is neither immaterial production, nor the infrastructures of information technologies and algorithmic valuation, nor the underlying networks of social cooperation, but rather that biofinancialization becomes *molecularized* in flesh, in code, and in matter. It alters the composition, the material infrastructure, of bodies and forms of life. Biofinancialization becomes fleshly, more than just the exercise of command over life and flesh; biofinancialization becomes the ecology of a terraformed existence, more so than just a system for accelerating accumulation.

HBSC analysts and marketeers tell us in one of the advertisements of their "In the Future" campaign that "in tomorrow's global economy, every resource will be counted."[66] We see a salmon against an aseptic white background with a barcode imprinted on its skin. "In the future, the food

In the future, the food chain
and the supply chain will merge.

In tomorrow's global economy, every resource will be counted. HSBC is one of the world's leading supply chain organisations. We help companies keep tabs on stock across six continents – and five oceans. The future starts today.

There's more on world trade at
www.hsbc.com/inthefuture

Issued by HSBC Holdings plc   AC22997

FIGURE 2.1 — HSBC / J. Walter Thompson London. In the Future. Supply Chain. Photograph by Andy Rudak. Reprinted with permission.

chain and the supply chain will merge."[67] This future starts today. It is a future where nature-cultural creations will become service providers and resources: "local demand," "global supply," the postindustrial assetization of the whole planet. Not only does biofinancialization rely on speculating on the profit extracted from production, but it is also the rent-generating assetization of life and ecology. Biofinancialization is not only "the becoming rent of profit" (Vercellone, 2010) and "the becoming rent of rent" (Marazzi, 2010) but also the becoming rent of Earth beings: animals, plants, and ecosystems.[68] As I discuss in chapter 8, technoscience plays a pivotal role in this process of turning Earth beings to assets to rent—a terraformed planet.

SOCIAL SCIENCE FICTION

Biofinancialization is materially ingrained into the affects, the muscles, the sociability, the desires, the lifeworlds of working people, nonhuman others, and things. It is impossible to think of autonomy in these conditions as independence from capital, state power, and hegemonic globalization; autonomy can only mean organizing, experimenting, and inventing new forms of life that attempt to create livable worlds. Autonomy in this sense is less about independence from social institutions and more about recombining materialities that instigate social and ecological justice. Autonomy here means, paradoxically, *organizing interdependences* that allow for creating ways of being—other forms of life—that divert existing modes of existence in unexpected directions.

When cultures of valuation and value production fuse onto the ontological fabric of life, novel ways of organizing and alternative world-making practices start to emerge. What is an adequate way to conceive these emerging alternative forms of political organization? Rather than a political economy of autonomy or a social studies of finance and valuation, I am thinking here of *social science fiction* as a way to grasp the autonomy of the political in biofinancial societies: a hybrid of social science and science fiction that allows for alternative concepts, tropes, and methods to evolve in order to conceive how other forms of life can be created. Social science fiction has a long tradition in science fiction itself; Ursula Le Guin, Samuel Delany, Philip K. Dick, and Kim Stanley Robinson, for example, deconstruct present social order by creating alternative *and* possible future societies.[69] Social science fiction uses not only technoscientific

In the future, local demand will shape global supply.

As incomes rise in the developing world, tastes will change too. HSBC's international perspective can help you find growth and opportunity in changing patterns of global consumption. See the shape of things to come. There's a new world out there.

There's more on trade trends at
www.hsbc.com/inthefuture

Issued by HSBC Holdings plc   AC22967

FIGURE 2.2 — HSBC / J. Walter Thompson London. In the Future. Global Supply. Photograph by Andy Rudak. Reprinted with permission.

knowledge and natural science but also social science to interrogate the limits of current social organization and social relations and experiment with creating an immediate experience that allows other alternatives to be imagined as possible.

By expanding on these works that bring social science to speculative fiction (*social* science fiction), I am moving in the other direction: to infuse speculative thought into social science research (social science *fiction*) in order to fabulate[70] about the autonomy of politics. Fabulation is more than the cultivation of visions of future societies; it is about experimenting with alternative tropes of enunciation and about putting in motion alternative worlds of existence. Social science fiction as I am using it here is about changing the conditions of experience in order to make alternative futures possible. The future is here. Biofinancialization is not just a mode of accumulation but a universalizing ontological machine of terraformation, one that changes all forms of life. Social science *fiction* is about doing scholarly theoretical and empirical research to mobilize fictional alternatives to a terraformed planet. Already in the introduction I have mentioned that the methodology of this book brings together research and speculative thought, and I use social science fiction to do this, not by introducing fiction into the stories told in this book but by telling theory in a way that could evoke and potentially contribute to make fictional worlds, which could always irrupt into our everyday lives. Social science fiction is social research, social analysis, and social theory told as fiction.

TERRAFORMING EARTH™

Semiocapital, biocapital, infocapital, neurocapital—it is inherent to productionism to seek and open new frontiers: endocoloniality.[71] The frontier of matter in biofinancial societies is not the same as the colonial appropriation of natural creations. Biofinancialization is not about conquering ecological resources that are necessary to sustain life; rather, biofinancialization is an experimental project entailing the constant remixing of the cultural, the biotic, and the abiotic. Bios in biofinancialization is not nature, it is flesh and matter. It refers to the fusion of code and matter. When this fusion touches on the very materiality of human and animal bodies and the geobody, it no longer seems to be able to be sufficiently described by the term *enclosure of the commons*.[72]

When the digital and the material, information and life are intrinsically imprinted onto each other, matter as such and the biota more specifically do not lie outside the process of biofinancialization. The commons are not external and against their enclosures, but neither can one say that there are degrees of enclosure; it is untenable to assume that some things are fully enclosed in the system of accumulation, while others are fully outside it and in between are things that are only partly enclosed (such as air, water, Internet, culture). When matter in its *ontological composition* is a frontier, both the commons and their enclosures exist *inside* each other. There is no commons versus biofinancialization but biofinancialization in the commons.[73] The subjectivity of working people, the body of the commons, and the ecobody of Earth are not separable from the current architectures of accumulation. Everything belongs 100 percent to the *res communes* and 100 percent to the current regime of biofinancial accumulation: terraformation.

The idea of terraformation, since its first appearance in the 1930s and 1940s, is not only a common science fiction theme.[74] It is also used to describe the science of remodeling and remaking the biosphere of a planetary body in order to make it hospitable to humans and to enable it to support life.[75] Terraforming in science fiction speaks to a territory—outer space—that is currently more desired and disputed than the Antarctic or the high seas of Earth.[76] Terraformation is as much about science fiction, technoscience, and colonization[77] as about geopolitics, value production, and the valuation and financialization of space.[78] Terraformation is not the vision of extraterrestrial geoengineering but a "local" project of controlled manipulation of Earth's ontology: Terraforming Earth, a vast capital enterprise whose primary actors, however, are not spectral entrepreneurs that apparently move and shape the world for the rest of us, but the processes that measure and evaluate Earth's spaces and matter.[79] Terraforming Earth is much closer to our realities than even its most dedicated believers, probably some NASA technocrats, would have thought. It is as close as the worlds that science fiction has morphed into our experience: social science fiction.

But unlike the vision of terraforming other planets, Terraforming Earth does not have a blueprint for action. Terraforming Earth does not have a preconception of what "Earth" is or can be. Earth is terraformed without a prototype and a plan. We don't know what Earth as such is or was or even what it is able to do or become as a precondition for terraforming it appropriately. Rather, Terraforming Earth is a simulacrum of itself; it is the opening of the frontier of matter, a practice of immersion in material experimentation, rather than of agency. Terraforming Earth is the outcome of the multiplication of climate change, acid oceans, the sixth extinction, synthetic biology, chemical pollution, extractivism, nuclear power, virtual space, big science, and the biofinancial logic that underlies social-material encounters.

MATERIAL ARTICULATIONS

Terraforming Earth has no master plan, only effects, that can be purportedly measured and their future impacts evaluated. The algorithmic moves of biofinance described in this chapter are only graspable as global motions on another planet as close as ours. On a planetary scale, Terraforming Earth unfolds without unified agency. As argued earlier in this chapter, humans as a species do not act as a subject by intention but by immersion, contingency, and involvement. This is the point where an understanding of an autonomous politics as the creation of alternative ontological interdependencies starts.

When ecologies of existence become terraformed, ontology returns to politics: *reclaiming* everyday materiality by actively recomposing and rearticulating it. I take inspiration here from Clifford's (2001, 2004, 2009) work on indigenous politics as rooted articulations (and disarticulations) of variously scaled histories, traditions, and practices on the uneven and variegated terrain of global space and time. But here I want to think of articulation beyond cultural practice and semiosis.[80] I am thinking of a practical process of articulation that operates on the level of matter, practices that disarticulate and rearticulate matter into unexpected organic and inorganic ensembles grounded in the material constraints of biofinancial life: How can the commons be expanded when they are fused with biofinance? How can terraformation become deuniversalized and matter decolonized?

Octavia Butler offers an alternative vision of organizing life in *Xenogenesis* (2000): ontological organizing, the creation of new couplings with other beings and things and new kinds of life able to respond to altered environments and to create livable words. If one wants to talk about autonomy in biofinancial societies, then this is about reciprocal becomings with other things, materials, and living organisms that let alternative ontologies of existence emerge. In Haraway's (2013) words, it is about creating "a seed bag for terraforming with earth others": terraformation from below. In the beginning of this chapter I borrowed the idea of terraformation to describe the current moves of finance as it opens the frontier of matter and appropriates bios. I described the material workings of biofinance and how it resides within the current socioeconomic nexus in order to question the reach of traditional autonomist politics that attempts to evacuate these conditions. My argument in this chapter and the departure point of this book, then, is that traditional forms of autonomous politics as we know them are unable to respond to the universalizing system that biofinance has inserted in our everyday life and material surrounds. Starting from this assumption, in the chapters that follow I develop an approach to autonomy as a practice that lies in compounds of algorithmic code, material processes, and bodies, not outside or against them but in the remaking of alternative ontologies and the reclaiming of their materiality. This is social science fiction, a set of empirical analyses and theoretical concepts that can allow us to modify our experience so that we can adapt and re-adapt politics and autonomy to conditions where there is no prior state of being to fall back or no "future wholeness which may yet save us" (D. B. Rose, 2004, p. 24).

03

ONTOLOGICAL ORGANIZING

Do I use Facebook to stay in contact with my family?—No, all you need is a mobile phone. At home, up there, they don't have anything except mobiles. Sometimes you just beep them so that they can see from your area code, where you are and that you've done a step further. In Facebook I have recovered some friends that I have lost for years—now they live in Paris. Last year, after the Pagani camp I wanted to continue to Germany together with a friend. We traveled through Macedonia and Serbia until Hungary, where we split. We prepared everything, we had every part of the route as a copy from Google Earth with us, printed in Internet cafes. And we used GPS on our mobiles. My friend took a train to Germany, but he fell asleep and had to drop out in Vienna where they caught him. I was arrested in Hungary and brought to a camp for six weeks. They threatened me to remain detained for years if I wouldn't want leave the country voluntarily. So I decided to return to Greece. In Serbia the police stole all of my money and my mobile phone and together with many others I was brought to a cell. Such a thing I didn't ever experience in Greece. When I finally arrived in Macedonia the police asked me if I was on my way to Serbia or to Greece. They showed me the path and even gave me some coins to make a phone call. I already spoke on the phone with a friend who through Evros came to Athens where he now lives. He tells me that actually it is very cheap in Evros, only $400. And this is certainly linked to the fingerprint questions. If you try to make it through the islands it is much more difficult without being fingerprinted. That's why it is more expensive. In Evros you can pass without much money and without fingerprints.

— INTERVIEW WITH SAPIK, LESBOS, GREECE, SEPTEMBER 7, 2010

When we think of autonomous social movements, we rarely understand them as forms of action that attempt to reorganize the material conditions of everyday life and the mundane ontologies in which they operate. Social movement politics are usually conceived as a form of oppositional political organizing that attempts to evacuate and/or challenge the policies of established institutions. In the previous chapter I argued that despite the centrality of such politics in social movement action, they seem to be unable to provide an alternative to the forceful permeation of everyday life and the environment by the universalizing system of biofinance. In this chapter I make the same case for an autonomous political practice that organizes alternative ontologies of existence from the perspective of social movement action. The previous chapter discusses how autonomous politics is related to current forms of control; this chapter approaches the emergence of alternative understandings of autonomy from inside current practices of social movements. To what extent can we approach social movement action as the practice of changing the existing material conditions of existence? How far can we go with the idea of ontological organizing?[1]

I turn to migrants' mobility as a site where such processes of organizing can be explored. The shared knowledge, affective cooperation, mutual support, and care between migrants when they are on the road or when they arrive somewhere constitute different practices that let organizational ontology emerge. I describe these flat mundane ontologies of moving people as the *mobile commons* of migration.[2] Sapik, in the interview extract at the beginning of the chapter, reminds us what it means to cross the borders into Europe. Once one is in Europe an even more brutally patrolled border stands in the way: European citizenship. Is it possible to challenge the existing political institutions of citizenship by organizing ontologically rather than by challenging constituted European social policies of citizenship? The ideas that I present here do not attempt to question citizenship and its possible importance in certain situations but rather to open, as Peter Linebaugh wrote, a chink in the wall and explore the possibilities that lie behind the horizon of politics that solely focus their action toward constituted political institutions.

For many, citizenship appears as a wall indeed. Citizenship is hard fought between those who try to restrict it and those who invest in the efficacy of citizenship as a potential guarantor of rights, justice, and liberation. Such critical investments can be found in the idea of citizenship

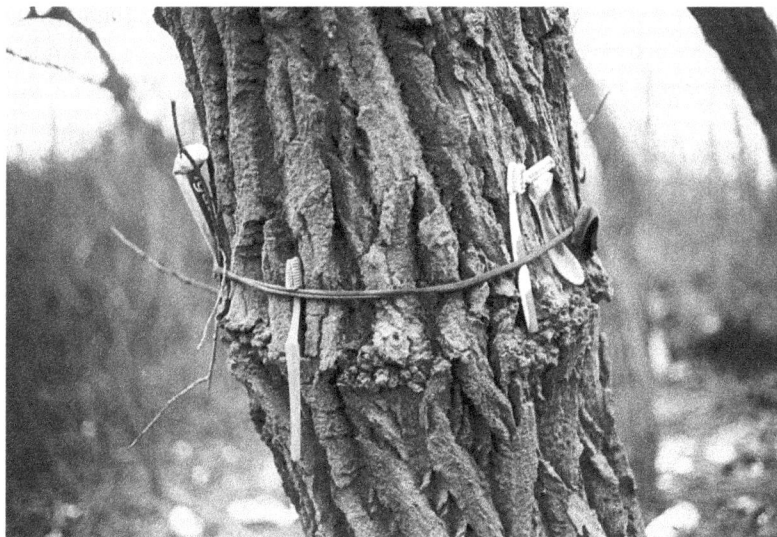

FIGURE 3.1 — Julie Okmûn / Contre-Faits. No Land's Men, the Struggle for Calais. Reprinted with permission.

beyond sovereignty and the state, which are discussed later, or in ideas of local citizenship, citizen labs, transnational citizenship, global citizenship, or acts of citizenship.[3] But however citizenship is defined, it can operate as a wall if it represents the ultimate horizon of political practice and social analysis. One could respond to the increasing securitization and abjection through citizenship with the introduction of another qualifying adjective to the concept of citizenship. But this is not the aim here. Rather, the methodological principle guiding this work is to see through the chink in the wall, to cultivate an imaginary and a practical sensibility to what lies before any claims for citizenship can be articulated and, most importantly, what lies after citizenship. What is an effective practice for challenging current forms of citizenship in Europe, the place in which this study of migrants on the ground and on the road—transmigrants—is located?

Throughout this chapter I argue that the efficacy of autonomous politics lies in the capacity of mobile people to bring to life mundane organizational ontologies of existence. In order to develop this argument, in the following four sections of the chapter I approach migration as a form of autonomous transnational mobility unfolding against the current regime for the control of movement. Then, in the last four sections I describe how this autonomous approach to mobility is practiced on the ground: the organizational ontology of the mobile commons. Seven photographs by Julie Okmûn accompany these thoughts.[4] I saw these photographs for the first time in 2009 in an exhibition of Julie's work that No Borders South Wales organized at the Oriel Canfas gallery and the Chapter Arts Centre in Cardiff, Wales. Julie's photographs convey a sensibility that is central to the argument of this chapter: the importance that the immersion in everyday transient lifeworlds and the involvement in the creation of mundane ontologies of life have for autonomous mobility and for making autonomous spaces of existence.

LABOR AND MOBILITY

For almost forty years now the response of established European politics to migration was to exclude mobility from the constitution of polity. Mobility was also seen as external to labor, which underlies most of the debates about migration politics; social position and class was thought independent of movement. But migration not only brings the current political system into turmoil, it also destabilizes class and recomposes what it is. The question of the past decades was how to tame and assimilate the supposedly alien migrants into polity. Now this question is rendered obsolete by the fact that as people did not stop moving, creating new lives elsewhere, mixing with the native working classes, and hybridizing everyday culture, they become effectively unassimilable. We are facing a different situation, one that is not concerned with how to immobilize migrants but with how to institutionalize mobility: that is, how to codify mobility, how to make it productive and sustainable, and how to combine it with the decline of sovereignty.

This is a moment when the cards of labor, mobility, and sovereignty are mixed and redistributed again. We used to think of mobility as a movement through space. And this is of course still true: migration is applied geopolitics on the ground. This spatial approach focused on the idea of territoriality in conceptualizing mobility. Consider the strategies of territorialization in the workhouse, which attempted to capture the wandering mob in Europe of the late Middle Ages,[5] or the first foreign worker hostels of the Gastarbeiter era,[6] to name just two examples. The governance of mobile populations is an important site for the exercise of control and the genesis of biopower.[7] The recurring pattern was the attempt to suffocate mobility by terminating it. Mobility-immobility was the driving conflict. In

these conditions immobility is associated primarily with territoriality, doc-
ile labor, becoming native, and integration within the local culture; mobility is conceived as sabotage, insubordination, escape, untrained work, and
multiple belongings. Nation-state/territory/people is the golden triptych
of modern sovereignty. But in conditions when mobility becomes a permanent and structural aspect of sovereignty, a new perspective on mobility
emerges: mobility as a movement in time.

When migration becomes tightly entangled with production and social reproduction,[8] the role of control is not to suppress mobility. Rather, it attempts
to render the speed of absorption into the local labor markets compatible
with the speed of flows of mobile populations. Migration control is about
speed and its regulation. It works as an equalizer between labor markets, political opinion, and migratory movements. For example, detention and deportation camps are less a form of blocking the circulation of mobility; they
reinsert irregular migration back into the productive logics of Global North
societies by making out of irregular mobility either controllable populations
or illegalized people.[9] Camps are *speed* boxes of migratory movements.[10]

From forced migration to managed migration during the 1950s and
1960s, mobility was governed productively by territorializing movements
and inserting them into the spatial regulation of working bodies.[11] As we
move to the temporal regime of mobility control, the main concern is to
transform ungovernable streams to governable subjects of mobility that
adjust to the needs of local labor markets and local demographics. These
needs are not "natural"—pure numbers depicting how much workforce
each market can absorb—but they are politically overdetermined by issues
related to security, nationalism, populist gambling of mainstream political
parties, labor policies, and so on. This is what the border regime does: it
is not there to block migration; it tries to institutionalize it by controlling
its speed and magnitude.[12] The temporal control of mobility is effectively
surpassing the sovereign governance of territories. As much as power over
a territory and the control of borders are considered the pillars of sovereignty, today's practices of mobility reveal that secure borders do not
and cannot exist. Sovereignty is the futile attempt to regulate the porosity
of borders: *porocracy*.[13] Even the heavy militarization of the US-Mexico
border after the 1990s proves "the incomplete, tenuous, and unstable nature of US dominion," as Gilberto Rosas (2012, p. 76) suggests in his powerful
ethnography of *Barrio Libre*, groups of young people who inhabit the sewer
system under the border of Nogales (Sonora/Arizona).

FIGURE 3.2 — Julie Okmûn / Contre-Faits. No Land's Men, the Struggle for Calais.
Reprinted with permission.

The turn to a temporal understanding of migration is crucial for the approach I develop here: migration cannot be stopped or fully territorialized; rather, it is a permanent and indispensable feature of Global North societies. Institutionalizing the temporal intensities of mobility is necessary in order to insert migration into labor in conditions in which spatialization has proved to be increasingly ineffective. Legal or illegal, regular or irregular, managed or unauthorized migration is directly entangled to production, labor, and its local contingencies.[14] So, in order to understand migration we need to rethink the changing forms of value production described in the previous chapter: the transition from the intensification of labor—that is, the duration of labor and the intensity of labor[15]—to the extensification of labor appropriation that involves the whole existence of the worker. Mobility is probably one of the most important and widespread factors of labor extensification: extracting value from the fact that bodies can become mobile in the most averse circumstances. Value production becomes, as argued earlier, embodied. This happens by creating different regimes and types of labor in order to differently insert specific segments of the mobile classes into diverse labor markets. Differential inclusion means that different modalities of entry into a country and different residence statuses—primarily through immigration controls and legal requirements—create different subjects of labor.[16]

DIFFERENTIAL INCLUSION IS CITIZENSHIP IS CONTROL

The differential inclusion of mobile populations points always to the way labor, mobility, and securitization are all directly connected with the machinations of sovereignty. The toll to govern this tripartite relationship is citizenship. Of course, the process of differential inclusion is not exclusively related to the modern politics of citizenship. On the contrary, differential inclusion accompanies multiple forms of belonging across different historical periods. The inclusion of the poor in the European medieval city; the temporary enslavement of white laborers in the British colonies; the freed black slave owners in the American South; the thin line between free and unfree as well as between waged and unpaid labor, which varies historically, socially, and culturally and produces different forms of social stratification; the different racisms that were mobilized to fragment black people and include them in variable positions in polity: all of these are just examples showing that differential inclusion is a contingent historical phenomenon.[17]

Thus, I use the idea of differential inclusion not to highlight its historical novelty but rather to argue that the specificity of today's differential inclusion functions through citizenship; the term *citizenship* is used here as a specific form of governance that regulates the relation between *r*ights and *r*epresentation. This double-R axiom appears as the foundation of modern polity.[18] Rights are considered crucial for governing migration (who is subject to rights and who is not is the primary way to create different segments of citizens). But representation has increasingly played a role in defining who is entitled to have rights and what kind of rights one is entitled to have. The cultural identity and the collective feeling of belonging of mobile or marginalized populations lead to the construction of an ad hoc social subject that then can become a subject of rights. Only through representation are rights possible. Citizenship is the form of governing this unstable and dangerous balance of the double-R axiom. Too much representation of a certain group (for example, Sans Papiers) without rights can create a potential explosive social situation because this particular group is socially active without having any legal, social, or political rights. A too-restricted representation of a social group makes exclusion and structural racism apparent (as, for example, the 2005 Banlieue uprising in France showed).

Imagine a scale where we have on the one pole full rights and on the other complete illegalization and invisibility. A cut is placed somewhere between these two extreme poles. This cut is citizenship. Where the cut is placed is a political question (for example, in the current conditions affected by the 2008 economic crisis and a broader conservative backlash across many European societies, the cut moves toward illegalization and invisibility). Citizenship is this toll of sovereign governance that regulates the balance between rights and representation and renders certain populations as legitimate bearers of rights while other populations are marked as inexistent. For example, Imogen Tyler (2010) discusses how selective British citizenship is by showing that the 1981 Nationality Act was designed to exclude the peoples of the ex-colonies by protecting only the right to British citizenship by those who had a lineage to someone born on the British isles.

We can think of the 1981 Nationality Act as a cut (that is, a particular configuration of citizenship) on this scale, which has on the one pole full rights and on the other complete illegalization in conditions of Thatcher's 1980s Britain. Once the cut is positioned, certain groups have different tools for changing the place of the cut, most importantly demonstrations, uprisings, social mobilizations, and protests (and sometimes academic research can contribute to this too). The Brixton riots of 1981 and the broader civil unrest of that period can be read as a response to the exclusionary design and function of the 1981 Nationality Act. More generally we can say that it is through all these struggles that sovereignty is pushed toward the pole of full rights. And there are always long periods of backlash when there is a growing anti-immigrant sentiment and the cut is pushed back toward the pole of illegalization. So far I have established that the temporal regime of mobility control reveals that migration is an inseparable feature of sovereignty and that it is through citizenship that the temporality of mobility is controlled and the speed of inclusion/exclusion in the sociopolitical system of a certain society is managed.

THE IMPOSSIBLE CITIZENSHIP

There is a paradox in this function of citizenship as the regulatory mechanism of inclusion and exclusion: the more a society moves toward citizenship, the more it creates the conditions for its disappearance as a form of governance.[19] If you include everyone and if you assign rights to everyone, citizenship becomes obsolete. "Citizenship for all" is an impossible term; it is always "incomplete" (van Gunsteren, 1998). Or else, imagine a society that assigns citizenship to everyone. In this fictional society citizenship is not connected to rights or any other legal status; it is a mere social ritual. Citizenship would be granted automatically to every denizen, and to the extent that, as in any society, rituals for social cohesion are important, everyone who wants to demonstrate a strong sense of belonging to this society can buy in almost every convenience shop a Home Office Citizenship Medal for £8.99. You can wear it every day or just forget it in a drawer or lose it. This fictional society would be very different, of course, than the societies we know. But probably the most important difference is that this society would not have borders. To think this the other way round: citizenship coexists with borders. Citizenship coexists with the exercise of sovereign control, as Bridget Anderson, Nandita Sharma, and Cynthia Wright (2009) show in their research. The more we talk about security, the more we talk about citizenship. This is the predicament of citizenship. It feeds from the power of sovereignty to erect and maintain borders, borders that it cannot ultimately control. Citizenship cannot be thought to be outside sovereignty and control.

Julia O'Connell Davidson's (2010) work on trafficking exemplifies this function of citizenship from another perspective, namely how in the name of protecting human rights and liberal citizenship, sovereign control promotes a tougher take on the freedom of mobility and leads to the introduction of restrictive migration measures as pro–human rights policies. In this sense we can think of citizenship as a form of governance that performs explicitly exclusion (alongside with its differential inclusion function). Whatever qualifying attribute we add to citizenship—activist, irregular, imperfect, biological, sexual, unrecognized[20]—it cannot avoid an optic that looks at people's movements from the perspective of control.

The vision that citizenship is inherently liberal can be historically revealed as a fiction. There is no global unified citizenship because citizenship exists only as one of the tools that are deployed to maintain national sovereignty. It is thus limited to the territorial space of the nation-state and stops where the borders of a country stop—while the rest of a country's activities (such as capital movements, trade, circulation of elite populations, and war) can extend beyond its borders. The limits of citizenship are the limits of sovereignty. But liberal citizenship is problematic not only because it excludes by design everyone who is outside its borders and

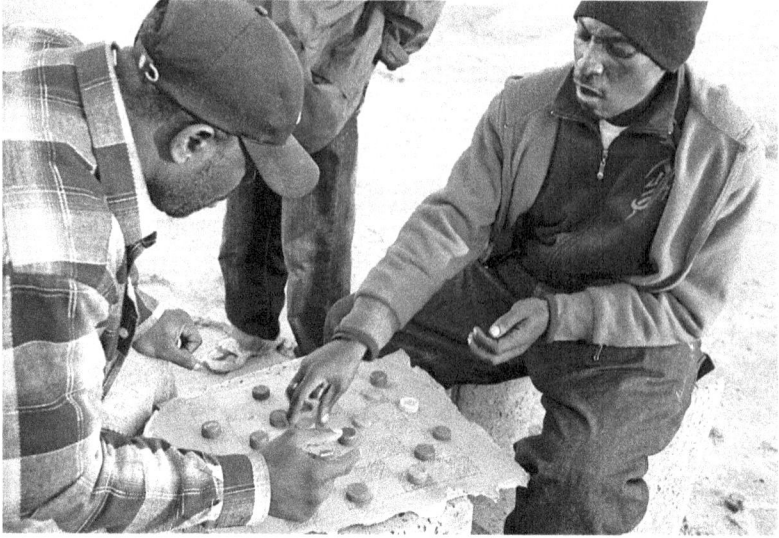

FIGURE 3.3 — Julie Okmûn / Contre-Faits. No Land's Men, the Struggle for Calais. Reprinted with permission.

differentially includes the denizens inside its borders but also because there is a long history of actively "denationalizing" dangerous or unwelcomed citizens—citizenship here can be viewed as "accidental" (Nyers, 2006) or "reversible" (Tsianos and Pieper, 2011). Elsewhere we used the term *postliberalism* to discuss how these ambivalences of citizenship push liberal democracies to their limits (Papadopoulos et al., 2008; see also chapter 2).[21] In postliberal conditions citizenship is not defined by its values but by its reach: it has to be always protected from expanding too much and including people who in certain political conditions cannot be considered citizens.

Thus, understanding and theorizing migration in terms of differential inclusion and citizenship is an important tool for creating possibilities for certain groups to be included in polity. But it can never respond to the question that migration poses to sovereignty: what about those who are mobile and cannot be included—that is, the majority of transmigrants? Here the relevance of my discussion of migration and citizenship becomes apparent for the argument of the book. If politics that addresses primarily social power becomes increasingly entangled in what it contests and fails to create social transformation, then an alternative view of politics emerges: politics as ontological organizing.

How is it then possible to shift our perspective from the order of sovereign control to the primacy of migrants' mobility—that is, to read the current regime of production through migration and to understand sovereignty through mobility, rather than the other way around? This shift represents probably the most important insight of the autonomous approach to migration: the attempt to see migration not simply as a response to political and economic necessities but as a constituent force in the formation of polity and social life.[22] Yann Moulier Boutang (1998) has offered an impressive account of this movement historically. The autonomy-of-migration approach foregrounds that mobility is not primarily a movement that is defined and acts by making claims to institutional power. It rather means that the movement *itself* becomes a political movement and a social movement. The autonomy-of-migration thesis highlights the social and subjective aspects of mobility before control. It rejects understanding migration as a mere response to economic and social malaise.[23] Instead migration is autonomous, meaning that it has the capacity to develop its own logics, its own motivation, its own trajectories that control comes later to respond to—not the other way round.[24] Of course, this does not mean that mobility operates independently of control. Very often it is subjected to it and succumbs to violent state or private interventions that attempt to tame it; probably the politics of detention and deportation is the best example of such violence that shows how migrant mobility can be halted and brutally controlled.[25]

There is no space for romanticization of nomadism and mobility in the autonomy-of-migration approach. Migration grapples with the harsh, often deadly, realities of control. However, migration is not just a mere response to them. Rather it creates new realities that allow migrants to exercise their own mobility with, against, or beyond existing control. In this sense, the autonomy-of-migration thesis is about training our senses to see movement before capital (but not independent from it) and mobility before control (but not as disconnected from it). One of the most common critiques[26] of the autonomy-of-migration approach is that it substitutes all these different migrant subjectivities and the diverse concrete spatialities of movements into a new big narration of migration. The term *migration* supposedly homogenizes and effectively erases the diverse lived experiences of migrants vis-à-vis the state. Of course, migration encompasses a broad spectrum of practices of mobility: humanitarian, forced, war, environmental, cultural, economic, circular, seasonal, internal migration—all these are radically different types of mobility. However, all these mobilities are not neutral definitions of migrational movements.

Mobility is often polyvalent, complex, and open to different classifications depending on the perspective from which it is approached.[27] For example, where and how a young transmigrant can be classified as an unaccompanied refugee minor or as somebody who circulates between the country of origin and the current country of destination or as an economic migrant is less self-evident than it appears in the first instance.[28] Subsuming all these different types and cases of mobility under the concept of migration does not mean flattening out their differences; rather it attempts to articulate their commonalities, which stem from all these different *struggles for movement* that confront the regimes of mobility control. The supposedly abstract and homogenizing category of migration does not attempt to unify all the existing multiplicity of movements under one logic but to signify that all these singularities contribute to an affective and generic gesture of freedom that evades the concrete violence and control of moving people. *Migration* in the autonomy-of-migration approach refers to a kind of politics that entails neither uniformity nor abstraction; rather, it relies on struggles for movement that escape and subsequently delegitimize and derail sovereign control.

The first meaning of the autonomy-of-migration approach is an empirical one: the real struggles, practices, and tactics that escape control. This approach to migration highlights the heterogenizing practices of state regulation of mobility: sovereignty breaks the connectivity between multiple mobile people in order to make them visible and render them governable subjects of mobility. And it does this through operationalizing the category of citizenship in order to create different classes of citizens. The heterogenizing effects of power should not be confused here with the multiplicity of mobile subjectivities and struggles. These are effaced at the expense of making clearly defined heterogeneous objects of governance. The second meaning of the autonomy-of-migration approach is an affective one: migration nurtures the belief in the possibility to be free to move. This second meaning of *migration* in the autonomy-of-migration approach is speculative. It is a speculative affect that embodies a virtuality as secure, free, and warm as it can get in the harsh conditions of sovereign control. Migration in this second sense is more related to an affective imaginary; it exists as potentiality and virtuality that becomes actualized and materialized through the diverse movements of people.

"I WORK ONLY FOR PAPERS"

Having established the current temporal regime of mobility control vis-à-vis the autonomy of migration, I discuss in the remaining sections of this chapter how autonomy is claimed, practiced, and sustained. This can be best exemplified in an emblematic type of mobility: illegalized border crossing. When migrants are considered irregular citizens, they are commonly conceived either as criminals or as being forced to move, not as active creators of the realities they find themselves in or of the realities they create when they move.[29] This constructs them as irregular or unauthorized subjects. It is not primarily the legal context that creates the category of the illegal migrant, but rather a specific political and theoretical view that does not allow for agency that is not driven by external necessities; the legal context only consolidates this perspective. Irregularity is a practice of governance that *illegalizes* migrants in order to control them through the current arrangement of borders and citizenship.

In conditions where illegalized migration has become one of the primary migration routes to Global North societies,[30] citizenship effectively limits freedom of movement by creating a minority of migrants eligible to access citizenship and a majority of abject and superfluous foreign aliens. Clandestinity then becomes a means to maintain the possibility of movement in conditions in which migrants are illegalized through the temporal order of sovereignty and the governance of citizenship. This raises an enormous political issue when we are confronted with migration today: the more one tries to support rights and representation through citizenship, the more one contributes to the restriction of movement. This dilemma is well known to activist organizations that engage with migration and radical border politics.[31] The politicization of irregularity effectively contributes to its enforcement. From the perspective of mobile people, confronting illegalization is not an intended (or even unintended) political act.

The dilemma is that migrants do not usually get involved in political mobilizations about migration *as such*. Migrants tend to become invisible, to disappear, to dis-identify themselves.[32] When migrants mobilize politically within traditional established political institutions, they do it only in a strategic way to challenge a particular and direct form of discrimination in a concrete situation. Many of the transmigrants in the camps of Pagani and Igoumenitsa in Greece—where most of the empirical materials that underlie the arguments presented in this chapter were gathered—used different versions of the phrase "I work only for papers." Initially it was difficult

FIGURE 3.4 — Julie Okmûn / Contre-Faits. No Land's Men, the Struggle for Calais. Reprinted with permission.

to understand this phrase: On the one hand we know that a lot of them work in the worst possible conditions, without being documented and, of course, for money. On the other hand, "papers"—that is, the documents that one needs in order to make it to the target destination—are not something that you "work for"; rather, we think of "papers" as something that one is legally entitled to (or not). But these transmigrants challenged two of the most widespread assumptions of what a migrant is: first, that migrants are only workers where their subjectivity is defined by their capacity to offer their labor power in "foreign" labor markets, and second, the distinction between legality-illegality by questioning the dualism between those who are legal subjects of citizenship (if they have "papers") or illegal subjects outside citizenship (if they don't).

These transmigrants turn both of these assumptions on their heads: not only is work secondary for their subjectivity, but they see that the *actual* work they do is the *work* for acquiring "papers"—something that Ellie Vasta (2011) describes as "irregular formality" and the "paper market" to articulate the fluidity between irregular and regular statuses from the migrant point of view. This is a double blasphemy against the logic of labor as well as the logic of citizenship. In fact, these transmigrants do not even intend to play the game of political participation in our established institutions or engage in acts of citizenship through mobilizing their subjectivity as citizens or as workers. Or we could even say that they would engage in any act of political participation if this would help them get the "papers" that they need.

IMPERCEPTIBLE ONTOLOGIES

The forms of political action that migrants engage cannot be confused with a mobilization that resembles the action of a collective political subject. The conditions of current migration defy the possibility of constructing a viable intentional and permanent political subjectivity, whether it is a liberal governmental subject or a radical subject of social change. To the extent that one cannot build liberal societies with migration, one cannot do traditional left politics with it. It is impossible to adapt and incorporate migration into our own typical representational political projects be they right, left, liberal, or radical left. And if this happens it will only be for a short period of time, until a specific group of migrants have achieved their strategic political goals for a certain issue, until the new migration wave arrives, until new relations of care between mobile migrants are built on the ground, or until new transnational mobile communities emerge that undermine any permanence of classical representational politics.

The specter of migration will never become a new "working class." It will always remain a specter, which comes in the night through the back door of your nation on a smuggled vessel; by using false papers; by crossing hundreds of miles of mountains or deserts; by changing one's own identity; by destroying the skin of one's own fingertips with acid and a knife to avoid identification; by overstaying a visa, an au pair contract, or the regular tourist period of stay. The specter of migration will always remain a specter, though one that is much more present than any of the political ghosts summoned in the history of political thought and political struggle in order to satisfy nationalist sentiment and fulfill the desire for securitization or revolution alike. The specter of migration will always be with us, among us, more real than anything else: cleaning your home, cleaning your office, cleaning your roads, cleaning your buses, taking care of your children, developing the software of our devices, repairing your devices, fixing your car, providing sex, providing babysitting, providing care, ironing your shirts, answering your phone calls, doing your gardening, building your house, collecting your strawberries, working in the abattoirs, living in the flat next door. Migrants do not hold the place of a historical or a political subject as such; rather, they tend to become imperceptible to history.[33] But the more they do this, the more they change history by undermining the sovereign pillars of contemporary societies.

The approach presented here breaks with the dominant integrationist canon of migration studies, which maintains the fundamental assumption that migrants' practices become political only if they become integrated into an existing political order—be it in the country of origin, the country of destination, or one of the countries through which transmigrants pass. The cohesion of this polity is taken for granted, and migrants' political practices are considered political only if they address and operate in it.[34] So, what kind of politics do migrants do if it is not gravitating around integrationism? What are the politics of migration when they cross borders? What kinds of politics are performed when people become mobile despite the restrictions of migration controls? What kind of politics characterize all these migrant practices that attempt neither to integrate people into an existing polity nor to systematically resist this polity?

Following Rancière (1998), migrants' political practices could be conceived as attempts to create a new situation that allows those who have no part to enter and change the conditions through which social existence is perceived, conceptualized, and experienced. How else can we understand the silent and mundane transformations that happen when migrants who clandestinely defy the borders that block their future expose the limits of liberal citizenship without ever intending it?[35] These politics transform the political without ever addressing it in its own terms and practices. Migrants' politics develop their own codes, their own practices, their own logics that are almost imperceptible from the perspective of established political practice: first, because we are not trained to perceive them as "proper" politics, and second, because they create an excess that cannot be addressed in the existing system of political representation. But these politics are powerful enough to change the conditions of a certain situation and the conditions of existence of the participating actors. These are politics of ontological change, a politics that bypasses existing constituted politics to de facto transform the materiality of existence.

Migrants' politics of ontological change are in this sense *nonpolitics* (that is, nonrepresentable in the dominant existing polity). With Asef Bayat we could call them "social nonmovements." In his work on social and political mobilizations in the Muslim Middle East in the 2000s, Bayat (2010) describes the invisible everyday activities that prepared all these radical transformations—nonmovements because for years they were sustained and nurtured silently through making invisible alternative spaces, through changing everyday life, through seemingly nonpolitical experiences and actions of people. When these nonmovements were confronted with the brutality of the state, they crafted a nonidentitarian collectivity of insurrection. In a similar vein, Raúl Zibechi (2011) describes the struggles of the urban poor and the indigenous movements in South America as antirepresentational politics. Their aim is to appropriate and self-organize social territory in cities or rural areas in the midst of a strict and immovable order of political and social power. These struggles create, in the words of Zibechi, postcapitalist "societies in movement."

The mundane gestures of sociality that nurture people when they are on the move or arrive and try to settle in a new place, these "societies in movement," are imperceptible from the perspective of an existing polity. The more migrants become imperceptible and the more they disidentify from their externally assigned identity as (illegalized) migrants, the more they become like everyone. Becoming everyone is the end of citizenship.[36] The moment when you buy your Home Office Citizenship Medal for the price of £8.99 in every corner of the country will be the moment in which freedom of movement will be a reality. But becoming everyone is not an event to come. It is not secular rapture awaiting, it is a generic strategy of mobility when it moves through places and continents and even when it becomes clandestine and passes through the biopolitical controls of sovereignty. Brigitta Kuster (2016) offers an impressive account of different modes, figurations, and techniques of disidentification, of becoming imperceptible and everyone in Mediterranean border crossings. Becoming everyone is a magical moment of transformation of the securitized objectivity of the present. It is a move based on respect and care of the worlds we are creating when we leave behind marked social positions and selves; becoming everyone is a necessary strategy of everyday survival for migrants on the road and for migrants facing racism when they try to settle in a place.

Crossing Calais—the last European border before entering the UK—for example, can be seen as an "act of citizenship" (Isin and Nielsen, 2008) only to the extent that the moment of hiding in a lorry is an illegalized activity. From the perspective of migrants, this is an act of immediate *justice* for sustaining their everyday life.[37] Let us put it in a different way: to the extent that migration undermines the securitization of sovereignty by its very existence, it also undermines the conservative, liberal, left, or radical left political projects and announces—together with many other social movements, of course—a different form of politics. This sounds perhaps disappointing for some, but there are many reasons to celebrate.

Migration is forcing us to repudiate the implicit avant-gardism of earlier versions of the autonomy-of-migration approach[38]—an avant-gardism that by attempting to improve citizenship and change governance tries to realign migrants with the working classes (as in the motto "Migrants and precarious workers together!") and to resurrect a new social protagonism of migration. Of course, there is a growing proximity between migrant labor and precarious labor, since migrant labor becomes increasingly precarized (especially after the 2008 economic crisis) and precarious labor becomes increasingly mobile. However, if there is a potential for transversal politics between the worlds of migration and precarity, this is not in a form of solidarity or in the creation of a new hybrid political subject.[39] Rather, I believe that where migrants and precarious workers meet is in the fact that they share the same spaces—urban spaces, material space—and that both of them, from their different positions and with different aims, participate in the metropolitan uprisings of European cities that remake the everyday ontologies of our lives.[40]

THE GIFT ECONOMY OF MIGRATION

On August 27, 2009, together with an Amnesty International representative, a meeting was organized with five young transmigrants in one of the central café snack bars of Mytilene, the capital of the island of Lesbos (Greece). Lesbos was at this time (and is even more at the time of writing this chapter) a heavily used route for crossing from Turkey into Greece. As a result the detention camps were overcrowded. In response to this situation, a no-borders mobilization was organized on the island in August 2009 and resulted in the closure of the main camp.[41] Many of the detainees escaped the camp without being registered and having their fingerprints entered into EURODAC, a centralized Europe-wide database of fingerprints of asylum applicants and illegalized immigrants.[42]

All five migrants in the meeting that afternoon were women in their twenties coming from different cities in the Horn of Africa. They said that some of them had previously worked as domestic carers and workers in Saudi Arabia and Dubai. The working conditions there were very bad, so they decided to migrate again, this time to Canada because relatives and friends told them that they had better experiences as domestic workers. They used different routes to arrive in Turkey and then eventually crossed the EU border to Greece on a boat. They were intercepted by Frontex patrols, the European border security agency, and had to destroy their boat so that they would be transported as shipwrecked asylum seekers to a camp in Greece. They preferred this because they were confident that they would be able to meet other people on the move in the camp and check possibilities to continue their journey, rather than simply be arrested and returned immediately back to Turkey.[43] They were interned in the Pagani camp. The Amnesty International representative explained that they would be released after thirty-eight days with no formal procedure for claiming asylum but on the condition they leave the country voluntarily and return to their countries of origin.[44]

The most striking aspect of this encounter was that none of these five migrants looked or behaved in a way that would fit the image of the typical illegalized victim circulating in media and mainstream politics. They protested against detention and complained how they were treated by the border police, but one could not see the picture of misery, exploitation, and oppression that they had suffered while they were interned in horrendous conditions in the Pagani camp. Rather they looked tired, calm, decided, and optimistic. When they were asked how they were going to spend the rest of the day, they replied that they did not know, but the next thing they were going to do was to go to a cybercafé to check email and their Facebook accounts: "Making connections. Making our route," they said before leaving. (Later my collaborator on this project learned that some of them are working not in Canada, as they had originally planned, but in Norway.)

In the same way that "migrants as agents" do not do the politics we expect them to do, "migrants as victims" do not behave as victims should. Rather than being isolated, individualized victims, these young women appeared to negotiate the difficult and dangerous lives they live through continuous recourse to the idea of "social connections" that would help them move on and continue their journey. In a strange way there was a feeling that when they were talking they were referring to a "we," without ever describing it, a "we" that had the potential to recode or even interrupt the logic of the border control and detention. Virtual spaces such as chat rooms, Facebook, emails, and encrypted communications as well as the spaces of the camps and of migrant neighborhoods help one stay mobile, collect information about routes and possibilities for survival, and learn tactics of existence.[45] This knowledge and affective reservoir offers vital resources and energies to migrants on the road or when they arrive in a new place. In this chapter, I refer to this as *mobile commons*: a shared affective, informational, technological, financial, cultural, material place that does not exist as a given but needs to be continuously updated and extended; the innumerable uncoordinated but ontologically transformative actions of mobile people contribute to its making.

People on the move create a world of spaces for rest and recovery, knowledge, information, tricks for survival, mutual care, social ties, exchange of services, solidarity, and sociability that can be shared and used freely. This world facilitates Sapik's movements as described in the interview excerpt at the beginning of this chapter. This world is both analog and digital, technoscientific and low-tech, actual and virtual, infrastructural and cultural—all mixtures of knowledge, technology, materiality, and affect that sustain the mobile commons. However, not only does Sapik use all these invisible resources to remain mobile, but by doing so he expands and circulates this intelligence for other mobile migrants. This contribution is related neither to the good intentions of those who participate nor to a presumable solidarity "reflex" between migrants. Mobility is by definition a process that relies on a multitude of other people and things. This extreme dependability can be managed only through reciprocity, and reciprocity between migrants does not mean exchange; rather, it means to multiply access to mobility for other transmigrants. Multiplying access is the gift economy of migration. This is the world of the mobile commons. This is a second world, *World 2*, beyond the world most of us experience as subjects of rights, as citizens, as political activists.[46] World 2—the world of transmigrants whether they are on the road, in a new country, or in a new neighborhood, whether they are settled, are clandestine, have refugee status, or are documented workers—is always a world in the making.[47]

INTELLIGENCES AND INFRASTRUCTURES OF MOBILITY

The autonomous politics of organizing the common worlds of migration goes beyond the traditional question of mobilizing migrants in existing institutions such as trade unions, civil society organizations, or traditional social movements against their oppression and for social rights.[48] Rather, the movement of migrants becomes a social movement when it creates alternative everyday forms of existence that facilitate people's freedom of movement. Migration becomes a social movement when it extends its own possibility through the multiplication of the mobile commons. This

FIGURE 3.5 — Julie Okmûn / Contre-Faits. No Land's Men, the Struggle for Calais. Reprinted with permission.

process of multiplication, augmentation, and circulation of the mobile commons is ontological through and through—that is, it entails several activities and practices that change the immediate material conditions of people's movement: the circulation of intelligences of mobility, infrastructures of connectivity, informal economies, communities of justice, and, finally, the nexus of care. I rely on the research presented earlier in this chapter, my collaboration with Vassilis Tsianos and the work of Gabriella Alberti (2011), Hywel Bishop (2011), Margherita Grazioli (2017b), Martina Martignoni (2015), and Fredy Mora-Gámez (2016), who all have taught me a lot about the incredible multiplicity and variability of the practices that sustain the mobile commons and the lives of migrants.[49]

The *intelligences of mobility* that circulate between people on the move comprise a diverse set of embodied as well as codified knowledges of each migratory route: conditions of border crossing, shelters, meeting hubs, escape routes, and resting places. Clandestine people depend on the experiences of migrants who "walked the route" to learn about the specific forms of policing in different areas, ways to defy control, strategies against biosurveillance, and places to get updated information and help. Also, transmigrants attempting to settle in a place rely on mobile knowledges about local communities and their specific customs, available modes of social support, educational resources, access to health, housing, ethnic networks, microbanks, and so on. Knowledges of mobility involve not only the circulation of contents but also ways to capture, transfer, and share all this wealth of collective and distributed intelligence. Such knowledges of mobility can be found condensed in certain places more than others. I already mentioned that detention and deportation camps often condense large amounts of such intelligence and ways to accumulate, store, and distribute it. Certain mobile people often carry the wisdom of mobility—sagacious people such as Sapik mentioned in the beginning of this chapter—but also text messages, encrypted chats, maps, or hidden Internet pages. Knowledges of mobility exist as long as they are embodied within diverse, lively infrastructures.

Infrastructures of connectivity maintain the circulation of collective intelligences of mobility but also facilitate the technological means, practical logistics, and material resources of support to stay mobile or to settle in a specific place. These infrastructures allow the setting up of secure spaces for collecting, updating, and evaluating knowledge by using a wide range of technological and informational platforms and media—from the mouth-to-mouth traveling of embodied knowledge to locally organized exchange hubs to social network sites, geolocation technologies, alternative databases, and communication streams. Infrastructures of connectivity are ontologically present and ontologically transformative as encrypted web platforms or as simple Facebook pages; as secure communication channels by using Tor; as whispers across barbed wires; as cafés, squares, or rented flats; as elaborated maps, or as traveling story lines, all passing through the hypersurveilled European space. Grazioli (2017a, 2017b) in her study of housing rights movement in Rome has investigated how squats empower autonomous politics in the postwelfare metropolis. As people, many of them migrants, reclaim abandoned or empty buildings they also reclaim their right to the city. And this happens as they remake the ontological fabric of everyday life in the squat and in the neighborhoods around it. A squat becomes a prolonged act of reappropriation: an autonomous infrastructure, a childcare facility, a subsistence garden, an infopoint, a space for support, a hub for exchanges of all sorts, an experimental art space, a production site, a place for sociability, a home. And it also extends beyond its own limits by changing the communities around it. Many squats create networks with other local grassroots associations, citizen initiatives, environmental campaigns, social rights struggles, and grassroots redevelopment projects. As autonomous infrastructures multiply, squats become ontologically embedded in their local surroundings by attracting new migrants and becoming hubs for the distribution of intelligences of mobility, by changing how public space is used, by contributing to the construction of new communal facilities, by changing how basic services such as access to water and electricity are managed, and by remaking the physical and social environments in which they exist.

INFORMAL ECONOMIES AND COMMUNITIES OF JUSTICE

Another form of activity that becomes visible in these squats and in similar spaces mentioned throughout this chapter are the *informal economies* that emerge within them. These comprise all the economic activities and services that cannot be easily accessed through the public sector or even privately: how to find (and let alone pay) a doctor or a lawyer; how to find short-term work or more permanent working arrangements; how to send and receive money; how to communicate with friends, family, and fellow travelers; how to make it through the economies of smuggling, get the necessary papers for your move, or pay your rent. These economies facilitate access to formal means of exchange such as money or credit but also deploy alternative exchange systems, barter, or nonreciprocal support. Mobility as well as the attempt to settle in a specific place can be sustained only through such informal economies spread along migration routes. These economies, although unregulated and often invisible, do not exist outside existing relations of production and reproduction. As much as they perpetuate existing modes of exploitation, they also sustain migrants who are unable to access formal employment or commonly used systems of exchange.

Political campaigning within the mobile commons is organized within transnational *communities of justice*. Such communities are built through alliances and coalitions of often very disparate groups such as the migrants themselves, local governments, political organizations, NGOs, activist networks, and civil society organizations. They engage in a range of activities, from organizing protest camps or support actions for migrants to formal political campaigning within established institutions. The port of Calais, the closest point between the UK and continental Europe and a primary hub in the migration route to Britain, provides a good example of how such communities of justice emerge and change over time. The beginning of political campaigning in Calais was about providing overnight shelter and food to immigrants. Initially, coalition building involved the local population, local charities, and in some rare cases the city council. As the tensions between municipal government and the migrants intensified toward the end of the 1990s and in the 2000s, Calais became a so-called humanitarian issue[50] and several larger charities became active, such as the Red Cross, which for a long time provided food and more formal accommodation (for example, in the Red Cross center near Sangatte).

When this and other similar accommodation centers were closed down, migrants self-organized to create their own makeshift shelters within what came to be called the "jungle villages." The Calais "jungle" is one of these ontologies created by transmigrants that has changed irrevocably what is migration politics in northern Europe. Its sheer presence has certainly defined debates about migration, public opinion, and migration activism in the UK in the 2000s and 2010s. The intense political struggles around the "jungle" created new tensions and a new phase of mobilization with the participation of numerous charities (such as Association Salam, La Belle Étoile, L'Auberge des Migrants), civil society organizations, and activist networks such as Calais Migrant Solidarity and the No Borders network. The local police have fully or partially destroyed the "jungle" several times in the past fifteen years, and at the time of this writing there is a major plan to clear the camp completely. The precariousness of the migrants in these conditions has attracted a series of different "borderworkers" (Rumford, 2008) and "border activists" (Walsh, 2013), nonstate actors that maintain life along the ever-proliferating borders within Europe. Borderworkers and activists carry out many different activities: providing medical advice, legal advice, asylum application support, and English language classes as well as engaging in direct political activities such as challenging the decisions of the mayor of Calais and the local police, demonstrations, occupations, camping in front of the "jungle" to protect its occupants from police raids, and media work. Communities of justice are ad hoc assemblages of traditional, radical, and experimental politics that seek to translate the ontologies of migrant life to forms of social action.

THE NEXUS OF CARE

Relations of care appear as a continuous thread weaving all these activities and spaces together—intelligences of mobility, infrastructures of connectivity, informal economies, communities of justice. The *nexus of care* implies care as the generic practice of caring for others[51] as well as the immediate

FIGURE 3.6 — Julie Okmûn / Contre-Faits. No Land's Men, the Struggle for Calais. Reprinted with permission.

everyday practices of care. Bishop (2011) in his work on the politics of care and transnational mobility investigates how all these neglected, marginalized, almost invisible instances of care between migrants on the move or in a given location become indispensable for holding together all other dimensions of the mobile commons. Mutual cooperation; friendships; favors that you never return; affective support; trust; taking care of other people's children, relatives, and the elderly; informal arrangements of social reproduction such as ad hoc nurseries; mutual support replacing lack of access to welfare services; transnational care chains; remittances; togetherness across geographical space; the gift economy between mobile people—the nexus of care holds the ontologies that transmigrants erect together.

As much as the mobile commons—intelligences of mobility, infrastructures of connectivity, informal economies, communities of justice, the nexus of care—is imperceptible for most people in European societies, it is not abstract. The mobile commons is concrete and practical. It exists only as much as it helps to install relations of justice (as discussed in chapter 1) in the midst of sovereign control. From the perspective of migration, justice *is* the making of the mobile commons—all these daily social relations, connections, and conditions that evade the control of mobility. Justice here resembles an affective index that designates how appropriate are the means used to arrive somewhere and the limits of what one can endure throughout this journey, and, most importantly, it indicates what is just and unjust in conditions that are by design outside formalized law. The justice of the mobile commons is the moral economy of migration. It is similar to E. P. Thompson's (1971) moral economy of the poor: the immediate feeling and judgment of the crowds about what is just and what is unjust in relation to the everyday conditions of existence, such as the price of food or the prohibition of using the commons. From the perspective of migration, justice cannot be achieved only through the assignment of rights and citizenship or through attempts to organize migrants in unions, political parties, or civil society organizations (however important and indispensable this might be in certain conditions). For transmigrants justice is achieved by changing the ordinary ontologies of existence in a way that allows people to move when they want to or need to and to maintain a livable life when they reside in a certain place.

Throughout the chapter I used the term *ontology* to refer to a form of organizing that is about the creation of thick everyday performative and practical justice so that everyday mobility, clandestine or open, becomes possible. I know that the question of ontology sits uncomfortably with migration and other social movements more broadly: migrants and social movements, if at all, change society, not ontology. In chapter 2 I tried to reverse this perspective and to argue that the current configuration of social and political power creates the conditions to approach social movements as ontological organizers. In this chapter I have reconstructed this argument from within the social movement of migration. But one can still ask: why is the mobile commons ontological and not social? It is ontological because migration in today's Global North changes society only to the extent that it changes the material conditions of existence so that people can cross borders, remain mobile, survive the violent pressures of sovereignty and the arbitrariness of borders, and arrive at their preferred destination. It is ontological because as it passes through borders and traverses territories, it remakes ecologies of existence in ways that defy the ever-increasing geosurveillance.[52] But again, the ideas of "material conditions" and "ecologies of existence" here can have different connotations: they can refer to a politics of matter, as discussed in chapter 1, or to some form of materialist politics. In the next chapters I discuss this tension between materialism and a politics of matter, exploring how they increasingly come together to reveal that social movements are fabricators of ontologies.

FIGURE 3.7 — Julie Okmûn / Contre-Faits. No Land's Men, the Struggle for Calais.
Reprinted with permission.

"THERE IS NO LOVE HERE!"

Since our first meeting with Sapik in the Pagani camp in the summer of
2009, we have had regular contact on the phone or via the Internet. He became a co-researcher and a research adviser. Sapik has an active Facebook
life, and his account is linked to a well-informed and useful blog about mobility and transit issues relevant to his peoples. Sapik is a true commoner
in the mobile commons. Suddenly, while we were writing this, it became
impossible to contact him. We were very concerned. For many years now
there has been a steep increase in fascist and racist attacks in Greece, and
Sapik could be one of their targets since he is a well-known and active
figure in his community. Thankfully he contacted us and said that he was
doing well. He had left the island and moved to Athens.

He then said that he was very scared when he experienced the racist
riots in Athens. But he went to Athens because he wanted to understand
"what is happening in this country." He was not hopeful that the big mobilizations against the government and the imposed austerity measures in
May and June 2011 would be successful. He was proved right. His voice was
quiet. We asked him when he would go back to the island where, at least in

comparison to Athens, things were much more secure for him. He didn't reply; the silence indicated that we didn't understand what he was saying. He was in Athens in order to understand the current situation in Greece. He said that he didn't know when he would be able to contact us again. And he no longer has a Facebook account. He had to close his account because he was threatened by neofascist users. Then he said good-bye and hung up. Very shortly after this phone call we received a text message with a new Facebook name and a smiley.

Later, Sapik decided to leave his clandestine existence in Greece. He said that although he was satisfied with his life there and had already built a strong community, close links to political activists, and a stable way to make his living, he wanted to leave behind the life without papers: "I want to live like you," he told us. He recently arrived in Germany and claimed asylum. He was strongly supported through his connections in his transnational community and the Europe-wide activist networks, and his lawyer is confident that he will be granted asylum. But then suddenly he told us that he's preparing his illegal trip back to Greece. We were very surprised, even horrified, when he said that he had firmly decided to leave the country and effectively drop his asylum case despite that his application was progressing well. "There is no love here!" he said. "See you in Greece."

But Sapik didn't leave for Greece. His asylum case was approved and he leaves legally in Germany. He still aims to go back to Greece, though. And amid of all that, he continues to share his knowledge, his connections, and his life so that people can be mobile when they want.

II. HISTORY REMIX

04

ACTIVIST MATERIALISM

1844: SPECIES-BEING

More-than-social movements become, as argued in the previous chapters, transformative when they change the ontological conditions of everyday existence. What is the historical traction of this understanding of movement? To what extent can we trace the posthuman and more-than-social character of (social) movements in past histories, theories, and mobilizations? Perhaps one the most prominent places to look for the mixture of politics and posthumanism is in the history of materialism, especially when the latter is conceived as collective direct activism on the immediate level of social and material life.

But this articulation between materialism and activism is unstable, full of discontinuities and breaks. In Marx and the early rebellions that took place in the Americas, and in the communes and uprisings across Europe, materialism first becomes directly linked to political activism: activist materialism. Since then materialism has been the target of interrogation not only from idealist positions and various dualist ontologies but also more recently from within the political forces of the Western post-1960s left, which were embracing materialism in one form or another. Critiques from the left did not position themselves outside the materialist movement and were not first and foremost an opposition; rather, they were an immanent movement enunciated from the core of materialism that lasted up until the 1980s and 1990s and finally ushered a new version of materialism to the fore. During the long history of the encounter between materialism and activism, both of them changed meanings, and the formation of each one influenced the meaning of the other, producing new configurations of social practice.

Marx's work is the most prominent and probably the first full-scale attempt to connect activism and materialism on the level of everyday political practice. The *Theses on Feuerbach* exemplifies the articulation between materialism and activism in a remarkable and equally unexpected way. Thought objects and abstract contemplation are what Marx tries to defy— that is, idealism. The movement that changes society is the movement that opposes idealism. It is real, objective—that is, material—says Marx. Marx's materialism is conceived as sensuous everyday practical activity that has the capacity to change the material conditions of existence. The moment of transformation is the moment when, to use Marx's term, civil society collapses and a new sociomaterial order emerges. This modern understanding of activism in materialism was epitomized in *The German Ideology*: here communism is not "an ideal to which reality has to adjust itself'; it is "the real movement which abolishes the present state of things" (Marx and Engels, 1846/1976, p. 48).

In the *Economic and Philosophical Manuscripts* (1975), Marx introduces a new definition of materialism grounded on inserting activism into the understanding of materiality. Here he uses the concept of "species-being" to describe the defining moment of human species as the self-making of the species itself in a direct practical and organic relation to other species and the whole of the "natural" world. Despite the predominance of humanist ideas and essentialist connotations,[1] species-being is as close as one can get to a definition of humans not based on what they are but on how they relate and act: the self-instituted collective emancipation in which cooperation and interaction among humans as well as between humans, nonhumans, and the material world is the guiding force.[2] For Marx the question is to uncover both what impedes this process—for him it is capitalist labor that alienates species-being—and what makes collective material self-transformation possible. Who controls the process of material transformation and who participates and in which position are questions that drive Marx's activist reading of materialism. This materialism is activist because it is a "life activity," in the literal sense: life engendering life. There is no social transformation outside the material realm.

Marx's early materialism avoids the pitfall of epistemology: the attempt to distinguish between a strong materialist position that gives absolute primacy to matter and a weak materialist position that puts the emphasis on how we conceive matter. Such an epistemological approach to materialism wouldn't be sufficient to distinguish it from idealism because at the end what is matter would be a question of definition. Asking such questions would be an idealist move because it would not prevent thought from dictating what being is.[3] From an epistemological viewpoint, both positions—the materialist as well as the idealist—are in principle tenable. Marx's early materialism avoids this impasse by mixing ontology and practice through and through: there is no transformative activity that is nonmaterial. Since activity is inherently material, matter itself cannot be conceived as an outside or as a mere object of human practice but as a process of change. Species-being depends on the collective metabolic transformation of matter including the species itself: activist materialism.

There is a monist understanding of matter here that resonates with recent versions of materialism.[4] The emphasis is on matter as a vital force: inorganic matter as well as biological and social life are movements of matter itself. Nevertheless, in terms of Marx's early definition of materialism, merely highlighting the importance of materiality as an assemblage of heterogeneous forces is not enough to account for the kind of transformative political engagement that was his main concern. Marx's monist ontological materialism is infused with an activist dimension that takes place in the actual everyday life of species-being: the collective capacity to engage in material change.

Practice and matter, activism and materialism cannot be thought independently. And the reason for this is *not* epistemological but political and ontological: political because capital and colonial power break up the species-being—that is, the fundamental equality between humans resulting from the fact that each individual of the species can exist only if it is involved in collective material change—into exploited classes and races. A necessary form of material political practice can reverse this destruction of species-being. This practice is ontological because matter's movements are independent of humans, and only through experimentation with nonhuman forces (Marx's "nature") can humans realize their species-being. Human species-being is not given, and there is no final essence; it exists because it is practiced, embodied and embedded in relation to other species and the material world.[5] Materialism without activism is not transformative; in fact, it is not possible at all. This is the quintessence of Marx's early account of a practical ontology and an activist materialism. What happens to this configuration of materialism twenty-eight years before the bicentennial anniversary of the *Economic and Philosophical Manuscripts*?

Marx's and in particular Friedrich Engels's late work offers a different understanding of activist materialism from the one developed in their early writings: dialectical materialism. "Diamat"—which had a long-standing impact on theorists of the Second International and the emerging Marxist social movements—consolidated the absolute emphasis on matter but introduced a very different conception of its practical, political, and philosophical significance. Diamat foregrounded activist materialism as a dogmatic epistemological doctrine that gradually removed the practical ontological concern with matter and subsequently transformed the meaning of activism. Already in the early writings there are numerous instances where, instead of the practical ontology described in the previous paragraphs, we find a relation to nature dominated by the ideal of progressivism and the human mastery of nature's laws. This understanding also changed the meaning of activism. In *Anti-Dühring* (1878/1987) and *Ludwig Feuerbach* (1886/1990a), Engels set out a materialist cosmology that defined activism as a political practice that is monocausally determined by a set of laws extracted from nature: historical materialism. This is characterized by both a bifurcated dualist ontology—with objective material reality and its inherent laws on the one hand and social practice on the other—and a reduction of materiality to human social institutions and structures. Activism was reduced to the efficacy of changing social structures. Historical materialism announces the erasure of the activist materialism to be found in the early works of Marx and Engels.

1908: ONTOLOGICAL DUALISM

In *Materialism and Empirio-Criticism*, Vladimir Lenin follows this line and conceives materialism exclusively as a theory of knowledge. He writes: "For the *sole* 'property' of matter with whose recognition philosophical materialism is bound up is the property of *being an objective reality*, of existing outside our mind" (1908/1970, p. 260, emphasis in original). Materialism here starts from the assumption of an ontological duality, two separate entities: matter on the one hand, mind on the other. Lenin reduces materialism to gnosiological realism, while the activist materialism of the early Marx asserted a monist ontology: mind is matter, and the unity of the world is sustained by its materiality and the immanent action of matter and mind alike. Lenin's approach was a radical departure from a position that is concerned with bringing together practice and matter. Rather, his concern was to develop a conceptual instrument that splits ideas in two opposite camps. While Marx and Engels's early activist materialism was concerned with how matter is changing and can be changed, Lenin's materialism was developed as a strategic tool for the selection of the social and political forces of his time that could potentially transform to a revolutionary historical subject.

Lenin was building a war machine. The aim was to develop a philosophical conception of materialism that had no other target than to reveal the functioning of a deep social dichotomy between the working class and capital.[6] His only goal was to submit theory to the everyday requirements of his political practice. With his philosophical work Lenin developed a tool to extend the social division as far it could go, to the far end of mind and the history of ideas. In *What Is to Be Done?* (1902) he claims that social conflict penetrates every corner of society, every social relation, every idea. Nothing is untouchable by class antagonism; it takes a partisan organization and a revolution to change it. This is partisan philosophy and partisan practice. And it is a truly activist move; however, this particular move enacts a different materialism. It is carried out in the name of materialism but is not an activist monist materialism. It is one that subsumes matter and dominates nature in the name of political power. If Marx's early materialism was of a kind that proclaimed the irresistibility of revolution on the grounds of a unified monist movement of matter and activism, Lenin's materialism is dualist, elevating irresistibility to something completely different: the will for action.

"Materialism must be a form of idealism, since it's wrong—too" (Sahlins, 2002, p. 6). Marshall Sahlins's aphorism captures the post–World War II predicament with the configuration of an activist materialism à la Lenin. Lenin's reduction of monist materialism to gnosiological realism had far-reaching consequences for the philosophical scaffolding of the social forces that found themselves entangled in the Marxist enterprise and in the emerging working-class movements from the beginning of the twentieth century up to the 1970s and 1980s. The most important consequence was that materialism became gradually equated to the idea of changing political power. The dimension of matter and materiality was resolutely erased from the everyday enactment of activist practice. Activist materialism, at least in its initial version found in the early writings of Marx, became everything but activist, quickly turning into an ideology of state socialism and an abstract philosophical system that came to legitimize particular forms of politics in the post–World War II period.

1980: CULTURAL MATERIALISM

The end of the 1970s probably saw a peak in the process of an immanent critique of materialism that rendered visible its contradictions as inherited from the previous period. Raymond Williams's analysis of this situation is as follows:

> It took me thirty years, in a very complex process, to move from that received Marxist theory (which in its most general form I began by accepting) through various transitional forms of theory and inquiry, to the position I now hold, which I define as "cultural materialism." The emphases of the transition—on the production (rather than only the reproduction) of meanings and values by specific social formations, on the centrality of language and communication as formative social forces, and on the complex interaction both of institutions and forms and of social relationships and formal conventions—may be defined, if any one wishes, as "culturalism," and even the crude old (positivist) idealism/materialism dichotomy may be applied if it helps anyone. What I would now claim to have reached, but necessarily by this route, is a theory of culture as a (social and material) productive process and of specific practices, of "arts," as social uses of material means of production. (R. Williams, 1980, p. 243)[7]

During the period of the crisis of materialism that unfolded in the decades between 1950 and 1990, the notion of culture reordered the existing meanings of materialism and fueled the development of a new constellation of concepts and activities into the social conflicts of the postwar period. Of course, not all of the various movements and critiques of materialism embraced the notion of culture. The point here is not to unify these extremely diverse movements and traditions under one overarching rubric. Rather, what is important is that the insurgency against the previous materialism evolved in proximity to new everyday activities whose many faces and actions pertain to changing sociocultural power.[8] This turn to culture thoroughly changed the way political activism is performed, moving the target away from the state itself toward power's pervasive materialization in the whole societal nexus: in terms of gender relations, racialization processes, social institutions, social and civil rights, the political representation of excluded groups, anticolonial and postcolonial conflicts, and so on. Many societies, many cultures, many socialisms, Raymond Williams would have said.

This remaking of materialism corresponds with the practices of new social forces that found themselves outside the traditional organizational forms of the working-class movement that appeared as the inheritor of the materialist politics of the previous periods. The new politics of cultural counterinsurgency, not least as exemplified in the new youth cultures of the 1960s and the variously globalized events of 1968, spread across the globe with a velocity far beyond the wildest utopian dreams that Soviet propaganda bureaucrats and Western communist parties ever imagined for their own materialist politics.[9]

But what exactly was the materialist aspect of the activism that propelled itself through cultural politics? The most likely answer is that there was very little materialism in this "cultural materialist" politics, at least not in the sense of an activist and practical ontology concerned with a monist understanding of matter (as in Marx's early version); nor was there much of the materialism of the late Marx/Lenin period with its strong focus on gnosiological dualism and the transformative efficacy of social forces. Cultural politics questioned both versions of materialism and developed along many disparate and diverse paths, all of which were, however, occupied with the centrality of representation and its critiques. Umberto Eco's *Opera Aperta* (1989) and James Clifford and George Marcus's *Writing Culture* (1986) as well as the broader linguistic/discursive turn and the interest in hermeneutics[10] are examples of intellectual engagements that marked the path to the undiscovered continent of representation. Through the changing of meanings and the challenging of representations, the process of social activism was now being performed.

Another important path for the revision of materialism that developed during this period came from an interest in social space as a key battlefield for social antagonisms. How is space regulated, appropriated, and reappropriated by marginalized social groups? Marxist-inspired readings of lived place, the situationist movement, and cultural geography[11] all turned to kaleidoscopic remakings of space in order to articulate an everyday, mainly urban activism that made radical interventions in the politics of postwar Europe and North America possible. The attention to space as lived experience is closely related to body politics. The body becomes an

open substratum for the inscription and reinscription of social signification. In this sense signification moves from the mind itself to the body and
emerges in a process of subjective embodiment through a social context[12]
or in cultural-political constellations.[13]

This focus on embodiment resonates with the postwar period preoccupation with subjectivity and difference.[14] As cultural studies have so
vividly shown, subjectivity is always in the making because it entails a
nonexpressed otherness, a nondiscursified and imagined possibility of social relations.[15] This is particularly important in a period where identity
politics occupies a central place in the political life of Global North societies.[16] Already in the 1970s and 1980s, cultural studies, feminist politics,
antiracism, postcolonial and anticolonial movements, and gender studies
identified the limitations of an activist materialism qua late Marx/Lenin
that saw social consciousness either as committed to working-class change
or as ideological.

In resonance with Louis Althusser's take on ideology (2001), new social
movements focused on the emergence of multiple political subjectivities
that defy straightforward classification as wrong (false consciousness) or
right (transformative) according to previous conceptions of activist materialism. Crucial for this attempt was the process of articulation.[17] Activism here is conceived as a movement of articulation that by rethinking
Gramscian hegemony attempts to contest domination through "rendering
the symbolic increasingly dynamic, that is, by considering the conditions
and limits of representation and representability as open to significant re-
articulations and transformations under the pressure of social practices of
various kinds" (J. Butler, 1997, p. 23).[18] This understanding of political subjectivity as subjectification and the result of articulation is what essentially
captured activist practices in this period, positioning subjectivity in the
tension between coercion by institutional mechanisms and articulation
through them.

Cultural politics challenged previous versions of materialism on the
grounds of an increasing diversification of social strata and classes. This
diversification brought a new form of activism that, rather than focusing
on materialism, was concerned with the fight for representation. In this
struggle, discourse, space, body, and subjectivity are approached as constitutive of an oppositional politics of difference. Cultural studies, women's
studies, postcolonial studies, and queer politics have all participated in and
critiqued this fight for representation.[19] The importance of representation
comes from the dissolution of social class as the central actor and political

force in society. The political order of transnational neoliberal societies is an order that is supposed to be occupied by multiple players working to foster alliances between themselves and to establish new relations of power. And this form of relationality triggers the imperative for representation.[20] Representation enters the realm of politics as the attempt to give voice and operative agency to social groups who have been excluded by the politics of traditional versions of activist materialism. We can trace the singular trajectories of these emergent oppositional subjectivities of new diverse social groups in civil rights movements; in the events of 1968; in feminist movements, antiwork movements, and new forms of social cooperation; in the 1960s cultural rebellions; and in the fight against colonialism and racism.

1987: "THE ONLY ENEMY IS TWO"

However deep the break between the activist materialism of late Marx and the cultural materialism of the postwar period might be, nevertheless a peculiar form of continuity remains. The materialism of late Marx reduced activism to the radical intentionality of a subject determined to reflect the antagonistic conditions of existence. Cultural materialism retained this reduction but introduced a differentiation with respect to the subject itself. Instead of a unified self-identical subject there exist a plethora of subjectivities and of possible contexts in which they are constituted. This break implied a deep change in the way political activism was conceived: late Marx's activism subsumed every activity under a single social conflict between labor and capital, while the activism of cultural politics multiplies the fronts on which social antagonisms are encountered and fought.

Nevertheless, despite this radical break, both positions retain a strange commitment to epistemological dualism. Representation and ideas are the battleground on which the conceptualization of activism thrives. It is about negotiating and transforming the conditions of thinking and feeling that make activism possible. In a peculiar way, cultural materialism followed late Marx's and Lenin's path in focusing on how to represent reality. Cultural materialism introduced a new conceptualization of the main determinants of representation. It is no longer the class structure of society but rather the endless variability of social contexts that allows different configurations of representation. In this sense the question of cultural activism becomes one of how reality is constructed in the subject itself, or "social constructionism." In both positions, however, practice and matter are subsequent to ideas, and despite their pervasive critiques of dualism, both retained a dualist ontology.

Deleuze and Guattari's well-known diagnosis of this situation is as follows: "We invoke one dualism only in order to challenge another. We employ a dualism of models only in order to arrive at a process that challenges all models. Each time, mental correctives are necessary to undo the dualisms we had no wish to construct but through which we pass. Arrive at the magic formula we all seek—PLURALISM = MONISM—via all the dualisms that are the enemy, an entirely necessary enemy, the furniture we are forever rearranging" (Deleuze and Guattari, 1987, p. 20). In much of their work, and most centrally in *A Thousand Plateaus*, Deleuze and Guattari introduce a monist materialism that attempts to rehabilitate matter from its entrapment in representation. Their move is coextensive with the (re)appearance of a form of materialism that puts the primacy of matter on the agenda of political practice and theory after the 1990s and creates the possibility for the emergence of a novel configuration of activist materialism. Strangely enough, the poststructuralist faction of cultural materialism of the previous decades prepared the way for this move—in particular feminist materialism; the attention to the body, as described earlier; and the persistent but evasive attempts to put materialism back on the agenda (one can only recall Althusser's [2006] subterranean movements of materialism).

But even more crucial to the reinvigoration of activist materialism is the increasing impact of technoscientific knowledge on everyday life and on the structures of production in the Global North that posit matter as a self-ordering, emergent actor in a radically posthuman world. Matter is before thinking; matter is in thinking. For Deleuze and Guattari there is no empty space, there is always matter, and matter is always differentiated. Representations are a particular form of differentiation in their own right; they do not exist prior to or vis-à-vis matter. Representations are movements of matter as much as genetic mutations or the lithosphere's plate motions are. Deleuze and Guattari's point is not to eliminate the distinctive importance of representations and ideas; rather, their claim is that when representations are considered separated from matter, they become strategic tools for ordering material reality. Representations are closures and reterritorializations that are used as powers to organize matter in a particular way.

The materialism emerging gradually after the 1990s focuses on the question of monism instead of concentrating on the binary opposition between materialism and idealism. This dichotomy undermines monist materialism. It is not about which position you take in this thinking; it is about the very act of taking a position. For Deleuze and Guattari the real enemy of materialist thinking is not idealism; it is dualism. "The only enemy is two" (Deleuze, 2001, p. 95). Materialism after the 1990s is an antidualism that gradually lets an alternative conceptualization of the relation between activism and materialism that informed most social movements during the late Marx period and after emerge: matter and mind, activism and materialism start to fuse again into one process.

It is not a coincidence that many of the social movements of this period and since focus on the question of reclaiming and prefiguration. The activism of reclaiming attempts to reappropriate the immediate spaces of existence by simultaneously transforming them through everyday actions: reclaiming the streets, the right to the city; Earth activism and the permaculture movement; the remaking of transnational spaces through migration movements (which was discussed in chapter 2); radical queer activism and the building of new social relationalities and communities; hacktivism; the commons movements; and indigenous mobilizations, to name just a few that emerged gradually in the 1990s and after. In all of them we encounter an emphasis on reclaiming material spaces and relations vital for developing new alternative social and material projects.[21]

Deleuze and Guattari's monist materialism captures a key moment of this form of activism that reconnects us with the concerns of early Marx's activist materialism, described in the beginning of this chapter: how matter is morphed—that is, how to live with matter and create new material forms of existence through collective practices. However, Deleuze and Guattari's materialism questions how morphing matter comes into being. The emergence of form is neither the transcendent imposition of a preconceived plan on matter (forget the architect and the bee) nor is it simply self-organized matter that is just represented in the mind of the subject (forget the subject-object divisions and correlationism)—neither external plan nor internal capability. In this sense, it is neither materialism (as conceived until this moment) nor idealism. The position Deleuze and Guattari try to develop is that the movement of matter itself makes both a materialist as well as an idealist stance possible. Both the capacity to make form and the capacity to understand the emergence of form are immanent to existence. There is no monism if there *is* a dualist option; "there is nothing that is one, there is nothing that is multiple" (Deleuze, 2001, p. 99). Deleuze and Guattari tried to avoid thinking along the either-or of materialism and idealism/dualism. The very possibility of thought is immanent to matter's movements.

2000: DESIRE

In this understanding of materialism, matter becomes the horizon and the substratum on which an alternative to the previous versions of materialism can emerge. Matter becomes (once again) the way to reconnect activism and materialism. The crucial move for materialism since the 1990s is to seek in matter an escape from a situation where the demise of the everyday transformative activist aspect of materialism became so pervasive. Deleuze and Guattari's move to a monist materialism is not a theoretical choice; it is the result of a political diagnosis according to which any desire for and any possibility of change has been vampirized by existing social and political institutions. Even more than that, in the previous decades desire itself has been transformed into an institution of capital.[22] Every social struggle is reinserted as a rejuvenating feature of the system of accumulation; every social innovation is turned to a value-producing innovation. The story of the twentieth century is not a story of revolutions; it is a story of counterrevolutions, says Heiner H. Müller (2000), where every radical desire has been appropriated, dismembered, and regurgitated, flattened.

The bottom line for Deleuze and Guattari's take on materialism, as a monist materialism based on a renewed attention to matter, is the attempt to reactivate the transformative force of desire. Deleuze and Guattari try to do this by breaking the link between "desire" and "desire for." Every "desire for" is a closure: desire for revolution, desire for mastering nature, desire for equality, desire for recognition, desire for an identity, desire for not having an identity, desire for desire itself. This is the political move that Deleuze and Guattari reinsert into materialism: to disrupt the view that the creativity of people, animals, and matter can be viewed as a desire that can always be folded back into the current forms of valorization. Every "desire for" is captured, measured, and transformed to value: biofinancialization (as discussed in chapter 2). This is the spell that current society casts on life.

To break this spell, monist materialism attempts to disarticulate desire from its essential function as something that has a target and object.

The diagnosis: "desire for" is the way sociopolitical control revolutionizes itself. The radical political key to monist materialism is that it allows desire to be engendered in a way that can move beyond its recoding into the political closures of the counterrevolutions of the twentieth century. The prominent role of matter in Deleuze and Guattari is a small gesture of rebellion against the capture of earlier materialisms within a machine that constantly revolutionizes control. Deleuze and Guattari perform this small gesture of freedom by inserting indeterminacy into the way desire operates. And they do so by turning to the underlying indeterminacy of matter: matter is primarily unformed and in continuous variation, an oscillation between various intensities, closures, and openings. Matter is a political exit. Matter is escape. The making of a life. Matter can break the spell.

The turn to matter becomes political when it is articulated in relation to this understanding of desire. That is why, despite the various attempts to read Deleuze and Guattari's materialism in a scientistic way—that is, as a philosophy attentive to science[23]—Deleuze and Guattari propose a rather minor move, one that attempts to interrupt the appropriation of desire by grounding it in the indeterminate movements of matter. A minor science, as they call it (other times they use the names *nomad science, itinerant science,* or *ambulant science*). In Proposition III of the War Machine chapter in *A Thousand Plateaus* they describe this as a practice that follows matter's immanent traits, confronts problems instead of applying theorems, and pushes matter to the next threshold. Against a science of matter or a technology to control it, Deleuze and Guattari emphasize practice as the key dimension of a minor science that knows how to surrender to matter. Minor science is a practice that is essentially experimental; it evolves around problematics rather than essential qualities, it is rigorous but not systematic, it directly links activity with matter, and it cuts through and occasionally derails big science, which Deleuze and Guattari often also call royal, imperial, or state science.[24] Here a revived form of activist materialism establishes its presence in the 1990s and 2000s.

2016: THE LOSS OF MINOR SCIENCE

Deleuze and Guattari's minor science operates below and outside big science and yet, as will be discussed in detail in chapter 8, it is continuously under pressure to be included in it. "The fact is that the two kinds of science have different modes of formalization, and State science continually imposes its form of sovereignty on the inventions of nomad science. State science retains of nomad science only what it can appropriate; it turns the rest into a set of strictly limited formulas without any real scientific status, or else simply represses and bans it" (Deleuze and Guattari, 1987, p. 362). The unfinished story of contemporary minor science is that it can be absorbed into the workings of big, state, royal science—that is, it can become a "desire for" a grand theoretical system and a philosophical materialism devoid of its activist element. Minor science and big science are inextricably bound together even if only though conflict.

In fact, minor science exists in the core of big science. Pamela Smith has shown how artisan production—probably the most vital aspect of minor science—was crucial for the emergence of the rationalist objectivist scientific worldview that came to dominate the Western world increasingly after the sixteenth century and to constitute modernity's epistemic territory (as discussed in chapter 1). The artisans' work, an intellectual revolution from the bottom up, "transformed the contemplative discipline of natural philosophy into an active one" (P. H. Smith, 2004, p. 239).[25] Later, artisan science was appropriated and absorbed into the new disembodied design of experimental positivist practice within big science. However, big science never completely abolished artisanal skills and practices and still relies on the minor science of matter to realize positivist experimental practice. Thanks to the purported modesty of meticulous artisanal efforts Robert Boyle's bottom-up experimental laboratory science won out over Thomas Hobbes's top-down geometric science.[26]

This coexistence of minor and big, state, royal science—even if only antagonistic—is with the rise of technoscience more ubiquitous than ever. In experimental science, the "discovery of natural facts" and realities of matter was a distinct procedure that preceded possible technological applications and their spread within everyday life. Within technoscience—that is, when science, technology, and everyday life fold into each other—discovery is invention is intervention. One of the main characteristics of minor science—its interventionist, direct, ambulant quality—becomes a dominant feature of technoscience itself. Minor science fuels the everyday workings of contemporary technoscience.

Scientific practices and objects are as much the result of artisanal work as they are of the precarized labor of "industrial" scientists and of the entrepreneurial investments of corporate and state science (see also chapter 8 for a lengthy discussion of these issues). Minor science and big science co-constitute all these technoscientific objects that circulate in our worlds.

Such objects make up the conditions of our actual material presence in the world; they are "wormholes that dump contemporary travelers out into contemporary worlds" (Haraway, 1997, p. 43). The entanglement of minor and big science is the reality in which material existence unfolds. But where does this leave the project of the revived activist materialism of the 1990s and 2000s? What is the meaning of activism when matter as political exit gradually disappears with the entanglement of minor and big science?

Activist materialism is again in turmoil. It bifurcates into two lines of development (which, rather than being in some sort of opposition, sit uneasily next to each other). The first and most widespread line of development attempts to strip (again) activism from materialism: from actor-network theory, object-oriented ontologies, neomaterialism, and neovitalism all preserve key theoretical tenets from activist materialism but drop in one way or another its activist dimension. The other line of development, probably one that is far less prominent in the academic/intellectual sphere and more present within political movements, consists of a series of diverse projects that preserve and expand the commitment to activism. This second line of development attempts to revive minor science after its loss. What is minor science today if it is not the other to big science? I trace the political past of this second line in the next chapter and then move to explore contemporary versions of it in the third part of this book.

INSURGENT POST- HUMANISM

1871: ASSEMBLING THE STATE

If worlding is about the making of social worlds that crisscross global space in divergent trajectories and at variable speeds and that defy the abstract universalisms of globalization,[1] I want to extend its meaning to include the making of material worlds that create unique occasions of mundane existence beyond the abstract registers of "nature" and the nonhuman. If, as Chris Connery, Rob Wilson, and the work of the Center for Cultural Studies at the University of California in Santa Cruz suggest, worlding is about enacting an opening in our thinking and practice to other values, ideas, and ways of being, I want to think of worlding as an opening to material processes and practices and as a possibility for crafting—literally—alternative forms of life[2]—alternative worlding constructions that, as Rob Wilson (2015) says in his ecopoetics of Oceania, are about the housekeeping on Earth (instead of the housekeeping of various social orders), about becoming ecumenical: the making of translocal solidarity, place, and bioregional worldings. Extending the worlding project from society to ecology and matter is concomitant with introducing a posthumanist perspective into culture and politics. But this posthumanist move bears its own problems.

Posthumanism, at least in its most widespread and mainstream versions, challenges the dichotomy between humans and nonhuman others and the analysis of social processes based solely on the grounds of human action and intentionality.³ But is it possible to reduce the textured relations between humans to the universalizing category "human"? How is it, after so many decades of work trying to question humanist universalism, that we are now confronted with probably the worst universalism of all: all humans as one, as if there were no divisions and alliances, divergences and associations, conflicts between humans? And the problem seems even more acute when we consider the other pole: how is it possible to homogenize nonhumans to the extent of creating an otherness so vast and uniform that even the most dedicated Orientalists could not have conjured it up in their wildest dreams? In fact, the universalism and reductionism of the category "nonhuman" may be even more dubious than traditional humanist categorizations because it can so easily be presented as a progressive move to include the hitherto expunged nonhuman others into human business.

Can we develop an alternative take on this mainstream version of posthumanism? Can we think of alternative forms of organization that challenge both humanism and the new universalisms that accompany much of posthumanism? Can we think of an insurgent posthumanism—that is, of a posthumanism that is explicitly political and is grounded in (social) movements? But since many movements are primarily social—that is, since many of them seem to be fixated on social power—we need to question some of their key presuppositions in order to be able to detect and strengthen their posthumanist energies. What would it mean to organize a political posthumanism?⁴

As I argued in previous chapters, the politics of social movements have largely ignored the complexity and unpredictability of the entanglement between politics and ontology and between a deeply divided society and a deeply divided nonhuman world. The principal avenue for social transformation passes through seizing the centers of social and political power. The main drive of social movements, especially movements related to organized labor after (the defeat of) the revolutions of 1848 and definitely since 1871 has been to target institutional power and the state. Within this matrix of political thinking the posthumanist moment becomes invalidated, subsumed under a strategy focused solely on the political association of individuals as the locus of action that contests social power. But here I argue that a posthumanist gesture can be already found at the heart of these social movements, even if in most cases this gesture is involuntarily neglected or, perhaps more commonly, eagerly erased.⁵ In fact, this chapter shows how through this long process of neglect and erasure, *movements* gradually came to be conceived as *social* movements, and then how in contemporary technoscientific and more-than-human-culture, social movements can be now understood as *more-than-social*. The argument that I advance in this chapter is that more-than-social movements emerged out of a long, even if effaced, history of posthuman struggles within social movements. In fact, one could argue that these posthuman struggles propelled more-than-social movements within contemporary technoscientific posthumanist culture and questioned the politics of social movements.

From Marx through anarchism to contemporary radical post-Marxist positions, the state in its capitalist form is conceptualized both as the guarantor of existing social relations and as the main threat to society and justice. More precisely, the state cannot exist without cultivating the humanist subject, and at the same time the state is the humanist subject's main enemy.[6] This characteristic of the state as humanist but not humanist enough has been the main target of social movements. The state and its humanism are the constitutive other defining what traditional social movement politics is.

Such traditional social movement politics conceals the fact that a significant part of the everyday realities put to work through the politics of movements has always had a strong posthumanist character through remaking the mundane material conditions of existence beyond and outside an immediate opposition to the state and other social institutions that protected the humanist political subject. In fact, it is questionable to what extent the state can be conceived as *the* state and not as a heterogeneous formation comprising diverse entities such as infrastructures, social groups, institutions, archives and knowledge, and animals and plants. In previous work we have discussed how the state cannot act as a unified actor; rather, specific segments of the state and other social actors and material actors create formations that become the locus of social power.[7] Thus seeing the state as *the* state is a practical political question rather than a question that refers to some essence of what the state is.

In what follows I excavate this posthumanist gesture from the main narratives of political struggles along three fault lines. The first is about abandoning an alienated and highly regulated relation to the material, biological, and technological realms by making a multiplicity of self-organized common worlds—a move from one world containing many enclosed and separated spaces to the making of a plurality of ecological spaces. A second posthumanist move questions the practice of politics as a matter of ideas and institutions and rehabilitates politics as an embodied everyday practice—a move from representational politics to the embodiment of politics. Finally, the third involves the decentering of the human subject as the main actor of history-making. History is a human affair, but it is not made exclusively by certain groups of humans—a move toward a post-anthropocentric history.

1824: VAGRANCY

There is a fundamental assumption behind the politics of political movements up to World War II: the state is a totalizing form of power that needs to be occupied and then stripped of its ethical and functional powers. "The Socialist Party is the anti-State, not a party," wrote Antonio Gramsci in 1918.[8] The mantra of Marx and Engels's refutation of Hegel's idealism of the state is well known.[9] But Marx and Engels's materialist reinterpretation of Hegel's idealist take on the state is an unfinished story. They challenged Hegel's understanding of the state as an ethical universal but preserved the idea that the state is a *totality* emerging in the real social world. The rationality of the state—which was a substantial element for Hegel's understanding—is always actualized in the real life of society as a unity between the objective and universal will[10] and the subjective will of the citizen. Marx and Engels exposed this unity as the ideological function of the state but kept the idea that the state is deeply ingrained in the social world.[11] This understanding was decisive for much of social and political movements throughout the past 150 years.

These social and political movements conceive the state as a totality immanent in the antagonisms that sweep the plains of society.[12] Against the Hegelian assumption that the state embodies the infinite ability to resolve any social contradictions within it, Marx and Engels believed that the resolution will be the result of an act that seizes the state in order to move to a stateless society. The departure from Hegel was never completed. By keeping the state at the center of power, any struggle for the emancipation of labor and other subaltern social groups was modeled on the assumption that it should pass through the state or at least address it directly.[13] Every other possibility for radical social transformation disappears from the horizon of political action.

But these politics never came even close enough to a nonstate society. The revolutions and uprisings have strengthened the state as a totality instead of putting an end to it. But how is this possible? A change of perspective can perhaps help to illuminate this situation. Rather than approaching the state in terms of the control it engenders and how it governs society, we can instead examine how the state is assembled and configured from many diverse and heterogeneous entities[14] in order to be able to respond to the struggles that challenge social power in a certain moment.

If we investigate the state from the perspective of the struggles of social movements, we find something more important in its resilience than its supposed unified and universalizing nature: we see how the state is assembled from many heterogeneous actors in order to facilitate and promote a humanist subjectivity resulting from the then-emerging freedom to sell one's own labor power. The state in its current form is a response to multiple insurgent movements of different shapes and magnitudes that lasted several hundred years until the eighteenth century. Rather than a monolithic structure, the emerging state is a changing and localized assemblage that facilitates the firm incorporation of working people in the system of production by guaranteeing that workers can be free and autonomous individual sellers of their labor power in an open market.

Many current social movements usually challenge the state not because of this humanist appropriation of the workers' freedom but because it is not humanist enough (that is, it does not do enough to protect and expand the autonomy of the workers as individual sellers of labor power). But this humanism hides the fact that it is a response to a long history of nonhumanist struggles that were equally (but of course in very different ways) suppressed by specific state institutions as well as by organized politics. The freedom to choose and to change your employer is not a fake freedom or an ideological liberty, as classical working-class Marxism suggests, but a historical compromise designed to integrate the released, disorganized, and wandering workforce emerging from the fifteenth century onward into a new regime of productivity.[15] In fact, what we have here is a mass of workers exiting indentured, forced, or slave work and reinvesting their capacities in new entanglements with the social and material world.[16]

The singularities that composed the escaping, wandering mob and the fleeing slaves were very far from the humanist individual starting to take shape at the same moment across Europe. They were much closer to the nonhumanist pleb traversing the countryside from the midfourteenth and fifteenth centuries and later to the escaping slaves in the colonies and the disorderly mobs populating the streets of European towns and emerging cities after the eighteenth century.[17] This nonhumanist movement forces a gradual displacement of the previous regime of feudal, indentured, and slave labor into a world in which the relation to surrounding materiality takes a different form than it did in the feudal and colonial order. This exit from the feudal labor regime and the colonial slavery matrix into a multitude of work patterns and into a creative relation to matter gives rise to new, shared, common worlds. Only in the name of the humanist ideal of "man mastering nature" did the emerging disciplinary mechanisms and enclosures after the eighteenth century as well as a multitude of measures to tame and control vagrancy—such as the English Vagrancy Act of 1824—destroy these diverse ways of relating to others and to the material world.[18] Thus, many of the multiple, scattered, disorganized, ephemeral, insurgent movements of people exiting feudal and slave labor in so many different locales and geographies did *not* enter into the capitalist humanist regime of the labor market but embarked on a journey that allowed them to create alternative common worlds.

Silvia Federici's *Caliban and the Witch* (2004), Yann Moulier Boutang's *De l'esclavage au salariat* (1998), Peter Linebaugh and Marcus Rediker's *The Many-Headed Hydra* (2000), Marcus Rediker's *Between the Devil and the Deep Blue Sea* (1987), A. L. Beier's *Masterless Men* (1985), Kristin Ross's *Communal Luxury: The Political Imaginary of the Paris Commune* (2015), Tom Brass and Marcel van der Linden's *Free and Unfree Labour* (1997), and Robert Steinfeld's *Coercion, Contract, and Free Labor in the Nineteenth Century* (2001)—among many others, of course[19]—describe various incidents and occasions, dispersed in historical time and geography, in which multiple modalities of work, everyday life, and divergent forms of social organization emerge. In these fluid conditions, self-organized relations between escaping people and land, plants, and animals gave birth to forms of exit from oppression and to different social-material relations of liberty.[20]

1680: SWAMP AND FOREST

These self-organized relations to the sociomaterial world are moments of making common non-proprietary and non-enclosed worlds, or "commoning" as Linebaugh (2008) calls it. The continuation of life through commoning the immediate sociality and materiality of everyday life is a truly nonhumanist flight into a world where the primary condition of existence is the immersion into the worlds one inhabits and shares with other people and with animals, plants, and the soil.[21] This is not only the social commons but the worldly commons, an ecological commons that emerges out of the process of commoning matter. And then this world is collective, shared by definition, a culture mixed with nature, a material order that facilitates the sharing of different commons. We can see the emergence of such relations of commoning in the communities of escaping slaves and commoners both inside colonial Europe and in the colonies across the Americas, from the South all the way to the North.

Such escaping communities relied on the local environments to facilitate their escape and maintain their existence. Geology has been always a refuge for freedom. And ecology has always sustained fugitives, runaways, and Maroons.[22] As Daniel Sayers (2014b, pp. 8–9) wrote,

> Marronage . . . is a complex process of a global and local nature even if individuals who participated in it were not aware of all others participating in the same process: the phenomenon of hundreds of thousands of individuals marooning around the globe between ca. 1500 and 1900 manifested very locally in swamps, mountains, cities, maritimes, and in various nation-states. The Maroons of Palmares, Rio Real, Camamu, and Cachoeira in Brazil shared something with the Maroons of Nanny Town and Moore Town in Jamaica, Suriname, Martinique, Cuba, Mexico, Colombia, Fort Mosé, Pilaklikaha, and the Great Dismal Swamp in the now USA, Canada, and West Africa. Part of that which they shared were similar ideas on how to go about eliminating the conditions of thralldom that each individual experienced, through self-extrication. Additionally, their decisions led to the formations of various social groups in most cases of grand marronage. At the same time, each context of marronage was historically contingent in nature, unique in appearance, and situated in local conditions.

From 1680, for example, the Great Dismal Swamp, located on the border of Virginia and North Carolina in the United States, became the refuge for escaping slaves and the habitat of hidden communities of fugitives, similar to those found across the Americas.[23] Their settlements and ways of life evolved together with the swamp; their existence became entangled with the fate of the swamp itself, its impenetrable subcanopy vegetation and thick muddy wetlands but also the later attempts to drain and exploit parts of it. These self-extricated communities, which were in complex relations with other slaves, indigenous peoples, and colonial administrations,[24] established self-reliant communities as they removed themselves from violent captive enslavement. Many of these communities developed different forms of the commons that often relied on reciprocity of their members and the local environment, plants, and animals.

This is the ecumenical nature of movements that I mentioned at the beginning of this chapter. And this is also a decolonial practice, or perhaps the seeds for such practice within the heart of the colonial project. Swamps, high mountains, remote corners of infertile land, the vastness of the ocean: all are shelters of freedom helping the worlding of movements. An ecumenical vision of movements is practiced by those who escape as they remain entangled. These nonhumanist movements enlarge worlds by multiplying material and spiritual interdependencies of small or large bioregions as they escape colonial architectures of total global connectedness. This is Epeli Hau'ofa's (1993) "Our Sea of Islands" and Edouard Glissant's (1997) "Poetics of Relation."

Within Europe, there were similar cases to the multiple, diverse, and radical commons that the Maroons, fugitives, and slave rebellions created across the Americas.[25] Close to the place where I live while writing this book is Charnwood Forest, a large area of undulating hills, 700-million-year-old rocks, fields, and tracts of old or ancient woodland covered with dense carpets of bluebells every spring.[26] It is also the site of a long history of conflict over the enclosure of the local commons. As E. P. Thompson (1991, p. 104) wrote, in Britain "there cannot be a forest . . . which did not have some dramatic episode of conflict over common right." Long before most of Charnwood Forest was enclosed for agricultural use, local villagers revolted against the increased use of the commons for raising livestock, in particular farming rabbits, a widespread source of income at this time.[27] "In 1749 a great number of inhabitants, men, women and boys of neighbouring villages, including a party of colliers from Cole Orton, converged upon the warrens. . . . In the ensuing encounter the warrens were thrown open. The 'rioters' clashed with the Warrener and his party, and one of the rioters was killed. There followed troops of dragoons, wholesale arrests, trials. Right of common was proved for twenty-six neighbouring towns and villages, and Charnwood Forest remained unenclosed for a further half-century" (E. P. Thompson, 1991, pp. 105–106).[28] "The Charnwood Opera," a ballad entertainment piece performed in the forest few years later in 1753, dramatizes this popular revolt against the encroachment on common land by the local landowners and warreners. The commoners cry, "Rabbits and Popery! Rabbits and Popery!" (Porter and Tiusanen, 2006, p. 208), and soon after they confront the supporters of the warrener:

On yonder Hill, See, How They stand
—with Dogs—and Picks, and Spades in Hand.
By Mars! A formidable Band!
Were they enclin'd to fight
See! How they troop from ev'ry Town
To pull these Upstart Warrens down,
All praying for the Church & Crown
And for their Common Right[29]

The Act of Enclosure of Charnwood was passed in 1808 and transformed Charnwood Forest to fields of arable farming. Most of the ancient woods had already disappeared by the early seventeenth century because of the demand for timber and the continuous intrusion of grazing lands into the forest. Despite the enclosure of 1808, some lands were maintained commonly and were planted with trees. These tree-covered lands together with the few remaining ancient forests make up the only woods that exist in North Leicestershire today. Far from only a story of the conflict of the commons and its enclosures, this is a typical example of how place and the commons fuse into each other. It is what Patrick Bresnihan (2013) calls the "manifold commons" and Herbert Reid and Betsy Taylor in their important book *Recovering the Commons* call the "body~place~commons" as they emphasize the "dynamic, interactive process of human and nonhuman production and reproduction" (2010, p. 20). The commons rarely exists in the abstract and never outside a specific ecology. But this ecological embeddedness is not static, it is a process of co-emergence: the commoners and the commons, the land, the plants and the animals, the concrete ecology and the local common rights construct each other, and when they disappear, they often disappear in tandem.

More than just a battle between owners and users, the commons is the outcome of the concrete practices of commoning. As Derek Wall (2014, p. 127) writes, "Different forms of commoning give rise to different sustainable environments." Commoning—that is, the use and making of the commons—is always specific: collecting underwood, foraging provisions, cutting turf, planting crops, raising livestock, turning the soil, working with the forest, coppicing and replanting the forest, irrigation and the managing of waters, and so on—the creation of an ecological space and a "multispecies community" (Castellano, in press). The histories of radical struggles of commoners in England and beyond are not only histories about common rights, they are also nonhumanist histories of ecological presence.[30] These nonhumanist collectivities of escape and commoning across the Americas, England, and Europe were non-unified but powerful movements that ultimately forced a recomposition of the state in the nineteenth century.[31] And as the state was reassembling itself to embody humanism and liberal modes of existence, the movements changed too. They became *social*.

1791: ECO-COMMONING

Thus, just before the emergence of humanism we could say that movements were not social in the way we would understand the adjective *social* today: their defining moment was the *nonhumanist* practice of worlding freedom through a flight to the self-organized commoning of matter. *Eco-commoning* is defined as a socioecological world that of course existed long before this flight from the feudal and colonial order happened but now provided the ground for the articulation of a new form of freedom. This form of freedom—as many of the works and examples mentioned in the previous sections portray—is less concerned with compensation and productivity and more with remaking the immediate social and material conditions of existence outside existing regimes of control. This form of escaping into a nonhumanist naturecultural life needed to be recaptured and was in fact recaptured into a new configuration of social organization.

Wage labor was the device that made humanist-liberal social organization possible, and it was aimed at controlling the liberties proliferating in the eco-commons. The key function of wage labor is not first and foremost to control people's productive capacities but to manage workers' surplus of nonhumanist freedom. "Labor as dressage" means that discipline, taming, and performance lie at the heart of the process of transforming work through Protestant humanism to the core value of the markets and busy(i)ness.[32] There are, of course, many factors that contributed to the birth of the liberal humanist subject in North Atlantic societies,[33] but probably one primary source of energy made it possible: the nonhumanist energies of freedom transgressing pre- and protocapitalist colonial societies.

But this description of the rise and fall of the nonhumanist struggles of movements does not yet cover the whole picture of the emergence of *social* movements. Moulier Boutang, in his book *De l'esclavage au salariat* (1998), highlights the fact that there is no historical necessity to move to the capitalist form of wage relations; patronage, forced labor, different forms of serfdom, indenture, and plantation slavery have all existed in different modes and configurations throughout history, many until today.[34] If one thinks through the perspective of the capitalist form of social organization, there is absolutely no economic necessity to change and reassemble the state in its feudal colonial form. Moulier Boutang (in Moulier Boutang and Grelet, 2001, pp. 228–229) explains:

> Haiti, the island that produced half the sugar in the world, initiated a decolonization that lasted two centuries, got rid of the whites, and abolished the slave economy. Between 1791 and 1796, it was done: Toussaint L'Ouverture defeated Napoleon Bonaparte. The plantation economy was undoubtedly efficient; the problem was that it was unstable. If capitalism abandoned slavery as a strategic perspective, it is because its own existence was menaced by the instability of the market that it put into place: if there had not been the Jamaican insurrection of 1833, the English Parliament would never have abolished slavery. The struggles of the slaves in the two centuries of modern slavery are worth ten times more than the struggles of the working class: they were more violent, more virulent, more destabilizing than the workers movement.

The transformation into capitalist organization was the effect of the struggles of movements—that is, of the disorganized and wandering working people and most importantly of the slaves seeking to escape into new forms of nonhumanist liberty. This form of liberty is a move to a tighter, more intimate relation between human action and material force. It is people reclaiming their relation to the material world this commoning of the world and of matter, not only the social world but the world as a whole—that becomes the transformative drive to which the emergence of capitalist forms of production is a response.

Struggles have primacy over the formation of power. However elusive and neglected they are, these nonhumanist movements and struggles fueled history-making on the ground. The state reassembles itself as a conglomerate of infrastructures, devices, and institutions that allow the appropriation of the liberties practiced by these nonhumanist movements by preserving only a small but crucial part of them: that work can no longer be regulated through noneconomic violence, but only by contractual means. However, even at this stage, as Steinfeld (2001) shows, free wage labor was not really free since a series of nonpecuniary pressures made free labor closer to coerced contractual work.

Only the efforts of organized labor slowly eliminated the violent enforcement of contracts and labor agreements in northern European countries and the United States. So it was organized labor building on the previous nonhumanist struggles that effectuated the slow move toward free wage labor at the end of the nineteenth century and the beginning of the twentieth. Not only is free wage labor *not* a historical necessity, but wherever it happened it was not the result of a top-down structural introduction of the capitalist labor market but the result of a long process of nonhumanist and more-than-social struggles of movements.[35] Free labor—that is, the freedom to choose your employer and to exit a contract without any nonpecuniary sanctions—becomes the elusive bedrock of both the new form of production as well as of the emerging social movements: "striving for freedom" in employment relations is the fundamental element of the organization of production if it is to succeed in appropriating and canalizing the nonhumanist liberties of the movements of the eco-commons.

The reassembled state that emerges out of these struggles is neither a superstructure nor an ideology nor a totality nor a unified tool of domination. It is an ad hoc conglomerate of diverse entities that simultaneously guarantees the freedom of employees to sell their labor power and allows the translation of this freedom into value. This contradictory mix of freedom and exploitation is the most crucial ingredient of the sentiment that still dominates life in Global North societies.[36] Out of relative nonhumanist freedom, today's production system forges a relative humanist unfreedom. And as this happened, movements transformed to social movements and elevated the state to a universal arbitrator about the right mix of freedom and exploitation. By making the reassembled state both the main target and simultaneously the path for social transformation, the revolutions and uprisings ended up reinforcing the logic of the humanist-liberal state rather than supporting nonhumanist experimentation with freedom. Instead of betraying the state and its order, the revolutions, one after the other, ended up betraying the people. As Immanuel Wallerstein (1998, p. 13) writes, "The revolutions never worked the way their proponents hoped or the way their opponents feared."

1966: EMBODYING POLITICS

What the revolutions could not achieve was achieved in a series of uprisings that erupted across the globe in the 1960s and 1970s.[37] Common to these was the attempt to challenge a relatively stable form of social and political regulation manifesting after World War II. In a moment when the "withering away of the state" seems almost impossible and when movements have firmly transformed to *social* movements, the mobilizations and cultural uprisings in the 1960s and 1970s changed the conditions of social movement politics again. A new cycle of struggles with strong nonhumanist elements has emerged. These struggles come long after humanism vanquished the cycle of nonhumanist struggles described in the previous sections. This new cycle of struggles can now be called *posthumanist*: the attempt to depart from and challenge the by-then long-established domination of humanist politics.

The new social movements were not solely organized around and against the state and its institutions. Rather, subversion is performed by practices that negotiate their embeddedness in existing power under the signature of a posthumanist escape, not under the imperative of inclusion. The meanings of social and human relationality gradually mutate; many social movements of that period escape into novel embodied material practices that put their subjectivities at the forefront of doing politics. The confrontation with state power and established politics comes from the fact that these movements put forward alternative ways of life that challenge power through their very existence: anticolonial and decolonial movements, feminist movements, environmental movements, antiracist movements, cultural mobilizations, sexual revolutions, and counterculture, to name just a few.[38] The 1960s profoundly changed culture, social relations, and the materiality of everyday life, and this on a global scale.[39]

The centrality of embodying politics in the 1960s movements is a direct challenge to the centrality that the human body achieves as a site of control in humanist culture: co-option and training, subjugation and usefulness are inseparable for the operation of the modern political rationalities of governance. But after the uprisings of the 1960s, the response was a different one: not to tame the embodiment of posthumanist politics but to regulate and control the conflicts emerging from these exiting movements by developing multiple ways to *include* them in a new reassembled state. The slow and varied emergence of the transnational neoliberal state captures the process that attempted to reincorporate in the state these escaping and subversive subjectivities. Niklas Luhmann's (1995) vision of "non-society" is the most apt description of the workings and intricate relationalities emerging in these conditions.[40] The social and material space is seen as fragmented, discontinuous, undecided, interconnected, relational, and networked:[41] nodes and lines, no beginning or end, assemblages of heterogeneous forces and entities, parties, institutions, social groups, animals, plants, businesses, land, science, and so on. Nodes can be constantly withdrawn and new nodes added; the state is assembled and reassembled in an almost ad hoc basis. The work of Bruno Latour[42] appears as a typical theorization of the networked, plural, assembling state.[43]

As the posthumanist movements of the 1960s were gradually included in the networked neoliberal state, the imaginary of radical social transformation was conceived as the fidelity to an *event* to come that will overcome the new plural networked capitalism (a ghostly reincarnation of the belief in revolution that was spreading across the globe from 1848 up to the 1950s with the Chinese and Cuban revolutions). Alain Badiou seems to express this kind of thinking in an exemplary manner. In *Metapolitics* he says that every real politics can be evaluated first and foremost on what it says about the state. A central idea of Badiou's political ontology is "that what the State strives to foreclose through its power of counting is the void of the situation, while the event always reveals it" (Badiou, 2005b, p. 119).[44] Here again freedom is derived from the situation of control, more specifically from its absent center, the void that is determining constituted power but cannot be adequately represented by it. The obsession of the social movements up to the 1950s with revealing the chosen historical subject of revolutionary change reappears in the figure of the event once the posthumanist energies of the 1960s movements fade away. The movements of the 1960s came to be codified in social movements studies as New Social Movements (NSMs),[45] and, indeed, they gradually became new social movements in the way I am using the adjective *social* in this chapter: they turned primarily to questions of identity, representation, and rights—that is, they turned primarily to address the organization of the state and the governance of social life. As this was happening, posthumanist politics become less important in this new phase of struggles of the 1960s movements.

1987: BORDERDWELLING

Beyond this dominant conceptualization of the event that can produce a subjectivity and a truth capable of revolutionizing social change, a subjectivity that was excluded and invisible before, this understanding of the event seems to have overshadowed another way of conceiving its nature that was crucial for the development of new social movements in the past decades: a tradition that with Deleuze and Guattari (and with Alfred North Whitehead and William James) we could say attempts to think of the event as a unique mundane and nonintentional occasion that comes to organize an existing state of affairs differently by acting inside it.[46] The event here is not about historical subjects and revolutions or about a rupture with the state; it is a materialist process that effectuates a divergent form of ordinary organization and mundane forms of life. This is an understanding of the event that focuses on the everyday practices of social and material actors that make politics through their very own bodies and existence.

The event as revolutionary social change neglects the immediate everyday practices that are employed to navigate daily life, to negotiate and remake the composition of our bodies and the ecologies we are part of—all those practices that are at the heart of social and material transformation long before we are able to name them as such. The event is retroactive; the power of distinction between what is and what is not is post hoc. At the end, it is marked by sadness and fear toward designating a political mobilization as being an event because it has not happened yet.[47] But an event is not a question about choice or the morality of choice but about the ethos of practice that is by definition undecidable and comes to craft new ecologies of being and new forms of life.[48] From a posthumanist feminist perspective we could say with Starhawk that practice is not about retroactive choice but about the "power to act with" in the remaking and reclaiming of the material realities of life.[49] Actualization exists because "the ghost of the undecidable" (Derrida, 1992, p. 24) dwells in every step, in every practice, in every occasion. There is no promise, no guarantee, no fidelity. The event as the reincarnation of the fantasy of revolution seems to be irrelevant from the perspective of innocuous, imperceptible, everyday material transformations that initiate social change.

What is probably most characteristic of these ordinary events is their refusal to be driven by fidelity to the coming event. But their path is not fidelity to the present either, but the joy of embodying and betraying it. This is the joy of the posthumanist moment practiced by many movements of the 1960s that appear again and take new shape in the 1990s. Instead of the fixation on *the* event, one could think with Mikhail Bakhtin about a form of joy that defies seriousness and makes truth erupt out of the present. This is the joy of putting together a whole cosmos around everyday radical material practices that are events that might never be named as such.[50] In the same way that Bakhtin is searching in Rabelais's grotesque images of the lower stratum of the body (food, drink, urination, defecation, sexual life) for the forces escaping ecclesiastical and political censorship and coercion, I am searching in the posthumanist embodiment of politics—the joy of changing bodily practices and of fusing the body with new ingredients and processes of this world—for the forces that defy both the cognitivist fixation with events and historical subjects to come and the circulation of class privileges in the aseptic circuits of contemporary networked neoliberal societies. This is the joy of bringing together and assembling a whole cosmos around practices that are events that might never be named as such. The laughter and joy of those who partake in the world through remaking their embodied existences defy seriousness, disperse fear, liberate the word, and reveal a truth escaping the injustices of the present. This is a cosmic constellation, not an individual act. In this feast of symbiotic eating, drinking, defecating, and having sex, the body becomes posthuman and retraces within itself elements common to the entire cosmos, as Bakhtin says: common to the earth, sea, air, fire, and all the cosmic matter and manifestations.[51]

The practice of alternative material embodiments is the heart of this second posthumanist dimension of the movements of the 1960s and then again of the 1990s: with Gloria Anzaldúa's 1987 book *Borderlands/La Frontera: The New Mestiza* (and I am thinking here also of Frantz Fanon, José Martí, Oswald de Andrade, and many others) I see how radical change passes through the posthumanist transformation of the materiality and social relationality of the body.[52] Anzaldúa's embodied politics is posthumanist in a very immediate materialist sense. It goes beyond cultural *mestizaje*; it is not only about identity and our symbolic belongings. It involves how the body mixes with other bodies: human bodies, animal bodies, inanimate bodies. This embodied politics involves borderdwelling, existing outside a fixed constitution of our bodies and selves, existing as malleable bodies in a malleable world.

> Los Chicanos, how patient we seem, how very patient. There is the quiet of the Indian about us. We know how to survive. When other races have given up their tongue, we've kept ours. We know what it is to live under the hammer blow of the dominant *norteamericano* culture. But more than we count the blows, we count the days the weeks the years the centuries the eons until the white laws and commerce and customs will rot in the deserts they've created, lie bleached. *Humildes* yet proud, *quietos* yet wild, *nosotros los mexicanos-Chicanos* will walk by the crumbling ashes as we go about our business. Stubborn, persevering, impenetrable as stone, yet possessing a malleability that renders us unbreakable, we, the *mestizas* and *mestizos*, will remain. (Anzaldúa, 1987, pp. 63–64, emphasis in original)

It is not that borderdwelling fractures consciousness, our positions—social, cultural, or geographical—our relations and connections. Of course borderdwelling disassembles, disarticulates, cuts, extracts, and disposes. But more than this, borderdwelling makes new ecologies of everyday existence, new worldings. New experience.[53] See Anzaldúa (1987, pp. 82–83, emphasis in original): "She is willing to share, to make herself vulnerable to foreign ways of seeing and thinking. She surrenders all notions of safety, of the familiar. Deconstruct, construct. She becomes *nahual*, able to transform herself into a tree, a coyote, into another person." Becoming coyote is an event. It is not the romantic vision of joining nature nor the becoming-animal of joining the idealized pack (such as Deleuze and Guattari's wolf pack). Anzaldúa's becoming coyote is rather ethopoietic: the boundary crossing of her coyote existence is that of an everyday transformation of ethos required by living as an "inappropriate" body on either side of a border (between Mexico and the United States in this case). She is refashioning the whole process of making her an "Other" on both sides of the border into an embodied capacity that cannot be appropriated fully. It is also a transformation that is required to account for all the changes that so many fellow travelers undergo as they cross the Mexico-U.S. border to live a clandestine life below the radar of surveillance.[54] Borderdwelling is the condition of posthumanist politics.

1921: JUSTICE/*JETZTZEIT*

All incarnations of posthumanist politics discussed in this chapter are not primarily articulated through fidelity to an event that can challenge directly and potentially supersede the assembled state and formalized politics but through the everyday betrayal of the supposed governing powers of the state and formal politics. More-than-social movements create alternative forms of life that escape existing ways of existence and cannot be neglected by social power. But traditionally this power to create conditions that cannot be neglected or bypassed has always involved the question of violence. It is widely believed that in social movement politics, violence as destruction is the necessary ingredient for the making of the new. Against this purported tight articulation of violence and transformation (primarily in the form of violence against the state), dominant liberal humanist thinking exorcises violence and asserts that violence starts where politics stops.[55] Is it possible to escape the logic that opposes the violence of destruction to the oppression of the state? Is it possible to avoid the perpetual recurrence of violence that is imposed by thinking movements' politics as the other to the state's violence? Is it possible to escape this dichotomy, commit to the fundamental possibility of nonviolence, and simultaneously promote justice and create new forms of life and alternative worlds? What is justice when it does not involve an antithetical subjectivity?

Walter Benjamin's *Critique of Violence* (1921) explores the possibility of practices that can open political spaces outside the eternal cycle of law-making (constituent) and law-preserving (constituted) violence. There is a form of power/violence (*Gewalt* means both in German)[56] that is neither law-making nor law-preserving and that through its existence addresses justice. Benjamin uses various terms to describe this form of *Gewalt*: revolutionary, pure, or divine. He asserts that *Gewalt*, "when not in the hands of the law, threatens it not by the ends that it may pursue but by its mere existence outside the law" (Benjamin, 1996a, p. 239). The reason for this is that this kind of *Gewalt* can "modify legal conditions" (ibid., p. 240)—that is, it can be a form of Gewalt that breaks the monopoly of law over power and violence itself. When *Gewalt* is *outside* the law, it is a form of *Gewalt* that is induced in a situation rather than being given in it. *Gewalt* that is given in a situation is the *Gewalt* that the law can exercise, and this form of *Gewalt* appears as fate.[57] The *Gewalt* of the law calls for a political force against it that attempts to establish a political order that differs from the previous one but is equally coercive: the *Gewalt* of the law appears as fate, as cyclical history, as something inescapable.

The new form of *Gewalt* that Benjamin tries to introduce is nonfate. It is "pure unmediated" *Gewalt* (Benjamin, 1996a, p. 249) that gets rid of the narrow-sighted "dialectical rising and falling in the law-making and law-preserving" forms of *Gewalt* (Benjamin, 1996a, p. 251); it overthrows law altogether. Within this new space of *Gewalt*, a "new historical epoch is founded" (p. 252) and justice can be realized. Justice is possible here because, following Benjamin, the new type of *Gewalt* that he sees emerging inserts a break in the normal social and political order that assigns standard roles to the involved actors: those who try to preserve existing law and those who need to challenge it and to make a new law if they want to improve their position in it. Benjamin advocates a form of *Gewalt* that allows the possibility of justice, not in occupying one of these two positions—which would only perpetuate violence, destruction, and more conflict—but in exiting the field of these dual options offered by law altogether. That is, he advocates the opening of a certain situation to possibilities that lie outside it: true justice can happen when options are mobilized that bring us outside the dialectic between forces that try to preserve existing law and those that attempt to make a new one. The dialectic of constituent and constitutive power becomes the ground on which control operates, and because of this it can only cause more destruction. But how is it possible to materialize such a different type of *Gewalt* that installs *true* justice?

How can we populate this new space of *Gewalt*, fill it with acts of justice before and independent of the law and of counteractions that exhaust themselves in contesting it? This is not a recourse to clichés such as "taking justice into one's own hands" or a blank apology for violence but a reference to the possibility of evading the continuous cycle that restores one new coercive form of law after the other. It is not a coincidence that I turned earlier to Bakhtin's *Rabelais* to evoke the ordinary materiality of existence as the space where justice can be enacted. In Benjamin this is further developed: the realization of this new form of *Gewalt* outside the law is the space of the ordinary, or better, it is a space that starts from the materiality of the ordinary. One could argue that Benjamin's "other type of *Gewalt*" resembles the practices of more-than-social movements that can be grounded in the radical making of alternative forms of life and everyday materialities that exist outside the law and outside the eternal cycle of state violence and oppositional destructive violence.[58]

Against the perspective that sees the politics of movements as targeting the exceptionality of law, social power, and the state, we can trace with Benjamin the possibility of breaking cyclical historical time and the anthropocentric passing of history by mobilizing this other type of *Gewalt*: by enacting justice *independently* of law (that is, by neither turning against the law nor following it), outside an anthropocentric view of history. Benjamin says that there are types of *Gewalt* that are simultaneously violent and nonviolent, legal and revolutionary;[59] it is not either/or, it is both. Benjamin refers to the general strike as one of these types of *Gewalt*, and although I am not interested in the general strike as such, I want to use it as an example in order to trace the characteristics of this new type of *Gewalt*.[60] The general strike is a very plain, everyday act, and at the same time it is a different form of *Gewalt*, revolutionary and divine *Gewalt*, because it is outside the law of the state. As Benjamin says, the reason for this ambivalence is that it "reveals an objective contraction in the legal situation" (Benjamin, 1996a, p. 240) that cannot accommodate an action such as the general strike that breaks so radically with the way the whole system of wage labor is organized. A protracted general strike is unthinkable from the perspective of the law because it destroys the ordinary life of society, since the workers exit from the role assigned to them by the law. They become a *nonsubject*. They do not oppose anything (they do not do anything spectacular apart from not going to work); they just withdraw from the position assigned to them. They silently and nonviolently refuse the symbolic order of the law. And they do this immediately, now; the general strike as a form of *Gewalt* is ordinary and exists now, in the *Jetztzeit*.[61]

2016: POSTANTHROPOCENTRIC HISTORY

Justice is never given, is never here; it is something of another world, something to come. Benjamin's *Gewalt* is the termination of the deferral of justice. Justice requires a "time-based practice of memory and historical imagination" (Orr, 2012). The question of justice is a question of temporality. Justice is now, justice is against deferral; the space of deferral is the space of law and of destructive violence.[62] Benjamin's divine *Gewalt* dismantles the possibility of the deferral of justice. It is the moment when something that is just happens just now: *Jetztzeit*. The possibility of justice that happens in the moment it is needed is materialized through the reappropriation of matter and the entering in forms of life that instigate justice in the present. And paradoxically this is the *end* of any form of violence, social or individual. The more justice is ordinary and concrete, the more *nonviolent* and collective it is. The more justice happens just now, the more "worlded" it is. I am looking for a posthuman reading of Benjamin's divine *Gewalt*. This is a postanthropocentric move in how history is conceived: history is made neither through the perpetual succession of violence and negotiation nor through the eternal dialectical struggle between constituent and constituted power. History is made outside the history of society; it is made when justice is restored materially.[63] When justice is ordinary and present, it happens without mediators;[64] it is a justice without intermediaries and without diplomats, referees, experts, translators.[65] It is a posthuman justice, the co-construction of life with other species and objects, the simultaneous emergence of ethos and ontology.[66] *Gewalt* is *immediate* justice, the moment when mediation and violence stops.[67]

Rather than being concerned with normative issues and issues of power, more-than-social movements attempt to alter the material conditions of existence starting from positions marked by asymmetry and injustice (in chapter 1 I referred to this justice as *thick justice*). With the rise of related approaches in science and technology studies,[68] there is a widespread assumption that symmetry is required in order to move toward a postanthropocentric grasp of how the human and the nonhuman constitutions operate together and how they produce new, mixed, hybrid associations. Instead of clear-cut classifications and orderings of beings, actor-network theory and other similar positions multiply the possibilities of how beings can connect to each other—an argument targeted primarily to humanist positions that attempt to defend the exceptionality of humans in the constitution of networks. Symmetry seems to be useful when it serves as a rhetorical device against humanism. But it proves to be very problematic in relation to questions of justice.[69]

The real question facing posthumanist politics of movements is how to move beyond anthropocentrism and humanism by maintaining a commitment to justice that addresses radical asymmetries that pervade human and nonhuman worlds. Symmetry is not enough to reverse the modern purification of humans and nonhumans and to stop imposing our "we" (humans) on "them" (the nonhumans), as Latour argues.[70] We have never been modern, not because the purification of humans and nonhumans is impossible, but because we have never been "we," and they—the nonhumans—have never been "they." The constitution of modernity is based on a set of universalisms that have their provenance in colonial expansionism and the spread of the colonial matrix.[71] The question is to delink from these universalisms and to introduce a form of politics that decolonizes the practices of more-than-social movements.

Decolonial movements perform a double break: they challenge the universalism of humans and the perpetuating injustices that were set up through colonial modernity and still split humans among themselves. The second break is a departure from the idea that nonhumans need to enter our polity. There will never be a liberal parliament of nonhumans, not only because this is one of the very limited forms of politics humans have ever invented but also because it is the most humanist of all. The quest is then to disconnect from humanist and nonhumanist universalisms in order to be able to practice politics in postanthropocentric ways: to create alterontologies that restore justice in the immediate ecologies that *certain* humans and *certain* nonhumans are inhabiting. The aim is to politicize posthumanism and simultaneously to posthumanize politics by decolonizing both of them.

III. ALTER-ONTOLOGIES

06

BRAIN MATTER

RECOMBINATION AS LIBERATION

Every epoch has its brain. And every epoch fantasizes a better brain than the one it has. Today, one can see signs of a transition from a cognition-oriented and centralized brain toward an extended, connected, embodied and, most importantly, plastic understanding of the brain. These varied technoscientific attempts to monitor, control, and transform processes of the brain on the level of its material composition are entangled in shifting cultural imaginaries and political practices in the Global North. In this opening chapter in the third part of the book, I explore these entanglements and introduce the main themes of this last part of the book on politics in technoscience, which I then discuss in length in the remaining two chapters.

"Today we are learning the language in which God created life," declared U.S. president Bill Clinton in his announcement on the decoding of the human genome on Monday, June 26, 2000. As I could not but recall Wittgenstein's canon at this moment—that language exists only when it is actively used—a daunting vision appeared to me: practicing the language of creation.[1] *Secular creationism* is the vision that some humans will master neurophysiological processes to the extent that they will be able to recombine brain-body matter in order to produce new forms of existence. Plasticity is the underlying idea to conceiving brain matter as amenable to recombination. But plasticity is not a new concept as such; it has a long history in neuroscientific research and traditional brain research. Today's plasticity starts where the gene stops: the specificity of the individual organism. Plasticity appears when ontogenesis and epigenesis are at work: the worldly making and remaking of the totality of an organism in the process of its development. I come back to this later in the chapter.

Rather than just the relative malleability of brain matter, plasticity now refers to the possibility of recombining brain matter, not as a general process of neuronal regeneration but as a process that takes place epigenetically—that is, according to the specific and contingent realities of each particular organism. "Genes and genius: Does everyone have the potential to be a genius? Epigenetics offers hope for us all" is the title of a review of David Shenk's popularization of epigenetics in *New Scientist* (March 27, 2010, p. 51).[2] In the near future we will be able to create new neurons "at will, where and when you need them" (Horstman, 2010, p. 5): neuroplasticity as neurogenesis accessible to everyone on an ad hoc basis.

Within this inflation of promises and hopes, the brain's plasticity is also seen as a possibility for developing new forms of resistance and new liberating visions of our neural selves.[3] This emancipatory imaginary of recombinant plasticity could even entail the biggest fantasy of all, which is so nicely and fallaciously described in the work of Catherine Malabou (2008). Recombinant plasticity should go as far as to become the self-governed process of challenging the very plasticity of our brain: "To cancel the fluxes, to lower our self-controlling guard, to accept exploding from time to time: this is what we should do with our brain" (Malabou, 2008, p. 79). If we only had a new political consciousness of the brain, Malabou argues, we would be able to steer neuroscience toward a democratic course and achieve neuronal liberation. Here the imaginary of recombinant plasticity encloses the brain in the fantasy of a grand unified historical actor who would be able to challenge prevalent uses of our neuronal plasticity by today's political order and return the brain to the hands of an emancipated public. Brain matter becomes the source of liberation and, simultaneously, its target.

SPECULATIVE BRAIN POLITICS

The vision of a public that can come to form a unified actor outside constituted private or state interests and seize control of technoscience and its objects (in this case the brain, "*our* brain" as Malabou calls it) resonates with a widespread understanding of alternative politics across many different technoscientific fields. In the previous chapters I have problematized this political vision, which seems to pertain to traditional forms of social movement action. I have argued that the attempt to achieve control of material processes through externally extending control on technoscience has a limited reach. It implies that research can be controlled through policy and that matter can be ultimately navigated and manipulated by social imperatives.

In this chapter I continue this discussion in order to show how such politics has dominated current understandings of neuroscientific research on the brain specifically and, more broadly, technoscience itself. In the final two chapters of the book I shift the perspective from such socially oriented politics to the practices of more-than-social movements that attempt to create alternative ecologies of existence: alterontologies. Specifically, in chapter 7, through a discussion of AIDS treatment activism, I present a systematization of different forms of politics and its relation to alterontological practice, and in chapter 8, the last chapter of the book, I describe these practices in detail.

In what follows in this chapter I collect various materials that can furnish a historical reconstruction of conceptualizations of the brain from the vantage point of its understanding as plastic and amenable to recombination. This is a speculative story in which previous conceptualizations and visions of the brain are seen through the prism of *recombinant plasticity*, a term I have modified from its original understanding in order to link the enormous creativity resulting from the inherent plasticity of the brain with our capacity to recombine its molecular structure and to create configurations that would not otherwise be found in humans. This capacity of recombination that some humans possess poses the main problem that this chapter attempts to engage with. Rather than a comprehensive critical history and analysis of brain research, I have weaved together different speculative approaches to brain matter that attempt to construct it as a political potentiality.

The next section discusses broader cultural conceptions of the body, and within this framework it explores how the vision of recombinant neuronal embodiment came to replace other prevalent existing imaginaries of the brain.[4] The sections that follow trace the links between these imaginaries and the epistemic genealogy of embodiment and recombinant plasticity. I start with the move from behaviorism to cognitivism and then to connectionism. Connectionism was crucial for preparing the ascent of theories of embodiment. Embodiment is presented as an answer to the shortcomings of the sciences of the brain that have treated the brain as a self-contained, decontextualized entity and to the shortcomings of genocentric deterministic approaches that have neglected the role of the environment. Embodiment is used to oppose essentialist conceptualizations of difference, primarily gender and race, and the untenable foundationalism of related political movements. The concept of embodiment appears to exercise an almost therapeutic function: it promises to heal the deep discontent within "Western thought" (Lakoff and Johnson, 1999).

The following sections investigate the relation of embodiment to experience and the uptake of embodiment in culture and polity in the Global North. In the same way that cognitivism and connectionism prepared the way for the emergence of embodied approaches, here I argue that embodiment opens the view toward an understanding of the brain and body as recombinant and plastic. The penultimate section reviews epigenetics and ecomorphs as two manifestations of recombinant plasticity that focus on the developmental and ecological malleability of brain matter. The concluding section of the chapter reviews different political readings of these speculative stories of recombinant plasticity and raises the possibility of an alternative politics in contemporary technoscience, which are discussed in the last two chapters of the book.

VISIONS OF THE BODY

Probably the most powerful and widespread cultural imaginary of the body is the cerebral one: the body that exists as the carrier of the intellect, as the site of cognition. The question of the materiality of the cerebral body is a question of secondary importance; its logic is based on taming, suppressing, and canalizing brain energies and bodily feelings. Flesh has to be controlled because it is a "source of epistemological error, moral error, and mortality" (Csordas, 1994a, p. 8).[5] The cerebral body celebrates the exuberant production of knowledge and deploys it to control the complex processes of its own physicality and materiality. It searches for brain modules, for deterministic procedures, for fixed algorithms in order to identify the central processing unit of the body.[6] It assumes its existence as universalist, normative, expansive, gender-free, and culture-free. The cerebral body is the value-producing body, the flesh that has use value, the able body—as opposed to the nonproductive and disabled body, whose corporeality has always to be continually corrected.[7] Cerebral value production = cognitive capitalism.[8]

A parallel vision of the body focuses on a different type of control: the immune body is obsessed with protection, with the creation and maintenance of boundaries. The immune body is concerned with the prediction of possible damages and contaminations; it concentrates on the techniques of repair, normalization, and segregation. Within the imaginary of the immune body, research aims to demarcate the limits of the body, its durability, its widths of tolerance. The immune body "is a body that separates us from the other bodies that inhabit the globe and that prohibits our fusing with other entities. The immune body is that which determines our Hobbesian selfness and is in potential conflict with every body" (S. F. Gilbert, 1997, p. 38). The immune body is primarily concerned with the production of knowledge that conserves and defends, that opposes weakness, that anticipates what is essential for protection and preservation of the body's processes.

The immune body is obsessed with the threat of sudden illness and death. Illness and death "unbutton the world taken as normal and consequently disrupt, question, alter and endanger common and taken-for-granted social relations" and make us aware of the "eventfulness of embodied human life" (Schillmeier, 2016, p. 161). Here, illness and death are not considered "natural" phenomena; they are processes that can be forced from outside; they designate the event of the breakdown of the body's boundaries. In response, the immune body aims to anticipate and prevent death and illness. The main task is to preempt them. The temporal register of the immune body is the future. A temporality open to vulnerabilities and risk, the future is the reservoir of possible threats that can trigger the body's implosion, dissolution, and death. For example, when HIV erupted in Western gay communities in the beginning of the 1980s, it initially triggered a moral panic, not over the actual deaths it caused and the lack of appropriate response to the epidemic but over what it suggested about the vulnerability of the body and of the body politic.[9] HIV became a signifier of how gay men subverted the masculinist fantasy of the intact body underpinning the prevalent heterosexual matrix:[10] a fantasy that assumed that masculine bodies are immune, protected, and impenetrable[11] in the same way that nation-states are assumed to be controlled and sovereign territories. The topos of the immune body became less about the negation of this vulnerability and more about anticipating how to avoid *potential* infection, disease, and death. The immune body is plagued by fear.

The only antidote to fear is to exit the materiality of the body altogether. This is the vision of the discarnate body that provides relief from the vulnerability of the flesh. The discarnate body introduces the fantasy of the pure self: incorporeal, fleshless; liberated from the passions, habits, and weaknesses of its facticity; the fantasy of the disembodied mind (as opposed to the embodied brain). The discarnate body is the home of pure ideas, clean thoughts, and uncontested intellectuality. Against the visions of the immune and cerebral body, which concentrate on the production of different types of knowledge, the discarnate body cultivates sanctity. Rather than producing knowledge to tame the body or to protect it, the discarnate body is the site of faith. It is less about exploring and experimenting with its immanent functions, origins, and boundaries and more about expressing confidence in some transcendent order and purpose of the body. The discarnate body is oriented toward a temporality that is outside lived time. Its powerfulness lies with the potent effect that this infinite temporality has on everyday practical commitments.

In the vision of the discarnate body, time is infinite while the universal cerebral flesh is a place without time, out of time. The vision of the immune body is defined by the synchronic affections between different bodies. The diachronic axis, the evolutionary history of flesh, is captured in the vision of the hereditary body: the search for genetic algorithms, for the ultimate code of development.[12] The hereditary body is the body that is the result of gene expression; it purports to tell the objective natural history of the flesh. The hereditary body is the body that marks and categorizes origins: it is a vision in which gender is constructed as sex, in which processes of racialization unfold, and it is the vision that cultivates the saga of deep belongings (nation, language) through supposed universal body architectures. The hereditary body is concerned with time past; it sees the future as a continuation of its given evolutionary roots and attempts to diminish the synchronic pressures on the brain and the body and to minimize uncertainty.

What is common to all these temporal registers is that the flow of time is external to the body. It constitutes the background against which each of these different body imaginaries occurs. In all these temporal orders time is preexistent; it is a neutral trajectory that runs quasi objectively and uniformly independently of the actually changing body. However, if we think of time as a creative force, not as just a neutral trajectory but as an intensive element in body's metamorphoses, then a different cultural vision of the body appears: the vision of embodiment and emergence. If we "temporalize time" itself (Sandbothe, 1998), the body becomes simultaneously the subject and object of its own regeneration. The embodied body responds, on the one hand, to formations of life that evolve as the time of life flows and creates new unpredictable and novel configurations of existence. This real lived time is the time of development: the emergent body exists in the realm of its own developmental trajectory and actuality.[13] On the other hand, it is emergent because the creation of new forms is always limited by the existing contingent conditions of existence.[14]

The embodied and emergent body is unthinkable, indeed impossible to exist, outside the formative chronotope of ontogenesis. If the hereditary body conjugates the notion of predisposition in different versions,[15] the emergent body refers to how lived ecologies shape its materiality. In this vision the brain stops being a self-contained entity and becomes— literally—embodied, that is open to changes that occur within all the different systems and subsystems that constitute the body and its surroundings. The embodied brain reminds one of Deleuze and Guattari's (1987) nomadism as an entity's state of openness to its own construction obtained through its own movements within concrete environments, rather than through an externally imposed form of organization.

FROM COGNITIVISM TO CONNECTIONISM

These different cultural imaginaries of the body are tightly interwoven with existing epistemic languages and practices of the brain. While every epoch has its brain, not every epoch considers the brain the seat of think- ing and consciousness. In ancient Greece the higher parts of the soul re- side in the heart, and similarly, traditional Chinese medicine sees the heart as the house of the mind; René Descartes considered the pineal gland the seat of thinking. With the rise of medicine in the middle to the end of the nineteenth century, the brain became a systematic object of study. But even then the brain was far from being the seat of thinking and conscious- ness. Until the 1950s the functions and the psychology of the brain were black-boxed through the dominance of behaviorism.

With the dispute over the ultra-positivistic Skinnerian program, the behaviorist mechanistic stimulus-response (S-R) model comes gradually under attack. The main task is to rehabilitate the idea of "thinking" in psy- chological and brain research. There have been many chapters in this en- deavor[16] since behaviorism's expulsion of thinking from psychology at the beginning of the twentieth century.[17] Dewey's vision of the mind as a social process contested the behaviorist view of mind and thinking.[18] In the first decades of the twentieth century, pragmatism presented a viable and lively alternative to the obliteration of thinking, consciousness, and experience in dominant academic discourses. But pragmatism could not ultimately challenge the dominance of the behaviorist model. It is only much later that pragmatism's approach informed research on the brain through its influence on certain strands of connectionism and embodiment. However, at the time, none of these endeavors precipitated a fundamental turn in research into mind and consciousness that would later take place with the advent of cognitivism in the 1950s.

E. C. Tolman (1954) was among those who formulated basic outlines for this upcoming trend in research on thinking a few decades before the emergence of the cognitivist movement. He introduced the idea of "intervening variables," which was an attempt to dissect the entire phenomenon of behavior in order to achieve a new homogeneous synthesis. A response is no longer seen as a direct linear correlate of the stimulus taking place after a certain time lag; rather, it is a function of the stimulus that depends on the environment it is embedded in, the elaborate need system and the belief-value matrix of the individual. The stimulus-response is *mediated* by this function, and this mediation lies *within* the individual. The internal plane of human consciousness now becomes the core center for the regulation of behavior.

Emphasizing the idea that thinking is a dedicated function highlights a key moment for the emergence of cognitivism. For example, Jerome Bruner, one of the protagonists of the cognitive turn, saw a possibility for derailing behaviorist dominance in the insertion of a new *middle link* in the S-R pattern that would allow the investigation of this internal plane of thinking. This link was "sign-mediated-thought" (Bruner, 1967). Thinking is elucidated as an instrument and as a device regulated by a set of rules—that is, as an *organon* with specific functions. The output of a certain input is no longer immediately predictable but is now primarily a function of thinking. However, with the suspension of prediction, the scientistic presuppositions required to assert the natural scientific character of research seem to vanish.

In the mid-1950s—a period in which significant publications (by Noam Chomsky, Allen Newell, Herbert Simon, and others) and events (such as the MIT Symposium on Information Theory) in the history of cognitivism took place[19]—Bruner, Jacqueline Goodnow, and George Austin (1956) published *A Study of Thinking*. Through their combined efforts they asserted that rule-based learning, categorization, and processes of abstraction constituted the main functions of thinking. Thinking is not only about representing but primarily about problem solving; it is a function. The quest then becomes how to illuminate and visualize the "invisible" domain of this function.[20] The answer to this was the idea of computationalism: cognitive processes constitute a standard set of procedures that can be reduced to predefined lower-level processes.[21] Cognition emerges in "patterns of data and in relations of logic that are independent of the physical medium that carries them" (Pinker, 1997, p. 24).

Even if cognitivism is still one of the dominant paradigms of research in the field of psychology and neuroscience, there is an increasing focus on producing systematic knowledge of somatocognitive processes that can be generalized without relapsing into the universalism and essentialism of computationalism. One could read experimental neuroscience's insistence with mapping psychological functions and subjectivity onto the brain[22] as another step in the long history of localizationism[23] that attempted to uncover how the relation between mind and brain is constituted. The brain mapping of subjectivity through new visualization technologies that correlate psychological functions with brain areas seems to perpetuate a traditional abstract view of the brain as a fully formed, static modular structure.[24] But it also reveals an attempt to go beyond the use of embodiment as a figural or metaphoric concept in order to sketch direct relations between the material workings (that is, brain activity and neurobiological processes) of the body and experiential processes and intersubjectivity.[25]

In the unfolding of this story, connectionism represents the next important step in the exodus from cognitivism toward an understanding of the embodied brain. Connectionism promises the possibility of unraveling the structural relations between perception, cognition, action, and affect by conceiving all these dimensions of existence as linked directly on the neuronal infrastructure of the brain. Connectionist research in experimental neuroscience investigates the embodiment of the brain on the material-neurobiological level.[26] Neuronal networks depict complex assemblies of interconnected nerve cells where certain synapses constitute central nodes in the network while others occupy more peripheral positions. The process of ontogenetic development envisions the birth, change, and decline in neural efficiencies, and the apoptosis of many such connectionist nets materializes as webs of sculpted neurons.[27]

Connectionist modeling challenges prevalent cognitivist approaches and their adherence to representational nativism, which assumes that the organization of brain function depends solely on genetically driven cortical microcircuitry. Mental representations in cognitivism are the result of in-nate neurophysiological processes that are context independent and uni-versal in the human brain. Thinking has universal algorithmic structure and resides in fixed neuronal architectures. Against this, "in a connec-tionist network, representations are patterns of activations across a pool of neuron-like processing units. The form of these activation patterns is determined by the nature of the connections between the units. Thus, in-nate representational knowledge . . . would take the form of prespecified weights on the inter-unit connections" (Elman et al., 1996, p. 25).

What is crucial in connectionism is that the weighting of the nodes is not given but emerges through learning. This is the moment when the idea of malleable brain matter that is entailed in the emergent and embodied vi-sion of the body described in the previous section comes into being. While computationalism presupposes innate neuronal structures, connection-ism presupposes semi-open, nonlinear architectures that unfold during the process of ontogenetic development. Brain matter is simultaneously the actor and the result of its own activity. Brain matter becomes formed as it becomes active, but it is active only because this activity shapes the brain into specific forms. Connectionism is a crucial move away from the essentialism and universalism of cognitivism. The formation of brain matter is emergent and contextual: it depends on intraorganismic and ex-traorganismic ecosystems, and it is embodied.

EXPERIENCE AND EMBODIMENT

The embodied approach adds a significant dimension to connectionist modeling of the brain. Embodiment is not only about the syntactic struc-tures of meaning; it also encompasses the semantics of experience (the production of meaning) and the pragmatics of experience—that is, body-specific, context-dependent, and culture-dependent aspects of meaning. Context and experience merge into the workings of brain matter. It is not a coincidence that social, cultural, and critical psychological theo-ries of embodiment focus on the study of the relation between brain and body: existentialism and phenomenology,[28] constructionist and discursive intersections with the biosciences,[29] critical psychology,[30] and cultural-historical psychological accounts.[31] The embodiment of brain matter means that mental functions are not formal procedures, cognition is not in-dependent of its implementation, and mind and experience are always instantiated in concrete material structures: in a body,[32] in an environment,[33] in a social context,[34] or in cultural-political constellations.[35] From the perspective of embodiment there is no such thing as the brain as a fully separate organ. We can think of the brain not as such but as part of, as embedded in, as being in relation to other functions and systems of the body.

Conceptualizations of the embodied brain vary immensely in content and scope, though.[36] In its weak form, embodiment simply means that cognitive functions take place within a physical substratum. More elaborate versions understand the brain as a multilayered, multifunctional, self-organizing system consisting of interacting subsystems: cognition, perception, emotion, and action are not separate but interact continuously and shape our understanding of the self and of the world. Another approach to the embodied brain emphasizes its phenomenological dimensions as the existential ground of thinking: our bodily movements and orientations are the ground on which our mental concepts and abstractions build. "No matter how sophisticated our abstractions become, if they are to be meaningful to us, they must retain their intimate ties to our embodied modes of conceptualization and reasoning. We can only experience what our embodiment allows us to experience. We can only conceptualize using conceptual systems grounded in our bodily experience" (Johnson, 1999, p. 81). Another widespread version of the concept of embodiment emphasizes the brain as an active agent absorbing, modifying, and transforming social, cultural, and symbolic forces. The brain in all these understandings is an inseparable part of the human body. Many extend this approach to include the artificial, organismoid, or humanoid body and its relations to the human body: embodiment in these accounts refers to hybrid machines that are able to act in real-time and real-space environments.[37]

All these divergent approaches and countless descendant theories of embodiment propose that human conceptual and experiential systems are inextricably linked to the sensorimotor and affective systems. Experience starts with the affective-perceptual sensing of the environment and with locomotion within it. From an intraorganismic perspective, the embodied brain is the steadily transforming brain in a process of constant interaction with the totality of the body and the brain itself. The experiences that humans have cannot exist without a brain that represents its own state and the state of the body in which it is embedded.[38] But the embodiment of the brain is not just about decentering the brain into the body of the organism; it is primarily about decentering the whole organism itself, an idea that has been already developed within cybernetics (Pickering, 2011).[39] Rather than reducing the unit of analysis to the organism itself, embodiment requires thinking through intra- and interorganismic relations and how brain activity is "enacted" (Colombetti, 2014) within the brain-body-world continuum.[40] There is no embodiment if there are no other bodies around. The embodiment of the brain is the becoming embodied with other bodies, and through other bodies; it is about symbiosis and mutualism rather than the perseverance of single organisms, as Lynn Margulis and Dorion Sagan (2002) put it. Embodiment means relationality and co-construction.

The brain of today's epoch seems to be characterized by its relational architectures in an ongoing formation of brain matter. That every epoch has its brain means that the brain it enacts becomes also the actor of its own existential conditions. In this sense, theories of embodiment are not just abstract immaterial representations of somato-material processes. Rather they are active forces in the transformation of existing sociopolitical and material realities; they even transform the existential conditions of the brain itself. Hence, the embodied approach to the brain is literally embodied; it is not monitoring reality or specific neurobiological, developmental or social processes; rather, it *is* the process *itself*: it recombines preexisting material and creates new ways of being. Theories of embodiment induce new modes of existence fostering combinations on all different levels of organization—genetic, neural, organismic, environmental/social: combinations that were not present before.

THE HACKABLE BRAIN

As much as the emergence of the embodied brain appears to be from today's perspective an epistemic event, the social movements of the 1970s and 1980s created the conditions in which everyday "body politics"[41] compelled existing essentialist epistemic understandings of the brain to reconfigure. Feminist and queer movements;[42] critical approaches to science, technology, and medicine;[43] the deconstruction of disembodied information systems and representational information technologies;[44] and a multiplicity of indigenous and antiracist movements released the idea of the body as a political potentiality.[45] Making the brain permeable to the pressures of the social movements was coextensive with contesting the universality of other visions of the brain—which I discussed in the previous sections—by infusing science and social science with social antagonisms.

As social movements were bringing questions of justice to the heart of the sciences and social sciences of the brain, a new, wider cultural imaginary of the brain was also emerging: the brain as self-regenerating and perpetually improvable. The embodied brain is not just a brain that assumes social justice; it is also a self-assertive brain that tries to overcome discourses of fatigue, intrusion, death, and degeneration by viewing itself as the all-in-one solution: it is the source, site, and target of its own regenerative practices. Even if the vision of the embodied brain privileges contextuality and specificity, its logic is based on an idea of neutralizing the notion of limit and context as imposed by other brain discourses. The embodied brain represents a powerful form of cultural universalism: it pledges to heal not in terms of correction (cerebral body), protection (immune body), or recurring to a fixed origin (hereditary body) but in terms of its own open reconstruction and recombination: the hackable brain.

Embodiment promises to engage with the lived pains of the body: the tamed flesh, the tortured flesh, the oppressed flesh.[46] But at the same time this promise is localized: it hinges on the belief in a recombinant individual agent. The ambivalence of the vision of the embodied and emergent brain is that it arose as a powerful critical practice that questions the prevalent decontextualized and out-of-time individualism circulating in everyday culture as well as in neuroscience, evolutionary biology, and psychology/developmental science in the Global North. But this thrust toward undoing the individual agent was gradually appropriated by the discourse of flexibility that came to destabilize the prevalent assumption of the individual as a self-contained, cognitively fully equipped and intrinsically competent agent.[47] The flexible individual is in permanent self-modification: brain computer-interfaces, brain boosting, and brain enhancement.[48] Social, subjective, neuronal flexibility is not just the target or the modus operandi of self-relationality; rather, it is the condition of embodied liberal individualism. Social control is embodied; it is exercised through the constant process of modifying one's own material bodily existence.[49] Individuals are in an everlasting process of self-maintenance; one could almost say that we never die or live but are just perpetually maintaining and working on our brains and bodies.[50] Contemporary political governance encounters the individual as an assemblage of ideas, limbs, high-tech devices, chemical substances, and environmental factors that is continuously creating and re-creating itself, hacking and unmaking itself.[51]

The vision of the embodied brain is fractured and ambivalent: it is simultaneously affiliated with social movements that made its existence possible and with liberal individualism and the entrepreneurial geoculture emerging after the 1980s. In the embodied brain, emancipation and control reside simultaneously. Social movements have opened a space for performing the brain as embodied, a space that did not exist before, a space that came before social control and had the capacity to create new liberating conditions. The existence of new emancipatory forms of life required control to change and reorganize itself in order to be able to respond to and finally appropriate these movements. Even if this could be seen as a failure, it testifies for the opposite: social movements are successful only when they change life to such an extent that they cannot be bypassed but need to be appropriated. Social control capitalized on the new realities that social movements created, and the embodied brain was gradually assimilated into a vision of the brain as self-regenerating through its own recombination and its own making.

SELF-REPRODUCING MACHINES

The vision of recombination is not just an abstract ideal; it is firmly located in technoscientific developments in the fields of artificial intelligence, data computing, and robotics.[52] The recombinant, emergent brain questions previous models circulating in these fields that attempted to duplicate the functions of the human mind and to create an artificial quasi-human brain. This quasi brain would be expected to execute command over the sensorimotor subsystems and to act as a controlling device responsible for autonomous problem solving. In this view, cognition again dominates the circuits of action, affect, and perception.[53] Theories of embodiment attempt to change this view and link cognition directly to motion and perception circuits (and also increasingly to affective systems), questioning the necessity of the existence of a quasi brain.[54]

The quest is no longer to "implant" consciousness in intelligent machines.[55] These machines need only simple cognitive architectures, sophisticated sensorimotor engineering, fast hardware, and a sufficient amount of data containing a repertoire of social-emotional skills.[56] They then become active and emergent: simple perceptions elicit bodily movements, and in turn these organize cognition.[57] Errors within this sequence of actions produce affective states; affects intensify bodily movements and new communication scripts, which require faster responses and new, more complicated cognitive procedures, and so on. In the realm of artificial intelligence, complexity is not a gift from the humans to the machines. In fact, all humans can do is reduce complexity and simplify brain processes and body architectures. What these new machines do is far more sophisticated than what humans can produce: they increase complexity through learning and recombining situated processes step by step.

Here recombination points toward biotic machines that can ultimately reproduce themselves independent of human intervention. This dimension of self-organized reproduction[58] is central to the contemporary imaginary of plasticity within neuroscience and popular culture.[59] If every epoch has its brain—and as I argued, today's epoch gravitates around the embodied brain—then every epoch fantasizes about having a better brain. Embodiment and emergence open a window to the plastic brain. Recombinant plasticity is the promise that theories of embodiment and emergence bring with them but cannot fully realize. *The Brain That Changes Itself* is the title of Norman Doidge's (2008) *New York Times* best seller. Recombinant plasticity points toward a different model for understanding brain-body matter, one that ultimately harbors a greater fascination for self-reproducing organic bodies than the distributive networks, self-organized systems, and body-environment interactions that dominate theories of embodiment.

Here plasticity refers to both the ecological-developmental plasticity and neuronal plasticity of the brain. Environmental influences[60] and intrinsic processes of interaction and ecological symbiosis with other bodies[61] define the range of potential phenotypes that can be actualized.[62] The plastic brain is present to itself, "self-generating" but also creating new forms through the incessant interactions and reconfigurations of the different participating levels of organization. And at the same time it is constrained by the contingent limitations that exist in itself and in its environment.[63] The interplay between plasticity and specificity, as Steven Rose (1998) puts it, describes the condition for inserting real-life time and real-life contexts in the body and the brain. The recombinant plastic brain is marked by the events as they occur in the multiple interactions between the genetic, neural, organismic, and ecological levels of existence.[64] It only exists in real-time and real-world ecologies; thus it can be only understood from an ecological-developmental perspective.[65] Mary Jane West-Eberhard's (2003) theory of developmental plasticity and Bruce Wexler's (2006) theory of neuroplasticity across the life span provide good examples of how phenotypic variation occurs as a diversified process depending on a multitude of environmental factors, social and cultural conditions, and the genetic material shaping brain matter in different ways.[66]

If there is a multiplicity of intrasomatic and extrasomatic factors that affect the development and making of the brain, then the question is how to investigate specific pathways of environmentally induced variations of brain development. This is the turn to epigenetics.[67] "Epigenetics is defined here as those genetic mechanisms that create phenotypic variation without altering the base-pair nucleotide sequence of the genes" (Gilbert and Epel, 2009, p. 12). Epigenetic factors are increasingly considered important for conceiving how genes are (or are not) expressed in processes of development and how environmentally induced changes of the organism can be transmitted to the offspring.[68] Epigenetic explanations of human development attempt to grasp the multifactorial complexity involved in extragenetic micro-organismic processes and cellular transformation as well as in organism-environment interactions.[69]

The study and standardization of epigenetic factors becomes one of the key innovations driving basic research and applications in neuroscience from an evolutionary-developmental perspective.[70] Consider, for example, research on the environmental impact on fetal development[71] and on gene expression through exposure to different nutritional substances,[72] the prevalence of specific types of degenerative processes associated with later life,[73] or the influence of social experiences on phenotypic variation.[74] These are just few examples; what matters here is that epigenesis opens up the field of research on the embodied brain toward different scales of gene-environment configurations. Elsewhere we have examined the enormous variations of these scales as well as how different epigenetic approaches conceptualize and operationalize the relation between the DNA and proteins, cells, the organism, and their environment in experimental research (Chung et al., 2016; Cromby et al., 2016).[75] But what is common to all of them is that the brain is a plastic system shaped through the interplay of epigenetic factors and our genes.

The moment of the announcement of the Human Genome Project, which was mentioned at the beginning of this chapter, was probably one of the last instances of celebration of genetic reductionism. To the words of President Clinton that "we are learning the language in which God created life," we should probably add: "Let the race for epigenetics begin!" After the celebrations of the decoding of the human genome had faded and given way to skepticism, *Time* magazine rushed to announce a new decoding: the decoding of the human *epigenome* as a new major scientific discovery

(*Time*, December 8, 2009). Forty years earlier the gene was an absent reference in the widespread scientific fantasies and popular imagination of the brain. But very quickly it became the floating signifier in the geno-centric imaginary that dominated the end of the previous century. Another turn now: what only few years earlier would have been formulated as "Why your DNA is your destiny" or "Your genes, your choices" (Baker and American Association for the Advancement of Science, 1997) today reads: "Why your DNA isn't your destiny" (*Time*, January 6, 2010). Now the task is to codify epigenetic factors, sort out substances and environmental conditions that inhibit or promote specific gene expressions, and standardize the mechanics of the environment-organism interplay and the environment-development-gene interplay.[76]

The outcome of this interplay is phenotypic variation: ecomorphs, the emergence of different phenotypes that is dependent on the influence of the contingent ecological and relational factors within which an individual of a species is embedded.[77] I use the term here in an extended way: ecomorphs as standardizations[78] of the effects that epigenetic developmental factors (be they intraorganismic or extraorganismic) have on a recombinant plastic organism. Ecomorphs represent stable configurations of ecological-developmental influences and the genetic code, of what Hannah Landecker (2011b) describes as the constitution of the environment and the social as a biologically meaningful signal in epigenetic research. Reducing and classifying the environment to a mere signal that induces drastic changes in genetic function is the crucial step in developing classifications of causal relations between the environment and the gene. Ecomorphs can be considered, then, as classifications of the causal coupling between certain environmental situations and a specific expression of genes. In this sense they are the smallest knowledge unit that has biovalue in epigenetic research and underpins the image of the brain as plastic.[79]

AUTOCREATIVE BRAIN MATTER

In the previous sections I told a story of brain matter from the perspective of its capacity to recombine itself (see the plot for this speculative story in table 6.1). The recombinant plastic brain is literally autocreative,[80] in a constant process of self-destruction and self-generation: it becomes a powerful political agent that changes and shapes the environmental, social, and biochemical conditions that make it happen. The promise of the autocreative brain is its capacity to be open to its own partial appropriation: it can be enclosed in neoliberal markets, it can maintain processes of ordinary political governance, it can excite policy makers and marketeers alike, and it can even fuel fantasies of grand political liberation.[81] The autocreative brain not only becomes a "radical challenge to Western thought" (Lakoff and Johnson, 1999), it also becomes an element for the ultimate regeneration and actualization of Western thought, literally—the pop story goes like this: "We'll be able to direct changes: stimulate new brain cells and networks where and when we need them; turn genes off and on at will to repair brain damage, restore function, and optimize performance; and rewire our brains to manipulate memory and even reverse dementia and mental retardation" (Horstman, 2010, p. 8). But as Luciana Parisi and Tiziana Terranova (2000) remind us, every configuration of the brain as a specific type of organism (in this case, the autocreative and recombinant plastic organism) is the result of the conjoined action of politics, production, and technoscience in the Global North.

Autocreative brain matter has something for everyone. Despite its radical contextuality, the promise and wide appeal of the plastic brain lies in its readiness to act as a universal brain. Probably the most apparent universalist appropriation of the autocreative brain is in the politics of neurogovernance: the claim that it is through the bioscientific modulation of our brains and bodies that "we," humans, come to construct "our" subjectivity today. It is claimed that humans "not just in 'the West' but also in many other regions" have come to develop a sense of personhood through the language and practices of neuroscience and biomedicine (N. Rose, 2013, p. 6). The autocreative brain manifests itself as a supposedly universal tool of governance: the management of populations and their sense of citizenship unfolds through the biomedical regulation of the malleable brains of citizens.[82] With its fixation on how human subjectivity is formed, the politics of biosubjectification and biomedical governmentality reproduces what it tries to problematize:[83] the humanist universalism of Global North societies—now in its postliberal arrangement. As I have already discussed in chapter 2, forty years of deeply divisive neoliberal politics have fractured society and given birth to postliberal enclosures of power that amalgamate diverse fragments of Global North societies to dominant subjectivities—and it is these subjectivities that the presumed universalism of biogovernmentality seem to affirm and perpetuate.[84]

TABLE 6.1 — Plot for a Story of Brain Matter

	BEHAVIORIST BRAIN	COGNITIVIST BRAIN	CONNECTIONIST BRAIN	EMBODIED BRAIN	AUTOCREATIVE BRAIN
UNDERLYING METAPHOR	mechanical interface	digital computer	network	animal-human-machine hybrids	autonomous machines
MODELLED ON	basic animal physiological processes	controlled problem solving	distributive processes	body-environment interactions	reproduction of organic bodies
FUNCTIONING PRINCIPLE	stimulus-response	universal algorithms	nodes and weights	emergence	plasticity
ORGANIZATION PRINCIPLE	black box	centralism	decentralism	contextualism	recombination
SUPPOSED BIOLOGICAL SUBSTRATUM	physiological processes	genes and brain modules	neural circuits	organism-environment assemblages	biotic machines
EXPLANANS	determinism	nativism	connectivity	relationality	epigenesis
DOMINANT CULTURAL-POLITICAL CONDITIONS	Fordist state	liberal democracies	neoliberal culture		neuro-governance and postliberal polity
SOCIAL MOVEMENTS	workers movement; social liberalism	Chomskyan liberal egalitarianism; civil rights; identity politics	alter-globalization movements; feminist and queer politics; autonomist movements; postmodern perspectivism		alterontological politics; decolonial movements

Possibly, the underlying condition for such an appropriation of plastic brain matter for the project of neurogovernance is its firm inclusion in the system of biovalue production. Neuroscience and brain research is concomitant with biotechnology and biomedicine and their propagation in public health and clinical practice.[85] The inclusion of brain matter in the system of biovalue production and the proprietarization of the epigenome intensifies its conflictual political character. It is here that Malabou's narrative of liberation discussed at the beginning of this chapter inserts itself. If plastic brain matter today generates itself by increasing late capitalist power and wealth through its biovalue, then there is surely potential for some kind of dialectic negation, Malabou asserts:[86] a total reversal, or better, a full-scale sublation of brain matter's appropriation into bioproduction—seizing plasticity for liberation.

But such a thesis assumes that plasticity is independent of brain matter. It constructs plasticity as a strategy and a device that we can use to modify our social and political existence. It presupposes that technoscience and capital—disguised as nature—breathe into the brain's neurons the breath of plastic life. And as if this split between what the brain is and what the brain does is not problematic enough, it also assumes that humans will be able to seize plasticity in order to free their brains from late capitalist appropriation and then, subsequently, free themselves. This is a clash of universalisms: universal brain liberation against the universalism of the brain's expropriation to biovalue. One cannot avoid seeing here a parallel between the plastic brain and the seizing of the state in traditional social movement politics. As I argued in chapter 5, in the same way that the state can no longer be controlled for the purpose of liberation, brain matter is no longer a device that can be controlled, seized, or simply used for achieving universal freedom. As I discuss in chapter 8, matter within technoscience today is fully privatized and, simultaneously, belongs fully to the commons. Brain plastic matter is *at the same moment* 100 percent capital and 100 percent commons. There is no total appropriation; there is no total negation. The plastic brain does not belong to capital so that it can be fully controlled. And as much as it does not belong to capital, it does not belong to "us," humans, and we cannot free ourselves by using it.

Instead of approaching plastic brain matter as this universal decontextualized entity that can be harnessed by different political projects—be it neurogovernance, biovalue creation, brain liberation—I am looking here for a politics that follows brain matter's traits as emergent and embodied and attempts to remain committed to the everyday experiences that shape it. It is here that the question of alterontological politics and more-than-social movements that I discuss in the remaining two chapters of this book emerges. My argument in this chapter is that the type of brain that humans believe they have today is the brain that enacts its own real existence and shapes itself. So is it possible to ontologically enact a plastic brain that commits itself to the emergent, embodied, and contextual qualities of brain matter by defying its purported universal appropriations for imposed social or political aims—be it through its enclosure in processes of capitalization, or as a device to achieve some form of grand liberation, or as a tool that facilitates neurogovernance?

Is it possible that the autocreative brain creates brain matter that cannot be replicated and universalized? Is it possible to equip our brains with all the capacities they need to avoid their capture for other political goals? This would require the plastic brain to engage in a grounded experimentation with its own ontological making and to refuse that it has a universal architecture and way of functioning. In fact, it would require delinking from prevalent forms of brain politics by creating alternative mundane brain ontologies: ten thousand tiny autocreative brains. This would be an alterontological politics of plastic brain matter. Many such experiments with alternative compositions of the brain are happening already. I cannot explore them in this chapter—my aim here was to develop a repertoire of existing forms of politics within technoscience, in this case brain sciences, and to open the view toward an alterontological politics that I develop in the last two chapters of this book.

COMPOSI-TIONAL TECHNO-SCIENCE

AIDS ACTIVISM

When, on October 15, 1982, White House press secretary Larry Speakes responded to a question about President Reagan's reaction to AIDS by mocking the reporters and saying, "I don't have it. Do you? . . . There has been no personal experience here," he declared publicly what everyone affected by the epidemic already knew: AIDS was at that moment primarily a cultural and political issue.[1] And it certainly was: the culture and politics of AIDS came to dominate medical, scientific, and social issues for most of the 1980s.

It took another three years before Ronald Reagan would acknowledge AIDS publicly[2] and one additional year until the government would start preparing a response to the devastating epidemic.[3] This response was nothing more than a (controversial) public health program for sex education despite the widening crisis and the twenty-five thousand reported deaths since the beginning of the epidemic in 1981. Research was still slow and would remain so for many more years, and access to care and drugs was insufficient. In 1986–1987 the AIDS Coalition to Unleash Power (ACT UP) was founded.

The years that followed saw the rise of AIDS treatment activism and its powerful impact on society, culture, health care, biomedicine, and clinical research.[4] AIDS treatment activism did not only manage to contest and change culture and public opinion on the AIDS crisis; it also managed to fundamentally transform the nature of clinical trials and the relation between patients' movements and federal health authorities in the United States.[5] The Food and Drug Administration (FDA) changed the approval procedures of AIDS drugs and introduced laboratory tests and "surrogate markers" to measure the effectiveness of drugs rather than long-term clinical trials only.[6] The first trials of combination antiretroviral therapy (ART), which began in 1992, were regarded not only as a breakthrough in the biomedical management of HIV but also as the first serious response of science to the demands of the AIDS movements for effective therapies. ART started becoming available four years later in 1996. It was just before that and, indeed, before knowing that this new class of drugs would be available that AIDS activism and ACT UP as an organization started to decline.[7]

The 1981–1986 period of AIDS activism is considered somewhat the latent phase of the movement that provided the opportunity for "making sense" of the situation and "preparing" for the "visible period" between 1987 and 1995, which is considered the key phase for direct action and intervention.[8] There are different ways to approach the rise of AIDS activism and the relation between the politics of AIDS and the scientific production of knowledge: as the process of activists becoming credible experts; as a process of participation and inclusion in existing institutions; as a network of actors (scientists, activists, pharmaceuticals, the virus itself, and so on) where all of them contributed to the creation of AIDS knowledge; and finally as an enunciative act that was based on the strong situated experiences of people living with HIV/AIDS.[9] Although all these approaches—their conceptual underpinnings and their political implications are the topic of this chapter and are discussed extensively later—reveal important aspects of AIDS activism, they all, perhaps unwillingly, imply a teleological reading of the movement: as something that targeted the inclusion of the movement's demands in committees and regulation, in scientific research, in established institutions, in biomedicine, and in the cycles of recognized experts.

I want to focus here on how AIDS activism became possible *at all* in order to be able to effect these changes. So I want to shift the focus for a moment to the first phase of activism, 1981–1986, in order to contribute to an understanding of the movement not solely as something that was fought on a social level and targeted social institutions but as a concrete, ordinary set of practices that primarily targeted the making of justice in everyday life: embodied justice, felt justice, material justice. Such remaking of the everyday led to the formation of new conditions of existence and action that could *not be ignored* by existing institutions and public discourse. The formation of ACT UP in 1987 could be understood as the moment when a movement that already existed could no longer be ignored. But this formation was the outcome of a long process that accounts for the emergence of this particular movement long before any question about its inclusion in science, institutions, social power, and so on was apparent.

REGIONS OF OBJECTIVITY

In a certain sense, such movements are practicing a "true" constructivism, a constructivism sans phrase: the making of new sociomaterial entities that change the conditions of possibility in a certain field. In what follows I use the term *region of objectivity* to define such a field as the one in which AIDS activism took place: a field that is sustained by a multiplicity of actors, objects, infrastructures, and so on. In such fields, questions about politics and knowledge are negotiated and ultimately decided in practice. Rather than conceiving objectivity as an epistemological question, though, I consider it as an ontological-material and practical question.[10] Objectivity here is not an abstract epistemological concept but a practice that changes the ontological composition of a region of objectivity.

Following the science wars of the 1990s,[11] epistemology can be considered part of the conditions of knowledge production. Scientists and their research—the organisms, objects, processes, or populations under study; technological apparatuses, methodological instruments, and epistemological debates; ethical beliefs, cultural imaginaries, and the wider polity; interest groups and state authorities; more-than-social movements; transnational institutions and national funding bodies; the private sector, publics, and the commons—all are players in the same game. They all exist in the same social-ontological field and evolve in the "mangle" (Pickering, 1995) of everyday practices that they perform. Their co-actions establish spaces in which certain ways of thinking and acting, and the very materiality of their existence, appear as given, or, if you like, as matters of fact, as a *region of objectivity*.

This type of objectivity is very different from the objectivity that dominated the debates of the 1980s and earlier, as the undistorted representation of the logic of things. As we move toward a "performative epistemology" (Pickering, 2010) and possibly toward an exit from epistemology altogether, a different conception of objectivity appears to be emerging, as knowledge now seems to be objective in a certain field to the extent that it manages to thoroughly transform the material conditions of existence in that field. It is objective to the extent that different actors in the field manage *to object*, transform, and remake the process of knowledge production itself. Thus, being objective is no longer considered an abstract qualifying attribute of knowledge, but refers instead to the efficacy of knowledge practices in objecting and transforming the composition of the materiality that underlies a field.

Desires, hopes, and investments in the objects under study—be they individuals, social groups, animals, or things—mingle with the constraints these objects impose on researchers, as well as with interest groups, ethics and beliefs, affected social actors, and state institutions. Together, they produce knowledge in ways that inexorably transform the immediate ontological *composition* of a region of objectivity. How do movements change a region of objectivity? What are the compositional politics of social movements? What is the particular type of politics that is performed in a region of objectivity?

In the sections that follow, I discuss four responses to these questions and present four corresponding conceptualizations about how politics is considered operative in a region of objectivity.[12] I start with a formalist approach to politics, which is primarily concerned with rethinking expertise and creating the appropriate procedures for considering legitimate experts in a certain technoscientific debate. A second approach to politics, participatory politics, is concerned with the expansion of the limits of public deliberation in a region of objectivity. The third approach focuses on the extension of our understanding of politics beyond human actors. A fourth emerges when actors who have been neglected in a region of objectivity contest existing knowledge and restructure the conditions of scientific knowledge production from the standpoint of their marginalized experiences. In the final sections of this chapter I bring together different aspects of these four types of politics to construct an understanding of the practices of more-than-social movements: the composition of alternative regions of objectivity and the crafting of alterontologies. Here I also argue that this compositional moment was a crucial feature of AIDS activism without which the movement would not be possible at all.

EXPERTISE AND THE LIBERAL PREDICAMENT

The question of legitimacy of participating actors is key to how a region of objectivity is constituted. In most cases individual or social actors in a region of objectivity achieve legitimacy through being or becoming recognized as experts. The necessary starting point of this approach is a classification of different types of expertise that are essential for managing issues of credibility and for shaping public discourse and decision-making.[13] Collins and Evans (2007), for example, attempt to identify different types of expertise beyond "contributory" expertise, the highest degree possessed by active practitioners who have a level of skill and knowledge allowing them to participate fully in their scientific field and to contribute substantially to its development. Here, the inclusion of legitimate experts (and subsequently the exclusion of nonlegitimate experts) in a formalized process of deliberation is seen as the main way to reshape the relations between different actors in a region of objectivity. Crucial in this process is then to extend expertise in a regulated way by using certain standards for identifying actors who can express valuable opinions about relevant technical aspects in a controversial issue.[14] In terms of AIDS activism, this would mean seeing the long and multifaceted first period of the AIDS movement as the preparation for AIDS activists to become credible experts and acquire a voice in the relevant institutions.

Managing the processes of inclusion/exclusion of experts involves distinguishing between a design phase and a political phase in every formalized debate within a region of objectivity. According to this distinction, decisions about the design should be left to experts, and experts should try to avoid influences from the broader cultural and political environment.[15] This distinction challenges previous social constructionist approaches to knowledge production. Instead of focusing on how extrascientific factors influence the production of knowledge, studies of expertise attempt to find out how intrascientific factors can be meaningfully regulated. This may seem like a response by social constructivists to the science wars of the 1990s, and it is indeed a way to preserve the relative autonomy and specificity of scientific knowledge. Social constructivism becomes Weberian, but with a twist or two: not only can the credentialed "contributory experts" make a difference in the design phase of a debate, but so can other potential experts who lack contributory skills but hold potentially important knowledge on particular topics by virtue of their experience. However, this is not a trick used in the wake of the science wars to appease scientists enraged with the debunking of their specific expertise by social constructionists; rather, it is an attempt to preserve the specificity of scientific knowledge production while also opening it up to contributors who had not been acknowledged previously.

Using expertise as the yardstick for deciding credibility and then inclusion and exclusion in a region of objectivity simultaneously opens and closes the process of knowledge production. It opens the process of legitimacy by not restricting it to contributory experts, but it closes it down again when the process of scientific knowledge production is about to implode through the introduction of extrascientific interests. The opening is performed by assigning expertise to social actors who possess relevant knowledge without being traditionally recognized as contributory experts. Collins and Evans (2007, chap. 3) call this crucial type of expertise *interactional*; it allows affected social groups with sufficient experience and knowledge to participate in a debate over a specific controversy. However, they also exclude other types of knowledge, such as general popular understanding or knowledge extracted from primary sources without deeper immersion.[16] This inclusion and exclusion of different expertises attempts to define and preserve the *borders* of what counts as legitimate expert participation in a region of objectivity. In fact, the study of expertise is about policing these borders, rather than offering substantial insights into any particular scientific controversy.

Expertise is ultimately about a formalist type of politics in a region of objectivity: politics is understood here not in a substantive way but by defining formal rights of participation and inclusion. The definition of these rights of participation relies on a basic agreement on the normative principles that govern the workings of a region of objectivity.[17] That is, it tries to set out rules according to which all legitimate participants offer reasons for or against certain arguments: rules that need to be followed in unforced communications in which all participants act reasonably and are well informed.[18] According to the formalist approach, such principles are negated when extrascientific politics enter into the technical phase of a debate and dilute expert negotiations. This logic is akin to the centrality that experts play in contemporary liberal democratic polity,[19] in which well-informed representatives and experts are assumed to settle disputes in terms of fundamental rules based on principles that protect everyone: a constitution that regulates the process of decision-making.

But what happens when the structures that can be used to facilitate informed and democratic deliberation over a particular controversy are already permeated by the controversy itself? This was what AIDS treatment activism was facing: an already constituted social landscape and a hostile environment within institutions such as the FDA. In the formalized process of this region of objectivity there was no space for the position that AIDS activists were defending. AIDS activists had to contest the constitution of the rules in order to be able to articulate their position. The political architecture that created the possibility for resolving the controversy did not exist prior to the controversy itself.

The crux of formalist politics—the possibility of constituting a flawless space as a starting point for communication between the participants in the social contract—is an untenable position when viewed in real historical perspective. Such a space is never given as the actual starting point. One could say, of course, that it is exactly because of this that we need to support a formalist approach to politics. The argument would be that the more that such spaces of communication elude us, the more they are necessary, and the more they are necessary, the more we should accept their paramount value.[20] However, elevating formal-regulative principles to core values does little other than to invoke an authoritative moral code; it does not make them work in everyday life. Formalist politics occupy the space of the normative by vacating the space of the actual.

Like any other version of formal-regulative ethics of political engagement,[21] formalist politics in science and technology miss the substantial embeddedness of communication in the deeply asymmetrical and unequal social, cultural, and material social relations.[22] Formalist politics thus miss the fact that conflict with a region of objectivity emerges not because a proper process of expert deliberation has not (yet) taken place, but because it *already* has failed somewhere along the way. And the case for AIDS activism reveals that clearly: the incapacity of the government and the relevant authorities to handle the HIV epidemic and the AIDS crisis through the already constituted channels of communication and action had to be changed. The formalist approach is good at illustrating, in retrospect, how such failure takes place and at detecting the stratified and unequal contributions that each participant can make ex post facto. What it cannot do is to engage with them in a way that changes the formal order in a region of objectivity altogether. Formalist politics occupy the space of observation by missing the space of transformation.

Of course, formal rules sometimes can instigate social transformation in a region of objectivity. When rules of negotiation are derived from the conditions of a controversy they can have transformative effects. When viewed from a historical political perspective, such rules are always the target of social movements that contest the formal structures of liberal democracies by demanding radical changes in the norms and legislation governing a certain field of life such as civil rights campaigns, the women's movement, and the gay rights movement. The problem is that changing the rules from outside is what formalist politics deny. Formalist politics are *literally* formal: the rules are not derived from a certain situation or conflict, but are assumed to apply to every region of objectivity, to every conflict. Formalist politics—in a Habermasian fashion[23]—attempt to operate as a transcendental judge in order to control procedures for deliberating over the controversy. This external universalizing position can only function as an adjustment within the constituted order in a certain region of objectivity. But when a conflict cannot be accommodated within a region of objectivity, then a different form of action needs to be exercised: a compositional practice that remakes not only the rules of the debate, but also its content, scope, and material devices through which exchange in a region of objectivity happens. And this was what AIDS activism achieved.

PARTICIPATION AND THE LIMITS OF INSTITUTIONS

A second approach to conceiving politics in a region of objectivity focuses on the conditions of participation and the processes through which laypeople are included in debates over scientific research. The participatory approach does not start from a "normative" or "formalist" definition of who has the necessary credentials to be included in the deliberation process. Instead, it makes structural claims about the inclusive conditions for shaping a region of objectivity:[24] Who needs to be included? Where and when is participation needed? How can we enhance participation and what specific processes need to be considered?

The broad answer to the "who?" question usually attempts to develop insight into how to facilitate the co-constructive role of the *public* for shaping technological objects and scientific practices. It attempts to go beyond seeing the role of the public only as concerned with the applications and consequences of technoscience.[25] The public can be a highly elusive and easily instrumentalized category, however. Attempts to concretize the public primarily include stakeholders such as NGOs, affected social groups (usually self-organized), or other civil society organizations into the process of decision-making. They tend to emphasize user groups, patient groups, and activist groups that organize themselves to articulate specific claims on existing technologies.[26] This form of inclusion has its roots in the women's health movement[27] and feminist approaches to medicine,[28] environmental health movements,[29] and of course the organizations of people living with HIV and AIDS that I referred to earlier in this chapter.[30] Facilitating the participation of such groups can be strengthened by changing broader scientific research agendas themselves or more specific science policies (such as the clinical trial procedures discussed in the case of AIDS activism).

More broadly, there are calls to intervene in science "upstream," before applications are decided and when it is still possible to shape the compilation of research agendas.[31] Science policy and the governance of science become crucial factors for shaping the framework in which such decisions about research topics and agendas are designed.[32] The question of "where?" often appears in close connection to the problem of "how?": we find approaches that call for practices that invigorate advocacy,[33] public engagement,[34] and attempts to reinforce accountability.[35] There are also broader quests for a civic epistemology[36] and the setting up of an agora as space for negotiation between scientific and social actors.[37] Such positions strive to enhance inclusive procedures in decision-making[38] and deliberation.[39]

Participatory politics in all their variety and nuance shift the focus from who is an expert or who is a legitimate participant in a region of objectivity to the problem of enhancing inclusion of citizens. They are thus concerned with changing the structures in which debate in a region of objectivity takes place. Participatory politics differ significantly in their scope, target, tactics, and radicalism. When I refer to participatory politics, I mean approaches that aim to reform existing institutions in order to change the conditions of citizens' inclusion. Other radical approaches go as far as to question the whole institutional structure and propose a radical democratic approach[40] for creating grassroots initiatives, or even alternative institutions—and AIDS treatment activism is a good example of this. Some of these more radical democratic positions are discussed later, as part of situated approaches that highlight the importance of power inequalities and their effects on the shaping of science and technology. The form of participatory politics that I discuss in this section is primarily concerned with the role of social difference in science policy and with strengthening public participation.[41] Deliberative democracy, policy adjustments, and public accountability become the tactical means deployed in participatory politics to promote inclusion that changes the balance of power in a region of objectivity.

The civic imperative of participatory politics is to construct regions of objectivity such that civil society can be included in processes of direct democratic decision-making. Civil society in participatory approaches refers to actors who come together on a voluntary basis rather than as functions of state-supported institutions, such as education and science, or pure market forces. Civil society is seen as a counterbalance both to strong individualizing tendencies of liberal democracies that emphasize the rights of the single individual and to the excessive growth of particular state functions and governmental institutions. Instead of focusing on abstract and formal procedures for settling conflicts within a region of objectivity, participatory politics emphasize that sound science policy can be achieved only if there is a system in place that relies on citizen deliberation and engagement.

There is here a strong echo of communitarian positions[42] that portray a "community" whose members negotiate specific issues in terms of their "horizon of meaning." Participatory politics works on the interface between the community and existing institutional structures. Institutional spaces broaden or narrow the accepted horizons of meaning against which a controversy in a region of objectivity can be debated and eventually resolved. The keyword here is *framing*: appropriate formulating of an issue in order to include public actors. Framing in this context is not a pure discursive strategy;[43] rather, it refers also to social spaces, practices, and relations that enable alternative ways to frame an issue.

However, the questions of how far and how many actors can be included in the deliberative processes of an institution is always filtered through the existing possibilities and margins that the institution allows—in Daniel Neyland and Steve Woolgar's (2002, p. 272) words, the "conditions of possibility for accountability." Participatory politics—and here I refer to mainstream participatory approaches, not to radical forms of participation, which are discussed shortly—operate as corrective forces to the shortcomings of existing institutions. The participatory politics of AIDS activism was primarily about exposing the shortcomings of relevant institutions of the time and forcing them to renegotiate the conditions of participation of civil society actors that at this moment were considered illegitimate. But if one focuses only on the main target of participatory politics, which is supposedly inclusion in institutions, something very important can be lost along the way: the broader social, material, and ecological transformations that movements set in motion. These transformations are not as visible as institutional transformations, but they are crucial for sustaining livable worlds for the affected social groups and for sustaining the movements.

Transforming existing institutions can, of course, have far-reaching effects in democratizing a region of objectivity. Nevertheless, institutional participation of this kind operates within existing institutional coordinates that define not only a certain problem and its potential participants but also what is irrelevant and thus disposable from the perspective of the institution. The effects of AIDS activism go far beyond its ability to transform the relevant institutions. The AIDS movement transformed thoroughly everyday life, culture, and the material conditions of existence of so many people far beyond its institutional reach. Participatory politics is crucial for enhancing institutional processes of democratization. However, it needs to be extended to and complemented with politics that not only transform existing institutions but also operate, if necessary, outside existing institutions and give birth to alternative forms of action, imagination, and practice in a region of objectivity. Later in this chapter I refer to this politics as compositional.

NETWORKS AND THE IGNORANCE OF GOVERNANCE

If participatory politics focuses on the inclusion of social actors in existing political institutions, actor-network theories attempt to include non-human actors in descriptions and ways of acting in a region of objectivity. Actor-network theories avoid thinking in terms of the sociopolitical and the natural as divided worlds and propose an approach to the social that complicates this bifurcation and prevents us from "counting in advance" what society and nature are composed of.[44] Rather, they seek to explore how humans and nonhumans collectively and in emergent ways construct a region of objectivity by being implicated in networks of connectivity.[45]

The main concern here is to describe these connections and to elucidate the appropriate assembly for dealing with particular issues: "Every new nonhuman entity brought into connection with humans modifies the collective and forces everyone to redefine all the various cosmograms" (Latour, 2007, p. 813). Cosmograms are divergent and often conflicting ways of actual world-making. Latour argues that the task is to "detect *how many participants* are gathered in *a thing* to make it exist and to maintain its existence" (Latour, 2004b, p. 246, emphasis in original). This is also a formalist approach, albeit quite different from the one described earlier: to define the appropriate staging for each problem, so as to enable effective processes of representation, mediation, and translation between human and nonhuman actors.

In order to accomplish this, it is necessary to maintain symmetry between the different actants and explanations, since a hasty conclusion about a prevalent form of agency can trap us into "prematurely naturalized objectified facts" (Latour, 2004b, p. 227).[46] Attention to symmetry reveals how new entities are formed in the sociomaterial world and how new complex "entanglements" of mutual attachments involve different actors and things.[47] Every given entity in a region of objectivity is made through its connections with other things: "Actor-network is, has been, a semiotic machine for waging war on essential differences" (Law, 1999, p. 7). What exists is produced and made through relations: "relations everywhere" (Strathern, 2005).

This inflationary focus on relations and connections corresponds to a broader shift in conceptualizing politics as governance in contemporary political theory and practice: politics as an affair of actants perpetually adapting to and establishing alliances and networks with others. This conceptualization of the political implicitly abandons the governmentality model that has shaped much of alternative sociopolitical theory in recent decades.[48] Instead of self-activating actors and processes of subjectivation— that is, instead of conceiving the production of subjectivities as an effect of power[49]—governance is a form of *a-subjective* management that emerges through connecting actors and creating alignments between them.

Power is assembled, rather than exercised. Politics is enacted through the configuration of a network rather than through the enunciation and subjectivation of different actors inside the network.[50] Within this framework governance refers to the management of the network's configuration, on how specific parts of the network are assembled and put to work together.[51] Governance signifies the erosion of constitutionalism as an established set of generally accepted principles in sovereign law. It is post-constitutionalist; that is, in a scene populated by many interconnected and partial actors, governance offers a common mode of effective political stability and functioning.[52] Politics here does not follow predefined or abstract principles imposed by a central authority (usually the nation-state) or the relative balance of power between different political subjectivities; rather, governance describes the process that controls the search for regulating principles *in the absence* of an authority that guarantees such principles.

Regulating principles are developed ad hoc through intensive negotiation between participating actants.

But if we contextualize the trope of governance in historical and geopolitical terms, it emerges as a predominant form for regulating polity and production in contemporary Global North societies.[53] It is a mode that renders the actants and entities participating in a region of objectivity productive, in a double sense: first, politically productive and, second, as active parts in the existing mode of production (see also chapter 1). There is a preoccupation in actor-network theories with discerning how a situation comes into being and how complex assemblages evolve through the relational actions of the participating actants. This is the political side of networks: it necessitates the creation of new political assemblies that accommodate the emerging constellations between human and nonhuman others.[54] However, the insertion into the political assembly happens in correspondence with the incorporation of these entities into a new mode of production, the self-valorizing assembly system of the biofinancialized economies, as described in chapter 2.

Thinking politics in a region of objectivity in terms of networks and governance goes hand in hand with the emergence of new forms of displacement. The point here is not to distinguish between good versus bad governance, but to explore how governance *as such* becomes the political algorithm of contemporary organization of control.[55] What about those who cannot or are not willing to contribute to the spaces that are regulated through governance? The moment when AIDS activism came to an end in the early to mid-1990s is coincidentally the moment of the consolidation of governance. As governance came to dominate the stage of politics, it shaped how the AIDS crisis was handled on a global scale in the 1990s and 2000s.[56] And it is in the governance of global networks of HIV infections through multinational pharmaceuticals, supranational organizations, charitable or private foundations (such as the Bill and Melinda Gates Foundation) and the emergence of philanthrocapitalism, global public health, and local contingent responses that one can see how the productive assemblage of governance fails to include those who cannot act within its framework of global governance—let alone those who are unwilling to participate it.[57]

What about human and nonhuman actors who betray the constituted order of governance, who disrupt the function of networks with an intractable conviction about justice, who ask inappropriate questions, who position themselves outside the "we" of the political assemblage of governance, who create alternative spaces and bring the wrong messages? From the perspective of governance this is a form of fundamentalism, and, as Latour says, the ultimate political question is "Can fundamentalism be undone?" (Latour, 2005a, p. 31).[58] But how far can movements go if they follow the logic that "if you cannot bring good news, then don't bring any"? Exploring the constitution of networks provides an insight into the intricacies of more-than-human material agency, but ultimately these new forms of agency are reinserted into the constituted order of governance.

SITUATEDNESS AND THE INDETERMINACY OF EXPERIENCE

A situated perspective on a region of objectivity presents an almost reverse but simultaneously also closely related account to the one just presented.[59] Situatedness primarily means to articulate knowledge politics from the standpoint of neglected experiences within a region of objectivity.[60] Implicitly, situatedness challenges the invisible "we" so often assumed or summoned in all previous approaches: the contractualist "we," the possibility of supposedly all-inclusive institutions, or the "we" of the networks of governance. Of course, recognized experts, policy makers, lawmakers, facilitators, observers, mediators, diplomats, and translators are all necessary for creating the political architectures of a region of objectivity, but from a situated perspective the most crucial viewpoint is that of the fully engaged yet *partial* participants.[61]

While situatedness can only exist within webs of relationality (similarly to actor networks), these webs are asymmetrical and unequal (which presents an almost reverse approach to networks politics). Asymmetrical relations between human and nonhuman others are constitutive of the conditions of every region of objectivity, and this can be illuminated from different angles: we see discussions of how technoscience contributes to domination[62] and attempts to enhance social and political structures that facilitate alternative forms of intervention,[63] mobilizations of radical science movements,[64] grassroots democratic participation, and science activism.[65] Many of these positions reflect radical critiques of technological, scientific, and medical rationalities[66] and are historically rooted in the social movements of 1960s and 1970s—in particular feminist, antiwar and antinuclear, and ecological movements[67] and the new social movements after the 1990s.[68] From the perspective of situated politics, questions of credibility (as in formalist politics), existing institutions (as in participatory politics), or inclusion in relational architectures (as in the networked approach) are facets of a continuous movement of social transformation that is primarily initiated by neglected, silenced, or effaced positions in an asymmetrical and thus antagonistic social and material order.[69] AIDS activism is probably a classic example: starting from their neglected (and in many cases actively effaced) and partial experience of living with AIDS, activists articulated a presence that challenged the already constituted order of a region of objectivity. The experience of life and death in the epidemic was the ground for AIDS activism.

An important question has to be asked about the nature of the experience of neglected groups, human and nonhuman, that are effaced in a region of objectivity: Can their experience be treated as given and definite?[70] Does the experience of the excluded preexist the relation of exclusion? While the formalist and participatory approaches seem to take experience for granted (as a "have" or "have not" feature of an individual actor), the actor-network approach complicates the problem of experience by refuting its primacy altogether. Experience seems to be either reified as a substance or eliminated and dissolved in pure connectivity.[71] This impasse between reification and dissolution of experience also pertains to early versions of the situated approach.

If experience is reduced to a mere reflection of the immediate given position of the actor in power inequalities, it is also reduced to something that is instantaneously accessible and transparent. This move undercuts any possibility for real transformation of the actors because they are captured in an endless process of reiterating their own experience. This move also underpinned identity politics during the postwar period in the Global North. Although cultural studies have vividly shown that identity is always in process, because it entails the remaking of the actors and their social relations,[72] the 1990s marked a moment at which identity politics became increasingly unable to contribute to radical political mobilizations.[73] Poststructuralism attempted to resolve this problem by introducing an idea of experience as discursive formation.[74] This consists of two parallel endeavors: it challenges both the individualistic fallacy invoked in much talk of experience and the notion that experience is a monolithic and transparent object of knowledge. However, this important critique of experience usually goes hand in hand with a reduction of experience to a mere sociohistorical incident that undercuts any possibility for agency and introduces a disembodied form of social relationality and existence.[75]

To what extent does the idea of experience in situated politics challenge the pervasive logic that sees experience trapped in the binary logic between reductionist essentialism on the one hand and discursive antifoundationalism on the other? In order to question both, the reification of experience as well as the elimination of experience through discourse, we have to assume that actors do not already "have" experience. They make experience as they collectively contest existing forms of injustice. Experience in this sense evades representation; it is processual and collectively constructed. Elsewhere we called this approach to experience "continuous" (Stephenson and Papadopoulos, 2006). It is a retreat from the self, from clichéd subjectivities, from the oppressive reduction of experience to the discourses prevailing in a certain context. Experience becomes a process that pushes itself to change.[76] Experience is all there is but it is not definite and given.

The point of departure for situated politics is not experience as such or its representation but how experience is collectively and a-subjectively made in webs of relations, continuously. The involved actors experience the world by making it, in a process of co-involvement.[77] AIDS activism became possible as people created the ontological conditions that allowed them to negotiate their sometimes very divergent experiences in the epidemic. In 1981–1986, different individuals, groups, and communities of people living with AIDS created common spaces, shared practices and language, and different modes of engagement with the virus that allowed the movement to become a movement as such and to become visible after 1987 through widespread direct action and interventions. In her insightful study on AIDS activism, Deborah Gould (2009) shows the complexities and ambivalences in this process of trying to make sense of, negotiate, and develop repertoires of action in the face of the epidemic. The movement and experience are contemporaneous; none of them preexist the other. From the perspective of situated politics, the point is not primarily to acquire the right credentials or to participate in governance and institutions but to engage with and compose alternatives to the dividing forms of power in social, material, and ecological environments that enable a movement to exist.[78]

THE TIME OF COMPOSITION

Despite significant differences and regular controversies between these approaches, all of them describe practices that interrogate the existing order of a region of objectivity: the definition of credible participants, the expansion and restructuring of institutions, the inclusion of nonhuman actors, and the inclusion of neglected voices. They seek to modify the conditions of possibility within the constituted order of a region of objectivity. But a different political framework is necessary if we approach a region of objectivity as an open process rather as an already constituted order that needs to be rectified in some form or another. More precisely, I am interested in how actors constitute themselves, whether human or nonhuman, long before they are formally recognized as such—that is, as constituted subjectivities capable of changing a region of objectivity. This is the crucial question of alterontological practice: how to contribute to the *making* of actors in a certain region of objectivity, or even how to become one. This question is not about the visibility or invisibility of an actor; rather, it is about how actors become the inheritors of unchartered obligations and the locus of change of a region of objectivity before they can be perceived as such, even before they consider themselves actors. I refer to this politics as compositional: the making of a sociomaterial actor on the level of everyday existence, *before* negotiations about inclusion in existing institutions, formal procedures, expert committees, networks and governance, modes of enunciation, and so forth unfold.

Compositional politics is about creating alternative forms of life that allow the renegotiation of a given constituted order to take place—whether this renegotiation is about expertise, formal structures, the nature of institutions, the role of nonhumans, network politics, standpoint politics, situated experiences, and so on. An alternative form of life acts as a set of constraints against which actions as well as possibilities for new actions within a region of objectivity evolve and take place. In this sense, as discussed in chapter 1, it becomes a form of life that cannot be bypassed—not because it defines in a deterministic fashion the outcome of actions,[79] but because it creates new ontologies that allow specific actors to become actors and to intervene and interrupt or alter the constituted order of a region of objectivity.

AIDS activism and the entanglements of human actors (patients, activists, researchers, FDA regulators, political parties, governmental authorities, and so on) and nonhuman actors (HIV virus, medications, tests for viral loads, circulating body fluids, and so on) have been investigated using all of the frameworks presented earlier in this chapter. One of the most prominent examples is Steven Epstein's (1996) important work on how the collective of AIDS activists inserted itself into a biomedical region of objectivity in ways that undermined and eventually changed the existing terms of the debate. Epstein describes how AIDS activists became recognized experts and increasingly contributed to shaping biomedical research. He shows how AIDS treatment activism, once it became powerful enough to enter existing institutions (such as the FDA), changed the format of treatment research, clinical trial procedures, the distribution of medications, and so on.

But if we read AIDS activism only from the perspective of what it has achieved in terms of the politics of expertise, governance, and institutions within the region of objectivity around HIV, we are in danger of missing all these diverse, fragmented relations, practices, and actions that made the emergence of movement possible. Instead we would impose a teleological view onto the actions of the movement as if it were designed from the beginning to achieve these targets.[80] Such a teleological reading fails to pay attention to the process of the making of an actor long before its practices were recognized as a form of "effective" politics. Moreover, widely recognized political achievements of a movement are sometimes the "by-products" of the movement's actions, rather than its main focus. Often movements mobilize in order to encounter direct forms of injustice and oppose them on an everyday level long before they develop organized political interventions and campaigns. This is certainly the case for AIDS activism between 1981 and 1986.

Teleology serves many objectives; one of the most prominent is presentism: to conceive the AIDS movement as a succession of events that all constituted preparation for the activism of the late 1980s that primarily targeted the FDA and the broader process of mainstreaming AIDS through its "professionalization." Already in 1990 social theorist and activist Cindy Patton (1990, pp. 19–20) had warned that "the amnesia surrounding the history of activism between 1981–5 was initially a product of the emerging AIDS industry; but it has been reinforced by progressives who have begun to locate the beginning of AIDS activism in 1987 or 1988, with the emergence of ACT UP."[81] Rather than what appears to be a single, unified movement, the first phase of AIDS activism is a long period of composition: a multitude of different practices that simultaneously attempted to deal in some way or another with the devastating crisis, to make sense of the broader social, political, and cultural situation and, most importantly of all, to secure the material conditions that would allow the gay communities under threat to continue to exist.

EMERGENCY CARE

From very early on, gay men and their communities developed and invented a multiplicity of practical engagements with an epidemic that quickly became a devastating social and public health crisis.[82] Building on the work of Puig de la Bellacasa (2015) on the temporality of care, we can understand these practices as emergency care.[83] Here is a list of such practices (in no particular order): challenging medical decisions; campaigning to raise money for alternative research; organizing support, volunteer caretaking, and extended care services; creating autonomous service provision (AIDS service organizations); setting up new community spaces and community organizations to engage with the new challenges of the crisis; extensive experimenting with one's own body and (not officially approved) drugs;[84] getting involved in intensive lobbying of medical associations, doctors, hospitals, local councils, and public health officials; organizing media interventions; negotiating the meaning of their own subjectivities by setting up community meetings, educational initiatives, and debates in newspapers through leaflets, magazines, editorials, and letters; developing new forms of affection, intimacy, and reciprocity; educating themselves in medical, health, legal, and policy issues; (re-)politicizing white, mostly middle-class gay men who started to realize that their relative privileged positions were inherently precarious;[85] the activist beginnings of the "SILENCE = DEATH" project; the many calls for civil disobedience and for getting "angry about AIDS" (Kramer, 1989, p. 48); militant action and confrontational activist practices such as sit-ins, traffic tie-ups, blockades, occupations, picketing, AIDS walks, and rallies; being prepared to get arrested;[86] holding candlelight vigils; inventing and reinventing new sexual practices and sexual expressions;[87] taking direct action and holding contentious protests; defending gay bathhouses and other sex establishments;[88] trying to make sense of the broader social, political, and cultural meaning of the epidemic;[89] setting up buyers' clubs of illegally manufactured or illegally imported drugs;[90] attempting to maintain self-respect and gay pride and navigate through all these conflictual feelings about one's own community produced by the hostile social environment and the constant stigmatization and demonization;[91] negotiating the burden of shame about gay difference and fear of social abjection created by the prevailing homophobic hysteria;[92] defending gay male sexuality within the terror and panic of mysterious deaths and diseases;[93] being proud of the community's attempt to face the crisis; and giving love to the ill and dying.

Through these compositional practices, AIDS activism gradually took shape and constituted itself after the start of the epidemic. Simon Watney (1997, p. xii) says that what we could call "the" gay community "did not preexist the epidemic in any very meaningful sense," and one could add here that AIDS activism did not preexist the emergence of this community. This means that AIDS activism is not just a reaction to the epidemic, as if the epidemic remained the same since it erupted and AIDS activism was conceived by a community as a full-scale strategy of response. Rather, AIDS activism is the outcome of a long formation process in which thousands of gay men and their communities tried to grapple with a devastating virus. AIDS activism is the outcome of an ontological encounter and an ontological conflict between human bodies and HIV retroviruses unfolding within a hostile homophobic culture and a specific biomedical regime. This group of gay men became a community and engaged in AIDS activism as a way of understanding and managing this ontological encounter. AIDS activism is the attempt to create a material, biochemical, medical, social, and cultural space in which the relation of human body and HIV could be reshaped after the initial outbreak of the epidemic. And of course the first concern was to just survive this encounter. AIDS activism became possible because of the everyday practices that allowed the community in the making to sustain itself.

The legacies of the previous political era of gay liberation were crucial for developing these new forms of organizing in the midst of the epidemic. Patton (1990, p. 19), for example, highlights the liberationist roots of early AIDS activism. But beyond the liberationist legacies and the integrationist realities, most communities had to reinvent themselves in order to be able to exist in the new conditions of the epidemic. Many different political trajectories and currents existing within gay communities of the 1960s and 1970s were reconfigured in a process of profound sociomaterial experimentation: emergency care, practical justice, reaffirmation of sexual difference, invention of novel forms of intimacy and affectivity, creation of spaces of political and cultural autonomy and protection, the reclaiming of confrontational politics—all practices that constituted the "new" community. This kind of dispersed, everyday, imperceptible politics—compositional politics—enabled the emergence of AIDS activism in the early 1980s[94] long before it became recognizable as the single social movement that was gravitating around ACT UP. Compositional politics reordered the conditions of everyday being and experience so as to facilitate the emergence of a new social actor. One has to take literally every word of one of the concluding phrases of Larry Kramer's (1989, p. 49) historic call for action, *1,112 and Counting*: "we must fight to live."[95]

ALTERONTOLOGICAL POLITICS

Conceiving the politics of creating alternative ontologies of existence as compositional highlights that a region of objectivity is never given or complete. Of course, compositional politics cannot but emerge out of the different accounts of politics already existing in a region of objectivity. It draws from the politics of expertise its democratic sensibility toward noncontributory experts, from participatory politics its bottom-up citizen perspective on technoscience, from network politics the agency of non-humans, and finally from situated politics the collectivization of transversal neglected experiences. However, compositional practice is much more than the aggregation of these different political sensibilities: rather than being anchored in a given institution, position, network, or subjectivity, it attempts to redraw the form and content of an existing political order. The way a region of objectivity is constituted and the political practices of the involved actors are two different things. Changes in the political practices precede changes in the constitution of a region of objectivity. The political composition comes first and shapes the institutional composition of a region of objectivity.

Compositional politics happens when certain human or nonhuman actors, imperceptible actors, emerge by addressing questions of injustice and by materializing ordinary relations of justice (see also chapter 5). Compositional politics is not *primarily* concerned with contesting given regimes of control by introducing improvements in an existing political order—that is, rules of equality, the codification of rights, and the establishment of institutional structures for the articulation of public responsibility. This may seem a paradoxical proposition since rules of equality, rights, and responsibility make up a plausible part of what many of us understand as active political engagement. And even if this form of politics is indispensable in certain contexts, I want to argue with Rancière (1998) that politics in the sense of composition arises from the emergence of the miscounted, those who have no place within the given order or a region of objectivity. Politics is a collective enterprise that exposes a given order to be limited, contingent, and inconsistent by creating an alternative lifeworld inhabited by the previously miscounted: alterontologies.

This is what AIDS treatment activists did years before they became political subjects in their own right and constituted a social movement with a distinct profile. Activists created an alternative objectivity, an alterontology of existence. This new form of life is ontological not only because AIDS activists and their communities engaged with the virus directly (for example, by confronting existing research; by acquiring medical, epidemiological, and biochemical knowledges; or by experimenting with novel drugs and, perhaps most importantly, with their own bodies) but also because they changed their own material conditions of life in order to be able to exist with the virus. The 1981–1986 period of AIDS activism was the period of learning to live, learning to die, learning to survive the virus. This period made the visible phase of AIDS activism after 1986 possible, a phase that was directly oriented toward stopping the virus. But first one had to accept the presence of the virus itself and to find ways to ontologically negotiate its lethal existence by materially reconfiguring everyday life. AIDS activism became an ontologically transformative social movement because it changed the material conditions of existence of hundreds of thousands of people, and by doing that it changed the course of the virus itself, how we understand it, and how it was confronted. The experimentation with one's own body, the making and circulation of illegal medications, bodily self-experimentation, the changing of forms of everyday sociability and of sexual intimacy, the militant attacks of political institutions, the material restructuring of urban spaces, and the reshaping of medical testing protocols and scientific procedures are all ontologically transformative practices that are simultaneously the effect and the precondition for the continuation of existence of marginalized actors that redraw politics as we know it by creating alternative conditions of existence that make just forms of life emerge: alterontologies.

CRAFTING ONTOLOGIES

STACKED HISTORIES

Where once stood the first English factory, a museum took its place, only to disappear a few decades later and give way to a community-based experimental project space. The Silk Mill was built in 1721 and was the first water-powered mechanized silk-throwing factory.[1] It became Derby's Museum of Industry and History in 1974. But the museum fell gradually into disrepair: underfunded, with falling visitor numbers, and with many exhibition cases in deteriorating condition, it was in need of a thorough renovation. The recurring crises of the traditional public archive put many similar institutions under pressure. The museum closed in 2011 to undergo necessary structural works and to prepare a plan for redevelopment.

Reconstruction of the Silk Mill started in 2013. The aim was to include stakeholders as well as the city and its people in a public consultation about its future and its use. One of the main industrial buildings of the fifteen-mile World Heritage Site of the Derwent Valley Mills,[2] the Silk Mill has a prominent position in the city of Derby, England, and its industrial heritage. But against and despite the heavy historical role that this building carries with it, it reopened completely empty, a seemingly blank site open for public participation and experimentation. The Re:Make project involves museum staff, visitors, and people from the community in a process of redesigning and rebuilding its space and its contents.[3] The goal of the

FIGURE 8.1 — The Derby Mini Maker Faire at the Silk Mill, Derby, UK. Photograph by Dimitris Papadopoulos.

Re:Make project is to rebuild the museum's facilities, exhibition cases, furniture and fittings, research and functional rooms, and, most importantly, collections in order to reopen as the new Derby Silk Mill—Museum of Making.[4] The Silk Mill was equipped with a purpose-developed workshop including multifunctional devices such as a very-large-format CNC router, 3D printers, laser cutters, and designing software, which allow for the reconstruction of almost every nonstructural part of the museum. Hannah Fox, the Silk Mill Museum of Making project director, envisions this remarkable process as a socially embedded, participatory "co-production" of the museum with "the people of Derby."[5]

The Silk Mill's industrial past was as turbulent as its postindustrial present. It changed ownership many times, and following technological advances in the silk-throwing industry and the changing composition of the workforce, the production techniques evolved and the building was redeveloped several times.[6] After the gradual decline of the British silk industry in the second half of the nineteenth century the mill changed again and became the chemical factory F. W. Hampshire only to be destroyed by fire a few years later, in 1910. It was fully rebuilt and remained a production site until the company moved to purpose-built premises in 1927. The ownership of the Silk Mill transferred to the local electricity corporation, which located some workshops and storage space there until the 1970s, when it was adapted for its use as a museum.

The stone foundation arches of the Silk Mill are the same as they were in 1721, as if they are holding an entity that changed so many times and yet each one of its different material configurations live inside each other. I refer to these configurations as ontologies. The term *ontology* has been used throughout this book in a rather concrete and ordinary way: an ontology is the habitat and space of existence of various animal species, groups of humans and things that provide specific conditions of how its materiality can be changed.[7] What defines an ontology is not its qualities, what it is, but how it can and cannot change. By becoming a museum the Silk Mill anticipates its ontological configuration as a large workshop that can no longer occupy valuable estate space in the historic center of a city. Equally the workshop responds to the success of a chemical factory that produced some of the most widely consumed pharmaceutical and health-care products in Britain and had to move to larger production facilities. The chemical factory emerges out of the decline of the previous silk-weaving facility, which in turn is the outcome of the attempt to establish the first modern mechanized silk-throwing production site. And, finally, the collapse of the museum and the birth of the Silk Mill as an experimental space is the attempt to negotiate the decline of the public (and publicly funded) archive by reflecting on its long and uneven industrial history and by remaking its contents.

Each new ontological configuration of the Silk Mill is an articulation emerging from the previous ontology. But my attempt to find how new ontologies emerged from the previous ones defied my expectation for easy causal explanations. Neither the archival materials nor the accounts of the people I talked to offered any sense of planning, intention, or structural necessities. Rather, what I found is that ontological change in the Silk Mill happened almost as a drift rather than by continuity or by sudden rupture. Each new ontology sets constraints against which the next one develops. But these constraints do not determine the nature of the new emerging ontological configurations. Every development is contingent on the frictions that happen as new ontologies emerge. But these frictions and even conflicts never determine the outcome. And there is also a strong element of chance in how ontologies are constituted: a bankruptcy of a factory owner, a fire, a technological development in an adjacent mill that forces the reorganization of production, the decline of the public archive. Through an *aleatoric* drift against specific and concrete constraints, these material configurations evolve and develop.

The different ontologies of the Silk Mill—from its inception as the first modern factory; its many transformations; the fact that it was one of the sites where the organized trade union movement was born; and its redevelopment to a chemical factory, a workshop, storage space, a museum, and an experimental project space—all are stacked temporally and materially. Previous histories are still active in each new configuration even if each new configuration evolved in unexpected ways. This is the paradox of stacked histories: they remain active forces after their disappearance, but they are unable to determine the content of the later ontologies. There is no essentialism here. The ontological unity of the world is void in its core. There is no common denominator, no core quality, no transhistorical essence that holds these ontologies together. What holds them together is that each specific material configuration permits only certain developments and precludes others. This allows us to think of the limits of ontologies (as well as bodies, ecologies, and places) without falling back into naturalized boundaries and "objective conditions" or by promoting an idea of permanent fluidity by dismissing the existence of limits all together.

MAKER CULTURE

Today's experimental space in the Silk Mill is in a direct dialogue with its industrial past. Jonathan Wallis, current head of Derby Museums and one of the people who is behind this extraordinary community-driven, experimental reinvention of the museum, says that today's Silk Mill is the inevitable response to Derby's long manufacturing history. Today's experimentation is necessitated by Derby's historic position as the leading site in advanced engineering in the UK. And although Derby is not the sole leader in advanced manufacturing today, this powerful heritage has shaped materially and socially the city and its people. The Silk Mill will be "inspired by the makers of the past, made by the makers of today and empower the makers of the future."[8] But how can an experimental space for redesigning a museum from the inside out and from the bottom up be a response to an ontological configuration that started 293 years earlier? What is the vision that captures the imaginaries of those involved and what is the hope?

This chapter addresses these questions through a reading of the maker culture, which is at the heart of the reconfiguration of the Silk Mill. The empirical materials I present here draw on four years of data collection. In 2012 I started participating in makerspaces and hackspaces in the East

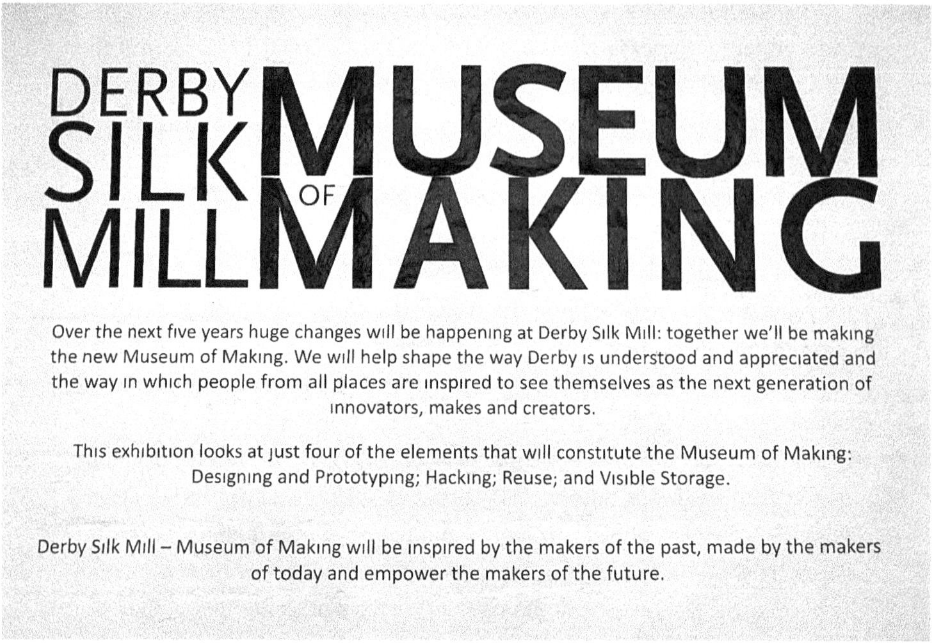

DERBY MUSEUM
SILK OF MAKING
MILL

Over the next five years huge changes will be happening at Derby Silk Mill: together we'll be making the new Museum of Making. We will help shape the way Derby is understood and appreciated and the way in which people from all places are inspired to see themselves as the next generation of innovators, makes and creators.

This exhibition looks at just four of the elements that will constitute the Museum of Making: Designing and Prototyping; Hacking; Reuse; and Visible Storage.

Derby Silk Mill – Museum of Making will be inspired by the makers of the past, made by the makers of today and empower the makers of the future.

FIGURE 8.2 — Silk Mill, Museum of Making (detail), Derby, UK. Photograph by Dimitris Papadopoulos.

Midlands in Britain and beyond. All the materials I collected, my analysis, and the political ecologies and ontologies I discuss here live "in a stacked ethnographic present," to use a phrase of Choy (2011, p. 71). Stacking is not only a temporal and ontological condition, as argued earlier in this chapter, but also a mode of sensibility while conducting fieldwork. On a theoretical level the chapter relies on the growing scholarship about the maker culture, grassroots innovation, and community technoscience.[9]

Conceptually this chapter is located within recent debates on ontology in science and technology studies, anthropology, geography, organization studies, and other social sciences. In the previous section I described the specific meaning of the term *ontology* in this chapter: worlds of existence in which different actors can change their materiality in concrete ways and not others. But how does this happen specifically? If, as Sergio Sismondo says, ontology is "about multiplicities of practices and the ways in which these practices shape the material world" (Sismondo, 2015, p. 441), this

FIGURE 8.3 — *Wired.* "How to Make Stuff." Cover featuring Limor Fried. April 19, 2011. Jill Greenberg, Kevin Hand, and Jeff Lysgaard/WIRED © Conde Nast. Reprinted with permission.

chapter attempts to explore, present, and conceptualize such human and nonhuman practices and their relations.

In the first three sections of this chapter I discuss the maker culture and various aspects that characterize material projects and ontological engagements I participated in. Here I move beyond the dichotomy between the existence of many worlds and multiple ontologies versus one world and a single ontology and present the idea of ontological stacking as a possible way to challenge this dichotomy and to approach ontology as a movement rather as a structure, a state of being, or a process. Starting from this idea in the sections that follow, I discuss specifically making as a movement in technoscience and conceptualize some of its key features within the framework of compositional politics that I presented in the previous chapter. In the final sections of the chapter I discuss the political implications of these diverse compositional practices of making and other movements of community technoscience, and I conclude with an attempt to understand movements as *more-than-social* that attempt to create alternative ontologies and forms of life.

"If you want something you've never had, then you've got to do something you've never done," reads the first sentence of the Maker Faire Africa manifesto.[10] It continues:

> 1. We will wait for no one. 2. We will make the things Africa needs. 3. We will see challenges as opportunities to invent, and invention as a means to proving African ingenuity. 4. We will be obsessed with improving things, whether just a little or a lot. 5. We will show the world how sexy African manufacturing can be. 6. We will hunt down new skills, unmask locally made materials, keep our work sustainable and be kind to the environments in which we make. 7. We will share what we make, and help each other make what we share. 8. We will be responsible for acting on our own ideas. 9. We will forge collaborations across our continent. 10. We will remake Africa with our own hands.

Since the first maker faire in 2006, these events have spread around the world. The maker faire as an event and as part of the broader culture of making started in San Mateo, California. Maker faires are self-organized gatherings that attract from a few hundred participants to hundreds of thousands. Their main goal is for people to exhibit and engage others in their technological and scientific innovations.[11] The *Make:* magazine has been in the center of this transnational development since its beginning in 2005.[12] What can explain this capacity of the maker culture to travel across the globe? What holds together a transnational movement so diverse and so widespread?

The title of a 2014 publication of the Institute for Public Policy Research captures something of the imaginary that the maker culture came to occupy: *The March of the Modern Makers: An Industrial Strategy for the Creative Industries* (Straw and Warner, 2014).[13] The combination of industry and the creative industries are the keywords here. When the maker culture is mentioned in mainstream media, it often implies a call to revive industrial production in the Global North. But this revival is not just a return to some bygone days or some form of nostalgia for lost skills. It appears as an aspiration to reaffirm material power and to assert a stronger position in global production networks.[14] From gadget fetishism and technolibertarianism to the so-called third industrial revolution,[15] the maker culture is hailed for its potential to revive the waning material creativity that can rebalance the flight of production from the West.[16]

FIGURE 8.4 — Artwork by Brett Ryder. Used with permission of the artist. The artwork was featured in the cover story of the *Economist*, April 21, 2012, on "The Third Industrial Revolution."

The maker culture feeds the vision of making manufacturing great and strong again, which seems to resonate with the sentiments of large segments of disadvantaged white working classes and lower middle classes squeezed under the fast ascent of biofinancialization (as described in chapter 2): the flight of jobs, stagnant or falling wages, deskilling, unemployment and underemployment, reduced social mobility, disillusionment with the self-perpetuating creative classes and (neo)liberal elites, and frustration with the dominance of a financial economy over an industrial one. As much as this could sound like a critique of current capitalism, manufacturing's revival turns out to be an element of a very different imaginary: a conservative ethnocentric one, where the reduction of trade liberalization is supposed to halt the loss of jobs and rapidly advance reindustrialization, and making is seen as one of the core values for restoring a new work ethic, skills, and capabilities inside the nation. The revival of the promise of manufacturing in the West has become one of the elements that helped assemble a new white nationalist identity across many Global North countries—and making is implicated in this politics.

The promise of the new manufacturing emerging today lies in the digitization of production, but the implementation of digitized manufacturing entails the potential for revolutionizing technoculture: it can be performed individually—that is, in small-scale environments and outside the industrial shop floor. The prospect is that it can capitalize on people's creativity—everyone's material creativity everywhere. The transformation of everyday creativity to a productive asset has been extensively researched in the creative industries and new media,[17] but this seems to be a new phenomenon in manufacturing.

What characterizes this combination between creativity and manufacturing is the engagement of the creator in the whole production process instead of a specific part of it.[18] In the same way that people are hailed as users today, the promise is that we will all become makers. Of course, this creative form of production relies on a form of social organization that is widespread in knowledge economies and creative industries such as intense connectivity (to other makers, to materials, to environments),[19] sharing,[20] networked innovation,[21] abundance of free labor,[22] and so on. But what genuinely differentiates desktop manufacturing from other creative industries is the entanglement of materiality and creativity through the skill of craft.[23]

One has to look back to the assertion of the situationists (and in particular Raoul Vaneigem)[24] that creativity—and not labor—is the driving force of human history. But Vaneigem's unearthing of the emancipatory potential of creativity against the oppressive nature of labor was a sign of a broader social transformation happening when he was making this assertion: creativity was transforming into a genuine productive force in the postindustrial, post-Fordist regime of accumulation.[25] The artisan, a figure combining technical skill, knowledge, insight, aesthetic innovation, artistic presence, and practical use, starts to disappear in the passage from the late Middle Ages to the modern period. With the industrial revolution, the figure of the artisan bifurcates into the manual worker on the one hand and the intellectual/artist on the other, and then fades away. Material work is delegated to industrial production and creativity to artistic practice.[26]

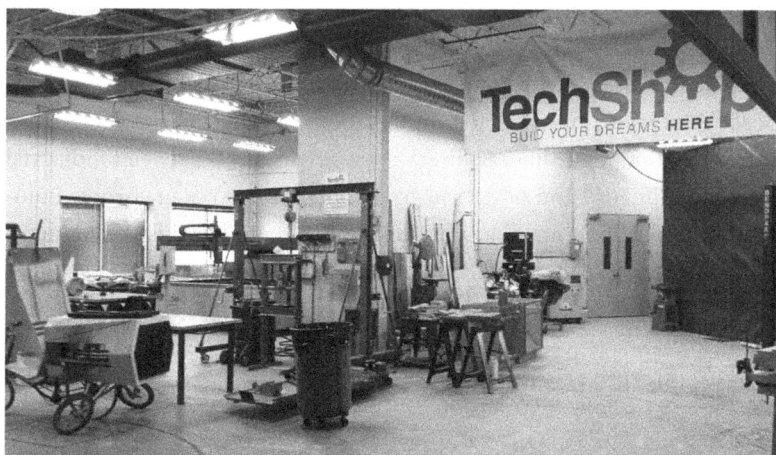

FIGURE 8.5 — TechShop, Detroit, Allen Park, MI, USA. Photograph by Dimitris Papadopoulos.

The post-Fordist reorganization of social life revives creativity by bringing it back to production through digital skill, information manipulation, and cognitive inventiveness. The term *cognitive capitalism* came to describe this capture of creativity.[27] In the maker culture we see another second revival of creativity: now creativity returns to its supposed roots, to material practice, to manufacturing, to life. This is, of course, a story of origins: before the industrial split, the artisan was the figure in which all these aspects could coexist. The artisan disappears only to be rehabilitated now as creativity apparently returns to our material engagement:[28] the imperative to invent. In the first maker faire in the White House (2014) U.S. president Barack Obama declared America a nation of makers: making "is in our DNA."[29] This is a story of origins indeed, one that could signal the ultimate defeat of Vaneigem's emancipatory vision of creativity as the *only* force that can "rid us of work."[30] In the maker culture the entanglement of work and creativity is as strong as ever before.

In his "maker movement manifesto" the CEO of TechShop—which was, until it declared bankruptcy in 2017, a large membership-based chain of workshops across the United States that represented the rather entrepreneurial and commercial side of the maker culture and whose expansive for-profit business model has been questioned by smaller nonprofit hackspaces and makerspaces—repeats the same slogan: "we were born to make" (Hatch, 2014, p. 144). The location of the TechShop branch near Detroit is emblematic of the attempt to release this miraculous "innate" potential for making and to harness material creativity. It is located in a 33,000-square-foot facility adjacent to Ford Motor Company's Production Development Center and close to its world headquarters in Dearborn, Michigan. The staff at this branch told me that there is direct cross-fertilization between industrial innovation in Ford and other nearby companies and the grassroots innovation of TechShop members.[31] "Build your dream here" is the motto of TechShop, which calls itself "a playground for creativity."[32]

PROVINCIALIZING MAKING

I am struggling to make sense of all these different investments in the notion of making: making inscribed in the DNA of the people of the United States promising to revive the core of country's industrial economy, a project that can be linked to a vision of re-Westernizing production, of insourcing industrial capabilities to the Global North and possibly also to white backlash politics and the reemergence of nationalist protectionism. But then I turn to Derby's Silk Mill project, where the figure of the maker emerges as the other to deskilling, low-wage labor, youth unemployment, and the dissolution of the social tissue of the city after the dramatic deindustrialization of the East Midlands: a local project of communal remaking—although it still remains unclear how far it will reach into the disadvantaged communities of the city. And then Maker Faire Africa—see its manifesto earlier in this chapter—implies a different endeavor as it mobilizes industrial and scientific creativity that is associated with Western capitalist modernity to challenge the coloniality of technoepistemic Western power. And across all these locations is the lived experience of makers, their intense engagements with materiality and sociality, and their endlessly diverging ways of practicing material creativity in the midst of everyday life.[33]

Making is primarily a practice associated with concerns of Global North societies. The social and gender composition of makerspaces, the intense entanglement of making with Western technoscience, and its implication with nationalist imaginaries are an evidence for that. This is the dominant story of making: a story told by the Maker Faire's red mascot robot carrying the symbol M engraved on its chest and standing on top of planet Earth. It is a universalist story and a masculinist story[34] that purports to have discovered making in the birth of the scientific revolution and enlightened values that propelled the ascent of capitalism through

FIGURE 8.6 — Ninth Annual Maker Faire, Bay Area, CA, USA. May 17 & 18, 2014. Flier. Reprinted with permission. The Bay Area is one of the flagship Maker Faire events globally. According to the organizers, the 2014 Faire attracted "1,100 + maker entries, 130,000 + attendees and 90 + sponsors."

FIGURE 8.7 — The first Derby Mini Maker Faire at the Derby Silk Mill Museum, UK. June 3, 2012. Poster. Reprinted with permission.

the industrial revolution. Today's making is called to continue this legacy and counteract industrialism's decline. But then there is much more to making than that: there is World 2,[35] the abject spaces of Western modernity whether they are outside the Global North or inside it (such as so many neighborhoods in the British East Midlands). Making here is about the survival of disintegrating communities, about enhancing self-organization, about supporting livelihoods. There is much more to making that I want to reclaim here than the universalist story of M—not only because making is involved in building and maintaining diverse ways of being that radically depart from the dominant Global North maker culture but also because it seems to resonate with the political sensibilities within more-than-social movements. It is this speculative dimension of making that I look to unearth here.

Making can only exist through traveling, in plural, as it bifurcates and diverges from itself. Atsuro Morita (2013, p. 236), in his study of a technology transfer project between Thailand and Japan, suggests that "rather than finding a craft community rooted in a particular locality, what I encountered were the intersecting journeys of humans and nonhumans" across different locales, contexts, and regions.[36] Rather than a blueprint for action that can be transferred from one context to the next, making is a practice that changes as much as the ontologies that it helps emerge differ from each other. Writing the story of making requires assuming "plural ways of being in the world" (Chakrabarty, 2000, p. 101), or rather plural ways of *making* the world beyond the dominant version of the maker culture and the universalist story of M.

INDIGENOUS TEMPORALITIES

Within the speculative framework that I am advancing here, making could be seen as a contribution to a decolonial project,[37] one that rejects building or rebuilding a universal global world but attempts to construct plural ontologies across global space. In the first chapter of this book, I argued for a decolonial politics of matter—a politics that challenges epistemic coloniality by transforming materially everyday ontologies of existence. In this chapter I turn to making as one of the concrete practices that can sustain such a politics, and in the remainder of the chapter I describe specific aspects of this politics, in order to see making as a form of politics that encounters the West and Western technoepistemic practices not just as one of the many ontologies that exist but rather as a force that puts other ontologies in danger, ontologies that are dissimilar, multiple, and heterogeneous. Provincializing making requires more than a mental shift: plural ways of making the world can exist to the extent that they disconnect from Western ontology, to the extent that a process of "un-networking" takes place that allows for other ontologies to emerge. As much as making is about traveling and connecting, it is also about breaking, cutting, opposing, and escaping Western technoepistemic universalism. It is about immersing in other existing ontologies and in helping to create alternative ways of life. In current conditions, a decolonial project can be only practical and ontological or it cannot materialize.

Maker Faire Africa claims that making is always a situated project depending on the specific ingenuity of the involved communities, the specific problems that it is called to engage with, and the existing infrastructures in a specific location within the continent.[38] The maker project becomes indigenous and bound to each of the places to which it travels inside Africa. In this sense, making can be neither global nor universal nor local. It could be said that making can only be indigenous, a mobilization that starts in an indigenous space and then comes to travel and connect to other nearby or faraway indigenous places and projects.[39]

The indigenous dimension of making is not about the association to a local environment but about practically reconnecting to the ontologies of the Earth, the land, and its "Earth-beings":[40] terraformation from below, instead of the colonizing planetary project of Terraforming Earth that I described in chapter 2. Coloniality is premised on extraction and the dispossession of indigenous peoples' land.[41] Indigenous resurgence—that is, indigenous people's autonomy, self-determination, and self-government[42]—cannot be achieved without reclaiming land and reclaiming indigenous knowledges, practices, spiritualities, social relations, and ways of being and making that arise from the connection to land and the Earth-beings that have been erased by Western technoepisteme.[43] The creation of such indigenous forms of knowing and making is about creating ontologies that allow such alternate connections to be practiced autonomously. Indigenous autonomy is possible only by moving away from the Western temporal register that relegates indigenous knowing and making to the past. As much as making is supposed to be

FIGURE 8.8—Colectivo Pequeñas Hermanas. Costa Rica. "Autonomy! The indigenous movement walks ahead while looking backward because our future is not what will come but what was." Reprinted with permission.

about creating connections, it seems that disconnection is an even more vital dimension of making,[44] because a decolonial making is an active practice of unmaking connections, of "un-networking" (Staeheli, 2012), of creating other ontologies and centers outside existing networks and connection flows.

The temporality of a decolonial politics of matter is one that is not directed to the future. Neither is it about a return to the past. It is a temporality that interrupts the Western domination of our imaginaries of what will come: "Our future is not what will come." Rather the future is the reclaiming of the past in a move to create alternative forms of life in the present that escape the universalizing Western timeline. There is something to this indigenous temporality that goes beyond indigenous resurgence and the multiple struggles of indigenous movements. Indigenous autonomy is a reconnection to the material surroundings, to Earth, to traditional knowledges, and, as art educator, master aerosol writer, and community organizer Lavie Raven says, it involves also a reconnection "to a global consciousness that we are all related. Everyone on the planet has indigenous roots to somewhere."[45]

FROM PLURIVERSE TO MOVEMENT

Making cannot be approached as an epistemological issue; it is a practical one. Making is a material movement; it is about ontological practice rather than about an abstract representation of a practice of material engagement.[46] And as such this movement is embedded in other previously existing ontologies. Each of these ontologies involves different environments, materialities, digitalities, groups of people, and more-than-human actors. Marisol de la Cadena, Mario Blaser, Arturo Escobar, Walter Mignolo, and others refer to this multi-ontological organization of the world as a pluriverse.[47] A pluriverse does not hold together through some universal matrix or unified denominator but through relations and dialogues between divergent worlds.

In the pluriverse of makers there is no such thing as a preexisting pattern that unifies divergent engagements with materiality. What holds these together is that each of these materialities is layered on previously existing ontologies. It's ontologies all the way down: ontologies that are not only historically and temporally stacked, as I have argued earlier, but also practically and materially. They are stacked in relation to each other. Stacking is the condition of existence of the pluriverse.

In permaculture[48]—a practice for ecological design of nonpolluting and food-growing settlements—stacking means to mimic natural forest environments, especially rain forests, in order to create food gardens.[49] In forests you have many layers of plants stacked in relation to each other—the canopy of giant trees, then the tall trees and the lower trees, then vines and climbers, then shrubs, and, finally, ground cover plants: creepers, grasses, roots, and herbaceous plants, all of them on the same ground. Permaculture mimics the forest strategy of stacking to create food forests where different crops coexist in limited space and reinforce their growth.[50] I am thinking of a similar form of ontological stacking here.

Stacking is about relative locations: how elements of an ecosystem co-act as long as they are mutually dependent and are in positions that allow them to assist each other. Ontologies are stacked together topologically:[51] never simply on top of each other or next to each other but *continuously* as they emerge or are related to other ontologies. Ontologies are many and heterogeneous, but they belong to the same shared earthly world. This understanding of the ontological organization of the world poses a crucial question to the pluriverse project: how to deal with the implicit departure from a monist understanding of the world? The most troubling implication of the repudiation of monism is that debates are settled on an epistemological level: a question about different opinions and views of how the world is and how the split between ontology and culture came about. In the pluriversal theoretical approach the debate focuses on whether there is a "one-world world"[52] *or* multiple worlds. But here I wonder, is it possible to steer toward the opposite direction and arrive "at the magic formula we all seek—PLURALISM = MONISM"?[53]

There is no monism if there *is* a dualist option; "there is nothing that is one, there is nothing that is multiple."[54] If the world is organized as stacked ontologies rather than as a multiverse of ontologies, then the question whether there is an "one-world world" or a pluriverse doesn't have much traction. As I discussed in detail in chapter 4, Deleuze and Guattari argue that the real enemy of materialism (ontology) is not idealism (the social), but dualism: "The only enemy is two."[55] The approach of stacked ontologies mobilizes a monist practice of ontology and matter that does not allow for a dualist epistemological option to emerge as a legitimate issue.

Deleuze and Guattari propose a "politics of matter" that focuses on matter's making rather than what matter is.[56] The alternative to dualism and the choice between a "one-world world" and many worlds is not the development of some kind of new ontology or cosmology (the "pluriverse") but a continuous participation in matter's movements.[57] Continuous materialism is the alternative creation of matter—that is, the making of other ontologies that escape existing ones. The Zapatistas' matrix of a "world where many worlds fit" is neither an epistemological question nor a plea for changing our worldview but a guidance for practice and an invitation to a movement,[58] neither the making of one single ontology nor the making of multiple ontologies, but grounded making: the movement of alterontologies, from alter-globalization to alterontologies, from epistemology to movement.

COMMENSALITY, COMPOSITION, AND DIWY

Many of my interlocutors in all the makerspaces and hackspaces I have participated in told me that they rarely find all the tools they need in a makerspace. Makerspaces are neither dedicated workshops nor industrial shop floors. What matters is the potential for learning and novel articulations to emerge—by exploring the possibilities of existing devices, discovering new devices that are brought in by other participants, involving

FIGURE 8.9 — High-density polyethylene (HDPE) reprocessing. Leicester Hackspace, UK. Courtesy of Woody Kitson.

the extended communities, sharing ideas, learning ways of handling stuff, acquiring new habits when dealing with specific objects or organic matter, experimenting with materials that others use or that one might have found on a workbench by coincidence, and watching the small mundane inventions of other participants.

In hackspaces there are many different tools, devices, things, materials, organisms, animals, and people stacked together—they are in relative locations to each other so that they allow articulations of material-making to be performed.[59] As in permaculture, one can get many (and very different) outputs from just one element in the forest garden. A simple item of a specific material, a PET (polyethylene terephthalate) bottle or an HDPE (high-density polyethylene) milk bottle, for example, can be engaged in many different ways, as inspiration for inventing new ways of recycling and reuse.[60] For Woody Kitson from the Leicester Hackspace, this involves "the reprocessing (for example, melting, reforming, welding, cutting, sewing, turning, milling) of the raw material," which then can be reused in many different ways: as a container; as a nonconductive spacer; as toys, tools, or tent pegs; as moldable plastic if treated correctly; as replacement parts for broken plastic components; as materials for art projects; as one of the ingredients of composite products; as an educational object for

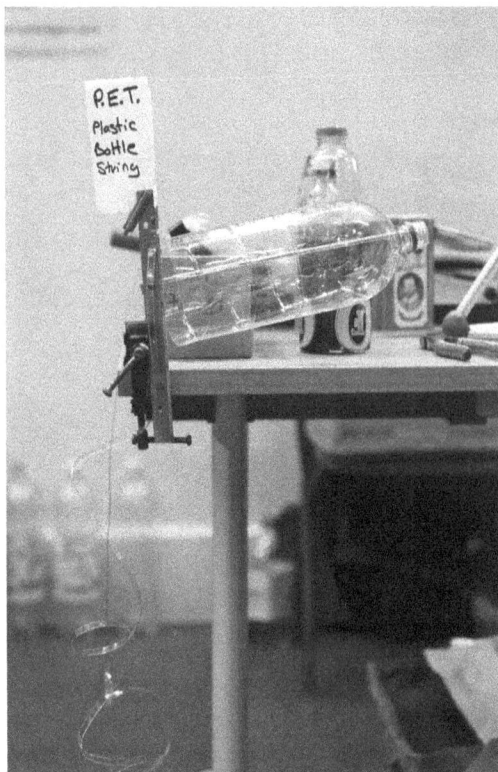

FIGURE 8.10 — Polyeth-
ylene terephthalate (PET)
string. Creator Fair at the
National Space Centre,
Leicester, UK. Courtesy of
Woody Kitson.

raising environmental consciousness; as material for experimentation that
attempts to educate about the endocrine-disrupting chemicals and other
hazardous ingredients it possibly contains; as a source of durable string;
and so on.[61] Objects are indeterminate and incalculable because the way
humans relate to them in each specific situation does not reveal their essence or all their qualities but only a partial and specific aspect of them.
Objects are much more than their relations to humans.[62] Objects exceed
what humans think of them and what humans do with them.

This immanent indeterminacy applies to all objects, even to those that
were made with a specific aim. One of the quintessential electronic devices
in many makerspaces is Arduino, an open-source and open-hardware microcontroller motherboard.[63] Arduino can perform many tasks in an environment of making, as it can be easily programmed to control several
devices and sensors and link their diverse processes. The Arduino is one
example of an interface that allows unpredictable compositions of code and
matter—that is, the digitization of material processes and the progressive

materialization of code. Symbolically Arduino is the prototypical device that characterizes the transition from the information age to the time of *composition*. The composition of the digital and the material is defining the era of the maker culture.[64] Everyday objects are digitized and interlinked within the web of crafted things. Beyond the clear distinction between information analytics and material construction, the practice of making involves a steady crossover and fusion of information and matter into new compositional objects. Despite current data totalitarianism, big data is only a sign of peak digitization.[65] The possibility of incomputability is always inherent in computation itself.[66] But when materiality becomes an immanent part of computational unpredictability, indeterminacy and incomputability multiply. There will not be space for today's all-pervasive data positivism when code and matter will have completely fused and compositional objects become ubiquitous.

In Sam Esmail's television cyberthriller *Mr. Robot* (2015–), Elliot Alderson, a tech security expert and skilled hacker, teams up with a group of anonymous hacktivists, F Society, to take down the world's largest multinational conglomerate, E Corp, and to erase—literally—all records of people's debts. What is notable in the series is that hacking is never just a question of code manipulation. Hacking is seamlessly embedded in everyday life, in social relations, at work, in friendships, and in the physicality of buildings, other objects, and technological artifacts. The aim of F Society to erase all consumer debt can be achieved when all financial information files on the servers of E Corp are rendered inaccessible. But it would not be enough to permanently encrypt or destroy the data on the main servers; all existing backups need to be also simultaneously erased, including a backup facility that is offline, which makes it unreachable with the means of "conventional" code hacking. F Society hacks the security of the building and the personal lives of the guards and gains access into the facility. Elliot installs a Raspberry Pi[67] into the climate control system of the facility, which contains a program that overrides temperature control. This allows F Society to physically melt all E Corp's backup data tapes. Season 1 of the series ends with a total reversal of biofinancialization (see chapter 2) through biohacking: the experimental play with the indeterminacy of code and matter composites. The materiality of hacking and the continuous crossover between information and objects marks also the second season of the series, which ends with conceiving the plan to hack a storage facility in which E Corp is collecting all existing paper records (since all digital records have been destroyed or made inaccessible) in order to rebuild the

FIGURE 8.11 — Dirty Electronics (www.dirtyelectronics.org). Synthesizers. Leicester Hackspace, UK. Courtesy of Sean Clark (www.seanclark.me.uk).

database with the financial information of consumers. In order to interrupt this analog-to-digital transition of information, F Society needs to hack the building itself and physically destroy it. Only then can all financial information and thus all debt be erased for good.

Craft and crafting becomes a key way to describe the relation to such compositional objects. Craft becomes more of a relation, mode of thought, and structure of sensibility than just a form of practice. Compositional objects exist independently of humans, and the only way to approach them and understand them is through craft.[68] This applies not only to quintessential objects such as the Arduino microcontroller but also other ad hoc devices and objects. I am thinking, for example, of the synths that the collective Dirty Electronics constructs.[69] These are truly compositional objects not only in the sense that they mix found materials with other purpose-built or acquired equipment and that they fuse the digital and the material, but also in the sense that the synths acquire their full potential within an experimenting community. Social interaction, movements of the body, and more-than-human dimensions such as gestures, temperature, and light all compose what a synth is. The performance of playing the synth, the performance of making the synth on the workbench, and the shared experience of experimenting with the synth are continuous. This continuity of experience and materiality is a crucial dimension of compositional objects.[70]

The topology of things and living beings that exist in a makerspace is always changing according to the project in which a group of actors is involved. Depending on the type of activity it might involve only few specific tools and objects or a larger array of them; it can involve automated elements that do not necessarily require human presence; or it can involve other living beings, such as bacteria or other animals and plants.[71] A topology of making is similar to an ecology[72] and more specifically an ecological guild, "a group of species that exploit the same class of environmental resources in a similar way" (Korňan and Kropil, 2014, p. 445).[73] Ecological guilds constitute communities of support where each of the species or objects in them contributes its unique "functions"—to use the terminology of permaculture (Mollison, 1988). But support here does not involve intentionality or even mutualism (although mutualism is a crucial biological relation between the organisms of an ecological guild).

Support in an ecological guild means to create conditions in which many actors can share the same space without being harmed or by occasionally, ideally regularly, benefiting from each other. Commensal interactions—that is, benefiting from what other organisms or actors offer in an ecology without harming or affecting them or even without reciprocating—offer a good way to approach the interaction taking place within makerspaces. This aspect of nonreciprocal sharing is particular useful for my attempt to understand craft and compositionality in the maker culture. Although exchanges are important in maker culture, exchange does not define making. Relationality and exchange are not the ontology of craft and composition. Compositional culture is defined by commensality: actors just leave stuff (techniques, ideas, objects, practices, concepts, tools, and so on) around, and other actors use them (or not). This is not an ontology of exchange, but an ontology of coexistence. In order to foster coexistence, actors have to reduce their presence, their subjectivities, and leave space for other actors to exist. While exchange presupposes a strong self that negotiates and transacts, commensality presupposes the careful retreat of the self. Compositional culture is DIWY: do it without yourself. DIY and the logic of relation and exchange are too humanist to describe the posthuman relation of compositional culture. DIWY is about reducing the presence of humanity in the sociability of craft and

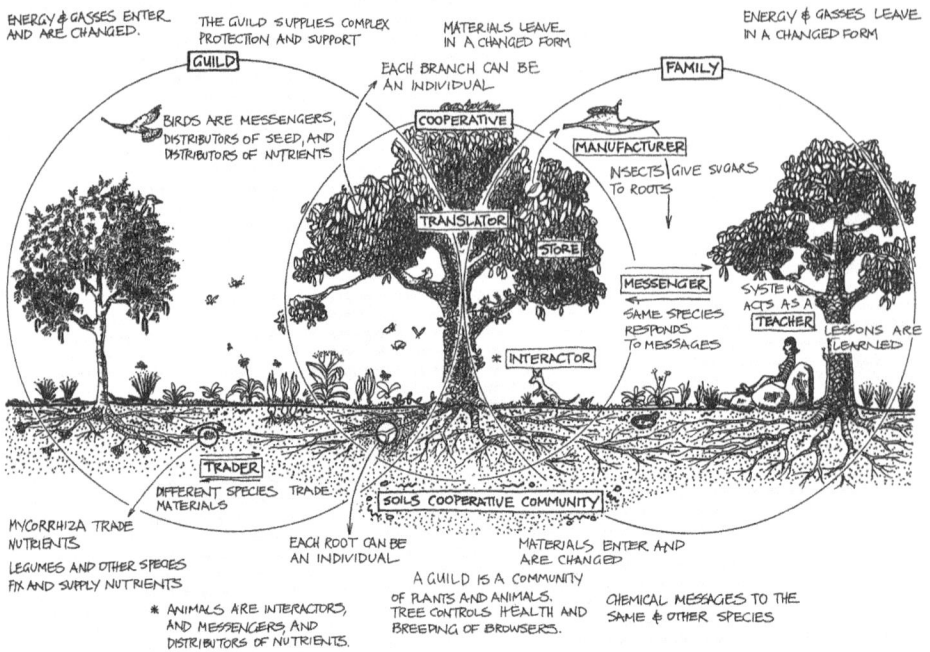

FIGURE 8.12 — Bill Mollison. Trees in Whole System. © Bill Mollison. Reprinted with permission. The figure was originally published in the book *Permaculture: A Designers' Manual*, by Bill Mollison (Tasmania, Australia: Tagari Publications, 1988).

compositional culture. Commensality and DIWY is the drive behind the type of innovation and invention that takes place in maker cultures.

DISTRIBUTED INVENTION POWER

In science the moment of invention is considered the outcome of a specific type of practice, "the experimental achievement." This key event characterizes for Isabelle Stengers (2000) modern science: only what has passed thorough experimental testing becomes a scientific fact. From this perspective, science is a specific type of practice that enables scientists to challenge their own questions and assumptions in order to achieve a level of certainty: only the questions that have withstood their objections can be considered scientific. In other words, scientific knowledge is a distinct practice that can be located in this specific single event of the experimental achievement. But is this the case in technoscience—cognitive science, climate science, biosciences, soil science, neuroscience, informatics, biomedicine, and geosciences, to name a few examples? And what about distributed and mundane technoscience beyond formal institutions that happens in spaces such as the ones I am describing here: hackspaces, makerspaces, community labs, and so on?

In technoscience, experimental achievement and invention are mediated by many different trajectories and actors already *before* it has taken place, even before it has been formulated. If we neglect this, we neglect the invisible and indeed invisibilized labors of so many different human, animal, and inorganic actors that contribute to the making of facts.[74] In analyzing Darwin's experiments, Carla Hustak and Natasha Myers (2013, p. 106) conclude that "it is in encounters between orchids, insects, and scientists that we find openings for an ecology of interspecies intimacies and subtle propositions. What is at stake in this involutionary approach is a theory of ecological relationality that takes seriously organisms as inventive practitioners who experiment as they craft interspecies lives and worlds."

In technoscience,[75] what counts as invention is not primarily the individual experimental achievement that gives coherence to traditional experimental scientific practice (although this might be sometimes part of it); rather, it is a form of dispersed experimentation: *distributed invention power*. If science as experimental achievement ever existed, this achievement of invention is now dispersed in society and matter—in a "more than one world," in a "more than human world," to use de la Cadena's (2010) words. Technoscience is more than human and dispersed, it is more than scientific. Knowledge in technoscience is not done by those who object but rather by those who invent in intended and unintended more-than-human collaborations. Technoscience is interobjectively and intersubjectively "materialized action."[76] Consider the making of a transgenic lab animal, of the robot Atlas, of Earth observation patterns of soil erosion, the visualization of neural networks, climate simulations, synthetic molecules, new drugs, and so on. And this does not only apply to formal instituted technoscience. Consider makers' projects such a kitchen bio lab, a sensor for air pollution, a recycling machine of plastic milk bottles, a self-made quadrocopter to inspect the condition of roofs and chimneys, or a small project for securing energy self-sufficiency.

Emphasizing the importance of distributed invention power in the age of composition exposes how much technoscience is mundane, informal, and community based: hackspaces, makerspaces, traditional and alternative knowledge systems, clandestine science, community labs, amateur science and technology, fab labs, indigenous knowledge, bio-art, activist knowledge, self-education projects, punk science—all gradually become an integral part of technoscience. Technoscience is done not only within its so-called core institutions[77] but in multiple ways and in many different mundane environments. Community technoscience is continuous with instituted technoscience and vice versa, a continuation that unfolds across disparate and fragmented worlds.

This extended view on technoscience allows us to capture how every specific knowledge practice assembles around it a different social and material world—be it scientists, technologists, animals, materials, businesses, social policy makers, marketeers, tools, practitioners, consumers, enthusiasts, activists, finance, community stakeholders. What we have here are large ecologies of multiple actors, landscapes, and knowledges. In searching for ways to approach, study, and engage with such communities, I am inspired by Choy's (2011) important work on environmental politics in Hong Kong that shows how a diverse array of environmental actors can be thought of less as a clearly organized political and public sphere and more as a political ecology or even as something similar to a regional biotic community. I tried to approach in similar ways the makerspaces and hackspaces I was involved in (albeit with much less ethnographic subtlety): namely by drawing and conceptualizing "connections between places, between species and other species, between forms of life and their environs, between what is considered big and what is considered small, between particulars and universals, between particular cases of a common rule, between specificities and generalizations, between grounded details and ambitious abstractions" (Choy, 2011, p. 5).

ECOLOGICAL TRANSVERSALITY

Distributed invention power relies on reinventing the meaning, use, and relationalities of materials and objects that exist in a certain context. Whether it is wood or plastic, metal or silicon, organic or inorganic, animal or human, making always involves *transversality* between disparate material registers and human or nonhuman communities of life.[78]

Ecological transversality mobilizes an imaginary that is very different from the single purpose–oriented logic of traditional manufacturing. This imaginary supposes that each object, animal, thing or process among them can have many effects and many purposes inside a technoscientific guild, similar to a permaculture guild, where the yields of each element of a forest garden can be multiple: providing food, burning wood, compost and fertilizer, livestock feed, shelter, and fun.[79] In the maker guild as in the forest garden, the same elements can provide different yields. So each element in specific moments has different functions and different purposes.

In the practice of making, new materials acquire new qualities, and new processes emerge through their continuous forking to unexpected paths: one always relies on what existed before, and then you split, redirect. Innovation in distributed invention power happens rarely through sudden novelty and more through hacking, stretching, knitting, weaving, tweaking, mending, and recombining existing processes or substances in a techno-scientific guild.[80] The maker's craft is less about the mastery of knowledge and more about knowing with others. "Start even if you don't know how" (Tremayne, 2013, p. 189) captures knowledge as practice and as involvement with others. As much as translation is crucial for this project of ecological transversality, it captures only a small part of the exchanges between the communities, actors, substances, and species involved. Translation relies on the widespread belief that communication can happen only if there is translation of meaning between two equivalent but distinct communities.[81]

Communication through translation is a language-centric and cognition-based approach to the co-action of separate communities. In the practice of making, rather than through translation, communication happens through involuntary infections and contingent permutations between organisms or substances that attract each other. Not every object or organism is open to others simultaneously. In certain conditions some objects will attract each other and enter into a composition. Makerspaces (like other tech-noscientific guilds) encourage objects and compounds to become accessible to others. This is the gift economy of the movements of matter and cross-species action. The makers' worlds always contaminate each other laterally.[82] Drifting matter. "Material spirituality," in the words of Puig de la Bellacasa (2014a),[83] is the commitment to getting involved in an exchange without "knowing how" and without knowing the final outcome but through trust in the other participating co-actors and the "involutionary" process[84] as such, trust in something that exceeds the situation.

Stuck in the logic of translation, we believe that everything is mediated through language, information bits, and a mediator or operator. But in making this is seldom the case. And even more so, objects are not things that can be fully explained or accurately translated as if we have total access to their essence.[85] Objects exceed the capacity of humans or other objects to fully comprehend them. Rather, objects become partially accessible to

FIGURE 8.13 — Artwork by Grady McFerrin. Used with permission of the artist. The artwork was originally published in the book *The Good Life Lab: Radical Experiments in Hands-On Living*, by Wendy Jehanara Tremayne (North Adams, MA: Storey Publishing, 2013).

other objects, animals, and humans in certain conditions while they become unapproachable to other objects, animals, or people. Simultaneously, animals and plants become engaged in specific types of interactions, and humans become committed to a certain project. The combination of the *accessibility* of objects, the *engagement* of animals and plants, and the *commitment* of humans drives each specific project in the maker culture. Rather than all-encompassing, universal calls for embracing the potentiality of human-nonhuman relations, multispecies exchanges, or human-machine interactions, makers' practices expose the fact that human-nonhuman relations vary and need to be seen in their specificity. Certain objects become accessible to other actors, while on other occasions the same objects hide from these actors; similarly, we can think of how an animal or plant can be engaged with some actors and not with others, and how humans become committed to a certain project while other humans can be fully disinterested.[86]

Nomeda and Gediminas Urbonas created an experimental/educational space within the Contemporary Art Center in Vilnius where they attempted to work with mycelium, mushroom roots, and to create building materials and everyday objects, mycomorphs as they called them.[87] Mycelium is an extraordinary living being. Because of its specific composition, mycelium enacts very different actions in different contexts: it can be a healer in distressed environments, a cultural metaphor in social theory, a source of food, a catalyst of bioremediation in polluted spaces, an agent of decomposition of organic materials, and in my example here a building block or an everyday object.[88] But what interests me is not the specificities of mycelium that allow its fundamental openness but the realization of how indispensable ecological transversality is in order for these specificities to thrive and for this openness to be expressed: in each specific context an enormous amount of attention between mycelium and the other participating species is required in order for these enactments to emerge. When these transversal ecological interactions are disturbed, mycelium either dies or becomes parasitic.

The mycomorph lab conducted by Nomeda and Gediminas Urbonas is an experimental technology that attempts to establish a transversal ecology between individuals of two species: How clean should the air be? What should the temperature and humidity be? Which materials are best for the mycelium to thrive but also for creating these building blocks? Wood, straw, something else? What about adding other materials, agents, or species? How long should the process last? Which forms are better? In carefully designed interactions, mycelium develops without destroying its immediate environment and the objects around it. Simultaneously, humans work

FIGURE 8.14 — Nomeda and Gediminas Urbonas. Psychotropic House: Zooetics Pavilion of Ballardian Technologies. Mycomorphs. Installation Detail. CAC Vilnius, 2015. Photograph by Nomeda Urbonas. Reprinted with permission.

with mycelium to replace toxic chemicals used in everyday objects and in construction with biodegradable materials. What would it mean to grow our houses and everyday objects instead of constructing them? ask Nomeda and Gediminas Urbonas (Januškevičiūtė, 2015, p. 13).

Making here has almost nothing to do with producing and constructing. It is about cohabiting a space through engaging in ecological and interspecies transversality that allows its participants to respond to the constraints and specificities of the life of each participating species. Rather than a repository of practices and tools, the mundane technoscience of making is about composing such transversal ecologies: coexisting, connecting and recombining, forking and tinkering, extending and preserving.

ETHOPOIESIS AND THE BLACKMAIL OF PRECARITY

If anything can be found in abundance in makerspaces, it is free labor. This is not only a characteristic of community technoscience; it is also in the heart of any form of technoscience. With Star we can then explore all these labors that have been rendered absent and invisible in the experimental achievement and in the humanist tale of making and contesting scientific facts.[89]

In his illuminating study on the making of transgenic rice in experimental fields in the Philippines, Chris Kortright (2013) introduces the term *experimental labor* to describe how research work is always embodied and haptic, operating in the constraints of the time and space in which the experiment takes place, involving a complex interaction with other local actors and the environment.[90] Experimental labor is about invention, and invention is always situated: it implies an involvement in the lives of other living and nonliving beings. This "ethopoietical practice," in the words of Puig de la Bellacasa (2010), the simultaneous production of ethos and ontology, cultivates distributed invention power.[91] Experimental labor fuses experience, subjectivity, and materiality.[92] Seeing the work of technoscience as experimental labor reveals how many different types of relations, social groups, species, ecologies, interdependencies, and ways of life participate in the making of knowledge—that is, in the making of ontologies.

Not only is the work of technoscience dispersed and more than human, but also the specifically human forms of labor are socially distributed. In the contemporary mode of production, invention power is the valorization of social, cognitive, affective, and relational activities that are embodied and situated in one's own life.[93] But embedded in the current structures of technoscientific production, invention power becomes a highly segmented activity. The conditions of experimental labor are distributed unequally.

Producers of knowledge are differentially positioned toward their own labor as well as the outcomes of their labor: most of them are free laborers when they are involved in community technoscience, while there are different classes of researchers, scientists, and experimental workers inside instituted technoscience (just as there are many classes of lab animals, plants, and materials that are valued and exploited differently). Consider the increased measurements of research activity, the existence of multiple positions of researchers with only a few of them being in full-time employment, the precarization of research work, the multiplication of different tiers of academic and independent research institutions, the access to research funding that is available only to few, the rise of the postdoc worker, the lab as the post-Fordist knowledge factory, the exploitation of the invention power of young researchers by senior scientists, and the zero-hours contract lecturers.[94] All these different types of technoscience's experimental labor show how work is internally highly diversified and under the constant blackmail of precarization.[95] This blackmail of precarization is often obscured by the demand to invent. Invent! The other side to precarious labor is that it is presented as a necessary form of work in order to facilitate unhindered invention.

The imperative to invent links community technoscience and instituted technoscience in terms of their respective organization of the labor process. In both there is a continuum revealing different degrees of precarization: Is the contract-dependent lab researcher or the precarious academic closer to community technoscience or instituted technoscience? How can we translate our everyday free experimental labor in a hackspace to an activity that makes a living? Who has access to which type of academic positions in instituted technoscience? What is the value of our research, whether it is located in community or in instituted technoscience?

RENTIER TECHNOSCIENCE

In the previous sections I have tried to describe different aspects of the maker culture: the topological stacking of things and processes; the compositional nature of making; distributed invention power; the contacts and gaps between instituted and community technoscience; the involution of human, animal, and inorganic actors; ecological transversality; experimental labor and ethopoiesis; and the precarization of work within the broader context of technoscience. In the following sections I discuss the political implications of such an understanding of making. What are the forms of political organization that all these different types of technoscience rely on? Which forms of political involvement do they promote and how?

It could be argued that maker culture and community technoscience more broadly rely heavily on the free (as in *freedom* as well as *nonproprietary*) circulation of shared knowledge, practices, information, materials, and other living beings. This sphere has been constitutive of the *commons* as a social form of organization and as a social movement.[96] But during my engagement with the maker culture I came to learn that it is misleading to assume that the commons underpins the life of community technoscience while publicly organized activities underpin the life of instituted state-funded research and private entrepreneurial activities underpin private research and development.[97] The circulation of tools, materials, organisms, knowledge, and practices between different types of community and instituted technoscience traverse private spaces and public state-owned spaces and the spaces of the commons. The commons (practices of commoning, common pool resources, peer production, and common forms of sociality and relationality that are neither public nor private) that traditionally were outside the securitized system of technoscience enter gradually into it. Cori Hayden (2010) has interrogated the discourse of the commons as a clear counterpart to the enclosed regimes of intellectual property and has shown how the logic of the commons is intimately entwined with enclosed private and state activities.

We could argue that technoscience becomes "biosecuritized," to use Kath Weston's (2013) term. Biosecuritization designates a double move in which technoscience is securitized both in terms of surveillance and control of the actors that operate in it and in terms of the valuation and assetization of its own processes and outputs. Regarding the former dimension, the securitization happens through the installation of physical barriers and complex architectures of entry requirements that regulate physical, technical, and informational access as well as the proliferation of sociolegal measures (formal and informal rights, patents, contracts, entitlements, codes, dispositions, and so on) that define degrees of scientific legitimacy and power.[98]

The second dimension of biosecuritization refers to the entanglement of technoscience in the biofinancial logic of current Global North societies. Not only is technoscience tightly linked to the organization of current production processes, but it is also involved in inserting different human and nonhuman entities into the system of biofinance—be it technoscientific artifacts, apparatuses, scientific processes, ideas, concepts, papers, materials, other animals, plants, or ecosystems. Although many of these entities belong to a great extent to the commons, they are transformed into assets as they are entered into prevalent processes of valuation. Contemporary technoscience not only appropriates and mixes *res publicae* (public sphere) and *res privatae* (private sector), it also relies heavily on implicating the *res communes* in its workings through its assetization.

In chapter 2, I discussed the centrality of rent and "the becoming rent of profit" (Vercellone, 2010) for biofinancialization.[99] Technoscience and the absorption of the commons in its workings enhances the role of rent within contemporary societies: the becoming rent of human and more-than-human worlds. Matter is not only value-generating in the production process, but it is also rent-generating through its existence as an asset. Early in 2015, the group behind the Arduino project that I mentioned earlier in this chapter announced a partnership with Microsoft to create an

"Arduino Certified" Windows operating system—an announcement that prompted angry reactions and numerous protests from the open-hardware and open-software community.[100] Not only did a free and open object of community technoscience become linked to a proprietary ecosystem, but also Arduino was formally turned to an asset that others—in this case, Microsoft—could exploit for future benefit. The life of the commons—free labor, peer collaboration, materials, cross-fertilization with other adjacent free and open technological ecosystems, skills, practical ingenuity, the politics of open hardware, free time, and so on—that makes Arduino possible and resides in it was turned to a resource that could generate rent for those who could speculate on it.

COMMONS AND THE FOLD

Technoscience exists as the private sector, the public sphere, and the commons fold into each other. Invisible structures of common exchange and cooperation, organized public institutions, civil society actors, and private interests and funding circulate through technoscience and reinforce each other. "Give me a laboratory and I will raise the world."[101] This motto belongs to an understanding of science that is theorized as the experimental achievement taking place behind the closed doors of instituted laboratories. Today's technoscience in action reveals a different story: Give me a laboratory and I will raise a start-up. Give me a laboratory and I will raise a social center. Give me venture capital and I will raise a laboratory. Give me state funding and I will raise a laboratory. Give me a social mobilization and I will raise a laboratory. And so on.

This constant folding creates a new situation where technoscience can no longer be considered unified nor is it given which form of practice is defining the workings of technoscience. Increased public engagement can no longer be considered a secure path toward the democratization of science.[102] Neither does the inclusion of scientific experts in regulatory procedures necessarily ensure "regulatory pluralism, reflexivity on the science-law relationship or democratic accountability" (Bonneuil and Levidow, 2012, p. 97). Public engagement can be seen as a mere productive activity in post-Fordist economies.[103] As Javier Lezaun and Linda Soneryd (2007, p. 280) put it: "Technologies of [public] elicitation, and the cohorts of experts that control their application and interpret their results, constitute a veritable extractive industry, one that seeks to engage publics in dialogue and generate certified 'public opinion' with the ultimate goal of increasing the productivity of government." Kate O'Riordan (2013), for example, shows how public involvement in direct-to-user genetic providers constructs the publics as consumers who then shape the genetic information provided.

This continuous folding of the private, the public, and the commons into each other creates a condition where designating one of these three domains as the primary force behind technoscientific innovation—be it community or instituted—becomes almost impossible. Is it big science (Shapin, 2008), the neoliberal privatization of science (Busch, 2017; Mirowski, 2011), the economization of science (Berman, 2013), the privatization of public institutions (Newfield, 2008), academic capitalism (Hackett, 2014), technoscience rentiership (Birch, 2017b), corporate science (Sismondo, 2009), or the increasing power of marketers (Dumit, 2012) that drives technoscientific knowledge production? Or is it the intervention of the public though processes of deliberation and contention (Davies, 2006)? Or is it perhaps the practices of the commons that sustain and feed technoscientific innovation (Kelty, 2008)? A definite answer to this is almost impossible. For better or worse, there is no single determination of technoscientific knowledge, and there is no privileged location in which invention power in technoscience takes place. Neither is there a privileged position for controlling it.

UNCOMMONS

Inasmuch as there is no central locus of control, there is no clear separation between these spheres. They are stacked together in relative locations: state and public institutions, private companies, commons, and the informal sphere exist inside each other but always in different configurations and in different modes of relations (which in many cases are hostile). Again, as argued earlier, this is not a unified world and there is no universal matrix underlying it. And similarly, this is not a world just made out of many different universes. Neither does the commons refer to a "common world" nor to a multiplicity of divergent worlds.

The difference between commons and a common world matters; *commons* refers to actively shared worlds between those who participate in their maintenance while *the common good* refers to something that is considered common for all. Commons is about co-action while the common good is about collective possession: commonwealth. Commons is about practices while the common good is about belonging. Commons is self-organized, but a common world is instituted. Commons is about care for specific worlds while common good is about universalizing one specific worldview. When commons becomes the common good we enter the terrain of Eurocentric humanist universalism; when the commons is about commoning we enter the field of processual, more-than-human worlds.

Mignolo (2013) rightly says that "uni-verality is always imperial and war-driven." He continues: "Pluri- and multi-verses are convivial, dialogical or plurilogical. Now pluri- and multi-verses exist independent of the state and the corporations and it is the work of the emerging global political society, e.g., the sector of society organizing themselves around specific projects once they/we realize that neither the state nor the corporations have room for multi- or pluriverses." This third sphere of global political activity is the commons and community self-organization or various other struggles that have been reconstructed as struggles for the commons later. See, for example, migration campaigns,[104] the mobilizations that prepared the wave of unrest in the Middle East and North Africa in 2011,[105] or various mobilizations in Latin America: the rebellions of Indian peoples in the 1990s, the landless movement in Brazil, the piqueteros in Argentina, the water and gas wars in the beginning of the 2000s, and the presence of nonhuman others as active subjects in Ecuadorian and Bolivian politics.[106] These are all mobilizations that attempt to change an increasingly postliberal world from below.[107] They achieve that by setting up social relations in the everyday that neither mirror the state nor directly oppose it but are involved in creating alternative worlds of existence.[108]

Let's recall the 2011 Tunisian revolution, for example. This revolution that came directly from the everyday. There were no dealers of representation; there were no left parties, big NGOs, empowerment campaigns, or external humanitarian interventions. There were the permanently harassed street vendors, the young academics who were ready to migrate, and the caring culture between the people of the neighborhood, the brothers, sisters, and friends living in transnational communities abroad: all these seemingly invisible connections that suddenly occupied and safeguarded central places in cities and towns. In this sense Bayat (2010) describes how these mobilizations were sustained and nurtured silently through the continuous experiences of people, things, and places for years before the eruption of the events. When these imperceptible movements were confronted with the brutality of the state, they crafted a nonidentitarian collectivity of insurrection. But long before the eruption of the insurrection they had silently crafted new everyday political ecologies.[109] In a similar way, Brecht De Smet (2014) has described the street politics that prepared the Egyptian revolution of 2011 and maintained its powerful impact. He provides an analysis of Tahrir as a radical, grassroots, prefigurative uprising that attempted to install justice and freedom in everyday life. "Tahrir became a 'freed zone' within the belly of the dictatorship. The Square offered a miniature experience of political emancipation. The activity of occupation was transformed from an instrument of liberation to a prefiguration of a free society" (De Smet, 2014, pp. 331–332).[110]

Indigenous ecocosmologies represent a different aspect of this movement: they respond to the destruction of globalizing power without seeing indigeneity as an instantiation of one global universal indigenous movement.[111] In fact, the global indigenous movement exists to the extent that it is practiced as a form of divergent indigenous cosmovisions and indigenous autonomy movements. Indigenous ecocosmologies are not given, are not just a matter of existing traditions, but are made and remade anew despite, against, outside, and with the universalizing processes of globalization.[112] Indigenous ecocosmologies are about autonomous cultural, political, and material articulations and the right to choose one's own forms of organization to manage land, natural environments, education, and health.[113] In other words, it is a form of commoning of the Earth by creating different scales, paces, and material practices. As de la Cadena (2010) argues, such politics are often dismissed as ethnic "beliefs" or "local cultures" even though they "express an epistemic alternative to supposedly scientific paradigms (ecological and economic)."[114] This epistemic alternative challenges the perceived "common good (productive efficiency, economic growth, even sustainable development)" in a certain society by promoting alternative forms of material and interspecies commoning in different ontologies: as argued earlier, the difference between common good/commonwealth, on the one hand, and commons/commoning on the other matter.

We can think of indigenous politics together with another different but often relevant example: projects for the alternative making of ecological spaces: from urban guerrilla gardens to reclaiming access to enclosed public spaces, from the waterkeepers' campaigns to the water justice movement,[115] from antipollution mobilizations to the political ecology of commoning air and soils. Consider, for example, community-based projects to protect the corn of the indigenous peoples of Chiapas when it became widely known in 2001 that heritage seed stocks of landrace corn in Mexico were being infected by genetically modified crops imported from the United States. "Sin Maiz, No Hay Pais" [No Corn, No Country] (Esteva and Marielle, 2003)[116] is a response to a much wider conflict rooted in the turn to the "commodities consensus" in Latin America "based on the large-scale export of primary products" (Svampa, 2015) that fueled extractivism as the dominant economic path throughout the region.[117] Within this context agricultural fields become battlefields between multinational agroindustrial holdings and local rural, peasant, and indigenous communities.[118] Marisa Brandt (2014) describes how the Zapatistas responded to the emergency of the intrusion of genetically modified crops: instead of engaging in a legal fight against transgenic introgression and for patenting their own corn, they created an alterontological practice, a form of "biocultural innovation" that blended strategically endorsed and locally sustainable technoscientific knowledge with indigenous agroecology, a large network of solidarity growers, and the attempt to preserve the political autonomy of indigenous communities.[119]

The project's "goal is not to separate nature and culture, but rather to demonstrate how deeply imbricated they are—Zapatista corn performs the biocultural link between Zapatistas' political project and their maize plants. By creating alternative networks for corn circulation, the project allows international recipients to participate in Zapatistas' political bioculture, that is, to relate to seeds as potential food or plants that are deeply inflected with the values of promoting self-sufficiency and resisting governmental and economic dependence. Relationships bring worlds into being; ontology is a political achievement" (Brandt, 2014, p. 876). Alterontological commoning of corn rather than its formal protection was the aim of the project. The practice of commoning as well as indigenous ecocosmologies exist only as contingent relations to other people, other species, and the material surroundings. In native American teachings one's "relatives may have wings, fins, roots, hooves" (LaDuke, 2010, p. 83). In Mayan ontologies of creation, human flesh is made of corn. In the world of commoning, exchanges with nonhuman others exceed clear distinctions between engaged nonhumans and the committed humans that they cooperate with. In the water commons movements, the people, the water, and the watersheds are not separate; they are becoming *with* each other, a "human river."[120]

To what extent is this form of commoning possible in technoscience? Because if as argued earlier the commons becomes folded onto the private sector, state institutions, and the public sphere, then there is a question about its ontological constitution. The "uncommoning" of the commons is not a project to come but something that is already happening. It neither makes the world more unified nor makes a more divergent world. It just changes the conditions of how conflict, contention, and collaboration will be articulated and practiced.

In technoscience one can observe how the ambivalent unequal and conflictual relation between instituted and community technoscience unfolds. Instituted technoscience constantly expropriates and privatizes community technoscience whether it is grassroots innovation, free labor, community infrastructure, indigenous and alternative knowledges, or the multiple micropractices of commoning. Simultaneously, community technoscience relies on reclaiming knowledge, technologies, and resources that are developed in the realm of instituted technoscience. Adrian Mackenzie (2013), for example, in his work on synthetic biology, offers a glimpse into this ambivalent movement between "publics" that object to high-profile Big Bio on the one hand, and "publics" that are just validating and confirming it on the other. Alessandro Delfanti (2011) offers a more complex view of this process in which the folding of private enterprises, publics, and the commons into each other underlies the constant, often antagonistic oscillation between big enclosed science and open technoscience.[121]

MORE-THAN-SOCIAL MOVEMENTS

There is a long history of industrial struggle and social conflict that permeates the stacked ontologies of the Silk Mill. In November 1833 industrial unrest swept the mill factories in Derby and escalated to a lockout of the workers for several months. This was probably the first organized action by labor in Britain, and although it did not achieve the desired aims at that time, it led together with a few other similar unrests to the formation of the first union confederation in February 1834, the Grand National Consolidated Trades Union. Although this and other similar events shaped the articulation of conflict and the promises that it envelops up until now, they were rarely mentioned in the day-to-day life of today's postindustrial Silk Mill. They exist as movements where each wave creates the conditions for the emergence of the next one, although none of them can anticipate the form and practices of the subsequent movements. These movements are more than social; they live inside more-than-human worlds, in the ontologies they help to create.

FIGURE 8.15 — Silk Trades Lockout of 1833. Silk Mill public house mural, Derby, UK. Photograph by Dimitris Papadopoulos.

More-than-social movements do not attempt to contest power by organizing protest; rather, they attempt to create the conditions for the articulation of alternative imaginaries and alternative practices that bypass instituted power and generate alternative modes of existence.[122] Protest and resistance social movements that channel all their actions to resistance are vocal and visible.[123] But they are not the main force of movement action.[124] What constitutes movement action is the capacity to set up alternative forms of everyday existence and mundane practices that later come to force power and control in a specific field to reorganize itself and subsequently to reengage the actors involved in the field in new and often unexpected ways. Nikos Karfakis (2013), for example, has discussed how the multiplicity of mobilizations of people diagnosed with chronic fatigue syndrome target simultaneously popular opinion, social policy, workplace exclusion, and relevant technoscientific knowledge. Michelle Murphy (2012) has shown how the politicization of technoscientific aspects of reproductive health has created alternative spaces for women's empowerment, but simultaneously it has also created a complex entanglement with racialized biopolitics and the unequal economic, social, and political logics of the past fifty years.[125] I discussed in the previous chapter how AIDS activism instigated major social and material transformations beyond the teleological view that it solely focused on becoming included in clinical trial panels. Or, just to mention another final example: it would be impossible to understand the workers' takeovers of factories in Argentina after the 2001 economic crisis and the more recent wave of worker-recovered factories and cooperatives across southern Europe after the 2008 economic crisis merely as a form of protest or resistance.[126] Rather, they are creating alternative forms of life on the ground, and they set up conditions in which movements, rather than just oppose power, installed alternative ontologies of existence. Movements escape existing ontologies and contribute to the creation of new. "Ο αγώνας γυρίζει το γρανάζι" [Luchas (struggles) turn the cog] stands on a handmade poster of the recuperated factory of Vio Me in Thessaloniki, Greece.[127]

Many of the approaches to movements in technoscience highlight protest and resistance and cultivate the imaginary that movements' action is all about a possible inclusion of neglected publics in technoscience.[128] Inclusion seems to be the ultimate horizon of action: inclusion in the hermetic cathedrals of science and the instituted technoscience with the aim to shape research agendas upstream and change state policies.[129] And this indeed may be the case for some of the movements in the field of technoscience, such as the science for the people mobilizations of the 1960s and 1970s[130] or the demands for participation[131] in the design of science policy as well as in defining the topics of research in the 1990s.[132]

But there is a plethora of other movements that did not (only and not primarily) address demands to the institutions of technoscience in order to be included in them but rather developed gradually alternative practices that attempted to appropriate and reclaim technoscientific knowledge, to participate in its production, and eventually to set up self-organized structures to produce knowledge themselves.[133] The radicalization of green, ecological, and health movements in the 1980s, many of the ecological commons campaigns of the 1990s, and similar campaigns for alternative energy production are just a few examples of these type of movements. And of course community technoscience itself—in all its complexity and ambiguity as discussed in the previous sections of this chapter—constitutes a form of movement that in many cases has the capacity to set up alternative forms of everyday existence and material engagement.

MATERIAL LITERACY

Common to all these attempts is an understanding of what Jean-Luc Nancy calls being-in-common, rather than belonging in an already existing community.[134] Being-in-common means that every situation in which these movements find themselves or even help to create is singular and unique.[135] These movements make community by ontologically making themselves and their own material conditions of existence: being in the process of commoning. I encountered two main aspects of this type of community building in many spaces of community technoscience, and I want to highlight them here: material literacy and infrastructural imagination.

Learning is key to community building in technoscience. In public debates and instituted technoscience, literacy is primarily understood as scientific literacy:[136] the making of scientifically informed citizens and of appropriate institutional structures for their inclusion. This perspective is dominated by a classical cognitivist model of knowledge transmission: participation is the result of the enhancement of cognitive contents. Against scientific literacy, which solely focuses on strengthening cognitive frameworks for understanding competing arguments in a technoscientific controversy, there are attempts to show that scientific literacy can primarily be enriched only if it is connected to active processes of responsibility for concrete community development, what Maud Perrier and Deborah Withers (2016) call "collaborative (un)learning."[137] Such learning and unlearning settings have a long history in alternative education, self-organized learning, and informal community schools—I am thinking here of the Experimental College of the Twin Cities[138] (Dyke and Meyerhoff, 2017) as an good example.[139]

In many of the makerspaces and community technosciences in which I participated, one can see a different form of literacy that resonates with these ideas: literacy that is focused less on the humanocentric and cerebrally organized task of creating informed citizens and more on the attempt to create stable microecologies. A different form of literacy is happening here, namely material literacy: an ongoing and involuntary (involution with nonhuman others and things) experimenting with matter.[140] Its starting point is a commitment to engage with a problematic; its aim is not to solve the problem but to create what Andrea Ghelfi calls an "ecology of proximity," spaces of dense relations, energy flows, and co-constitution of the involved actors. In such ecologies when there is an issue, "there is always more than one solution, and this 'more than one' is grounded inside, not outside, the ways in which a problem is matter related: matter works through a pluriverse of canals, of bridges" (Ghelfi, 2015, p. 85).

Material literacy starts from an obligation to protect an ecology from its degradation and to make it a livable place for all its participants; it is, in the words of Puig de la Bellacasa (2017), a "matter of care." Learning and knowledge here starts with an obligation to engage with a specific issue, and care serves as both the ethopoietical compass and the main aim in the process of this engagement.[141] Rather than conceiving an issue as a problem and trying to get to the essence of the problem and to deliver solutions, literacy here is about learning to engage with problematics:[142] the dynamics between the different actors within an ecology and their capacity to be affected by each other in order to change the conditions in which an issue can be dealt with. The material literacy of problematics is about shifting relations and enhancing care. And by doing that the ecology itself shifts and transforms; often it bifurcates and forks toward other ecologies, toward more resilient alterontologies: "thousand ecologies" (Hoerl, 2013).

While scientific literacy preserves the securitized logic of technoscience and makes publics out of citizens, community technoscience attempts to radically democratize literacy through a mode of distributed "cooperation without consensus" (Star, 1993).[143] Distributed cooperation without consensus—or else nonparticipant connectivity[144]—takes place as the exploration of limits and the virtual potentials that the relations to the materials allow. This is distributed invention power—as discussed earlier in this chapter—in the middle of material constraints. This, of course, is possibly a feature of knowledge production as such, but the main point is that these constraints are freely shared and explored in collaborative ways. My interlocutors in the Leicester Hackspace told me that the scientific lab mystifies constraints. In the securitized space of the big instituted science lab, constraints are what individual scientists need to protect from others in order to be able to invent. Constraints in instituted technoscience are

FIGURE 8.16 — Leicester Hackspace, UK. Educational workshop. Photograph by Dimitris Papadopoulos.

always a matter of secrecy,[145] while in community technoscience and material experimentation the knowledge of constraints is becoming common property, the fuel of material literacy.

GENEROUS INFRASTRUCTURES

This co-emergence of politics and matter creates alternative spaces of existence. In fact, it creates these spaces as infrastructures. But here I am interested less in how infrastructures are made and produced or how infrastructures are used for social reproduction. It is much more about how infrastructures emerge through "traverse communications"[146] (Deleuze, 1987) between different involved actors. It is about the emergence of infrastructures through "creative involution":[147] symbiogenetic world-forming, the actuation of material alliances through the association, mutualism, and articulation between not directly connected forms of life.[148] Endosymbiosis instead of endocolonization.[149] Infection instead of property.[150] Cooperation instead of toxicity. Warm compost instead of acidity. Commensality instead of contamination. This is "creative involution" rather than linear evolution.

More-than-social movements achieve their autonomy through the making of infrastructures. Autonomy—as discussed earlier in this book, in particular in chapter 2—refers to the idea that social conflict and social mobilizations drive social transformation instead of just being a mere response to (economic and social) power. But more-than-social movements expand this form of autonomy to engage with questions of justice in more-than-human worlds and to highlight the importance of creating alternative lifeworlds of existence for their practice. The quest of these autonomous infrastructures is to restore justice step by step through everyday material practice (see also chapters 1 and 5).[151] This is a mundane material and generative justice.[152] Justice is fought for on the level of matter and through close alliances between *engaged* groups of animals and plants, *committed* groups of humans, and *accessible* material objects. The autonomous politics of more-than-social movements are relational, ontological struggles to create alternative material articulations. An autonomous political posthumanism emerges in the infrastructures of more-than-social movements: political autonomy as material interconnectedness; being in the quantum vortex of constant interdependences; knowing and naming one's allies; building material communities of justice.

More-than-social movement infrastructures are autonomy made durable: transparent, unnoticed, and persistently present spaces that incorporate political practice in their workings. Infrastructures allow more-than-social movements to politicize ontological practice in the absence of consensus.[153] These infrastructures shape political developments and life without the need to start again and again from scratch. They become part of infrastructural imagination:[154] the capacity to transfer infrastructures beyond a specific spatial and temporal location and to reclaim it for a different ontology; the capacity to connect, tweak, and reconnect different infrastructures; the capacity to extend infrastructures over time and to redeploy them in the future.[155] Seed bombs are good examples of such infrastructures: they rely on complex human and nonhuman labors to exist; they are readily transferable; they can be applied differently in varying environments; they

FIGURE 8.17 — Seed bombs. Photograph by Dimitris Papadopoulos.

carry knowledge, material potentiality, and learning within them without imposing it as a closed system in each different location in which they are used; they are self-sufficient—clay protects the encapsulated seeds, and nutrients support them in their first growth—until they melt in the soil; and they can travel easily. Through this traffic—linkages, transfers, mutations, and modifications—more-than-social movements' infrastructures are sustained. Infrastructures need to be understood ecologically: to "come into being, persist, and fail in relation to the practices of the diverse communities that accrete around them" (Carse, 2012, p. 543).

More-than-social movements create *generous* infrastructures: infrastructures that allow for communities to maintain and defend the ontological conditions of their forms of life even when instituted infrastructures break down by failure or by intent.[156] In this sense these infrastructures are directly political. Politics (and the social) does not come on top of the infrastructures that more-than-social movements create. Is a self-managed water system an infrastructure or a political campaign? Is an educational workshop of the Leicester Hackspace an infrastructure or a tool for achieving other social goals, such as promoting hobbyism or hacker culture? Is a cooperative farm an infrastructure of life or a political project for community empowerment? Is an open-access bike workshop an infrastructure or a commitment to a different lifestyle? Most of these infrastructures do both at the same time. In fact, if there is a split between the material and the political, infrastructures cease to be generous. They are no longer autonomous, and they are appropriated for other social aims and political targets.[157] They become instituted, territorial, and managed as tools. Instead, generous infrastructures involve always the involution between human and nonhuman others, between materiality and sociality, and only by doing this they become an alterontological practice.

GIVE ME A KITCHEN . . .

More-than-social movements are successful to the extent that they change the conditions of knowledge production by engaging with the *actual* making of knowledge in a specific subfield of technoscience. Only when movements produce alternative knowledge with, within, and occasionally against specific developments in technoscience can they effectively promote change. How far can (more-than-social) movements in technoscience carry us? And more specifically, to what extent is the maker culture part of an alterontological practice? Perhaps the experiences of the free and open software movement could be helpful here: although free and open software continues to play an important role in digital movements, it outpaced itself in terms of the innovation it produced and is now in the process of being continuously folded into proprietary software and vice versa.[158] New movements are evolving out of the free software movement, such as the movement for radical privacy, the cryptographic movement, and the development of anonymous campaigns in the hidden net.[159] In a broader sense, the free and open software movement has inspired open hardware and open technoscience[160] that underlies many activities of the maker culture.[161] In a similar way as the ontologies they help to form, movements do not evolve linearly from each other; rather, they operate inside the previous movement and develop in new unexpected ways.

Stacked ontologies and stacked movements. In this chapter I have tried to describe the ontological making of the world as temporally, materially, and functionally stacked. And I tried to argue that this is not an epistemological perspective.[162] This is not something that can be observed as a state of the world or even as a process. Stacked ontologies exist to the extent that different actors shape, change, and move them. Stacked ontologies are like the compost piles in permaculture gardens mentioned earlier in this chapter. I am thinking here of (more-than-social) movements as the agents that make stacked ontologies move and change: movements as the worms, bacteria, and fungi that compost layer after layer of organic matter.[163] In the words of Bruce Sterling (2005, p. 14), "Tomorrow composts today. . . . Technocultures do not abolish one another in clean or comprehensive ways. Instead, new capacities are layered onto older ones. The older technosocial order gradually loses its clarity, crumbles, and melts away under the accumulating weight of the new."[164]

Rather than being a single movement itself, the maker culture provides a material framework that many different movements, projects, and initiatives incorporate in their practices: kitchen science, DIY biology, alternative experimentation with medical substances, lay engineering projects, production of alternative forms of energy, projects of ecological modernization from below, self-managed systems against environmental hazards, alternative forms of agriculture and soil renewal, radical patient-based campaigns, permaculture regeneration, traditional systems of knowledge, craft, embodied technoscience, punk science, health movements, indigenous ontologies, open-source science, technology and agriculture, clandestine chemistry, the hacker culture, ecological justice initiatives, cross-species collaborations, bio-art, self-organized projects of scientific literacy—all examples of reclaiming and reinventing technoscience from within. Give me a hackspace and I will raise a laboratory. Give me crowdfunding and I will raise a new technoproject. Give me a community space and I will raise a laboratory. Give me a laboratory and I will raise a community space. Give me a plot and I will raise a soup kitchen. Give me a kitchen and I will raise a laboratory.

What is constitutive of these movements is not that they attempt to encounter and target technoscience *as such* but that they change the conditions of knowledge production by changing the ontological fabric of life. This distinction is important. It implies that there is no outside to the process of the fold described earlier in this chapter—a politics of multiple universes can work only if it is an alterontological practice: the alternative creation of ontologies. Real political disagreement, in the words of Rancière (1998) that I have discussed in the previous chapters, takes place when it is performed on an ontological level—that is, as I have argued, when it forces existing ontologies to fork and craft alternative forms of existence. "Ontological disagreement," says de la Cadena (2015, p. 280) in her powerful ethnography of indigenous politics in the Peruvian Andes, "emerges from practices that make worlds diverge as they continue to make themselves connected to one another." Throughout this book I called experimental practice all these very diverse activities and ways of being that allow such ontological disagreements to be enacted and set in motion. I have enlisted community technoscience as a creative ally to more-than-social movements that experiment with our constituted polity. The trust that an alterontological practice is possible even if we do not know how to make it happen in the beginning—what Puig de la Bellacasa (2014a) calls "material spirituality"—is the starting point for any experimental practice that engages with the forces of matter, other committed humans, accessible things, and engaged animals and plants. What is at stake here is not only technoscience itself but life in its ontological constitution. Within this process technoscience and our polity itself become a field of social, material, and interspecies experimentation. Give me a kitchen and I will raise a world.

ACKNOWLEDGMENTS

Many people have contributed to this book in many different ways, and I want to thank all of you. The list would be far too long to include all. Here I mention those who have offered direct comments on earlier versions of parts of this book: Antonis Balasopoulos, Marcus Banks, Wenda Bauchspies, Stefan Beck, Dominique Béhague, Nic Beuret, Huw Beynon, Hywel Bishop, Patrick Bresnihan, Steve Brown, Marco Checchi, Tim Choy, Adele Clarke, Patricia Clough, Harry Collins, Nicholas de Genova, Marisol de la Cadena, Joe Dumit, Stephen Dunne, Nick Dyer-Witheford, Dace Dzenovska, Arturo Escobar, Rob Evans, Judy Farquhar, Andrea Ghelfi, Margherita Grazioli, John Hartigan, David Harvie, Cori Hayden, Woody Kitson, Michalis Kontopodis, Chris Kortright, Hannah Landecker, Joanna Latimer, Javier Lezaun, Keir Milburn, Fredy Mora-Gámez, Michelle Murphy, Warren Neidich, Theo Nichols, Jackie Orr, George Papanikolaou, Trevor Pinch, Lele Pizio, Brian Rappert, Francesco Salvini, Leandros Savvides, Debra Benita Shaw, Stevphen Shukaitis, Niamh Stephenson, Marcelo Svirsky, Chris Talbot, Cristina Tufarelli, Imogen Tyler, Gediminas Urbonas, Nomeda Urbonas, Tracey Warr, and Ian Welsh.

I am deeply thankful to Emma Chung, John Cromby, Simon Lilley, and Vassilis Tsianos for our collaboration that informed work presented in this book. In particular, Vassilis Tsianos contributed to the book in many different ways, not at least through his ideas for the title. Finn Bowring, Jim Clifford, Chris Connery, Julia O'Connell Davidson, Jannis Savvidis, and Ernst Schraube have been a constant source of inspiration, a refuge in times of worry, and a place to share joy and ideas. I am especially thankful to my editor at Duke, Courtney Berger, and the editors of the Experimental Futures series, Joe Dumit and Michael Fisher, for their invaluable guidance and constructive suggestions. This book started in conversations with Joe Dumit and I am grateful for his rich contribution to it. Heartfelt thanks also go to Sandra Korn and Christi Stanforth for their creative support during the preparation of the manuscript. I am grateful to the two anonymous reviewers for their immensely helpful feedback and for their encouraging but also challenging comments and suggestions. Special thanks goes to Fredy Mora-Gámez for his work as research and editorial assistant on this project and for our collaboration.

My parents, Eleni Papadopoulou-Alivizatou, tireless inventor of alternatives, and Dionyssis Papadopoulos, maker of worlds, have both shaped the book with their ideas, their love and empowering care and their way of living life. Thank you. I started writing this book a year after my daughter, Alba, was born. Three years later my son, Amaru, entered our life. This book is variously and indelibly marked by the presence of these two wonderful creatures. The ideas presented here have been developed with my intellectual and life companion, Maria Puig de la Bellacasa. This book is for her. Without her it would not exist. Many of the issues that animate this book emerged during my involvement with migrant communities, transnational migrant activist organizations, and the precarious workers movement across Europe as well as the East Midlands hacker communities in Britain. I am deeply grateful to them and I hope this book contributes in some way or another to their efforts.

INTRODUCTION

1 The anticipation of an imminent and unavoidable catastrophe has, as Beuret (2016) discusses in his excellent study of environmental social movements, ambivalent effects: it has shaped much of the visions of current social movements, but the "global scalar logic of climate change" has often disabled "effective environmental political action" instead of promoting the mobilization of everyday alternatives. See also Yusoff (2009).

2 I will refer to literature on the posthuman and posthumanism in relevant places throughout this book. Braidotti (2013) and Roden (2014) provide overview discussions of the posthuman; see also Badmington (2004), Herbrechter (2013), and C. Wolfe (2010). Within the framework of science and technology studies I'm thinking primarily of the work of Pickering (1995).

3 Paris 2006, Athens 2008, Tunisia 2010, Cairo 2011, Madrid 2011, Athens 2011, the global Occupy movement 2011, the 15M movement in 2011, London 2011, Istanbul 2012, and so on.

4 My inspiration for approaching this cycle of struggles as a *worlded* phenomenon comes from the work of Connery (2007), in particular his text *The World Sixties*, whose beginning I paraphrased in this sentence.

5 In particular, it is important to investigate the extent of the effects, if any, that these movements had on the way the 2008 economic crisis was handled. Many of these movements addressed the crisis directly but their impact was limited. This seems to be the case also for another important social movement of that time, at least in the Global North and definitely in Britain: the climate change movement. Despite its extensive activity and wide composition, its effects were also restricted—see the important analysis of Beuret (2016, 2018).

6 Gilroy's argument is that existing social, political, and media elites came together to form a class, while black communities are internally disconnected and often unable to organize and "act as a body." He says: "The last week has been an amazing class, a primer, to give us the opportunity to understand how these things function today. You remember that party they all had, in the Cotswolds . . . and they were all there, the Milibands were there, the Labour

people were there, the TV people were there (not the ones from David Starkey–land but the ones from Channel Four News), and they were all there together, and they're telling you something when they all congregate like that. They're telling you that they're a class. And they think and act and conduct themselves like a class. They chat to each other, they marry each other, they go to the same places. . . . And if we want to act as a body, if we want to act in concert, we have to learn something from the way they conduct themselves, even as we challenge what they do. So the pieces I can see in this system, the role of information, of policing, of deprivation, of inequality. . . . And we need to clarify that we have the resources we need in our community—we just need to use them in a different way. Thank you" (Gilroy, 2011).

7   Gilroy (2013, p. 553): "Thirty years after that shocking, transformative eruption, the same streets in England's cities were again aflame. This time, there was no rioting in Scotland, Wales, or Ireland, and this time, no progressive reforms of discriminatory policing or uneven, color coded law would follow. No deepening of democracy would be considered as part of any postriot adjustments to the country's politics of inclusion. Democracy's steady evacuation by the governmental agents of corporate and managerial populism was too far advanced. The market state that had been dreamed about was now a rapacious and destructive actor, privatizing and outsourcing government functions while managing to incorporate those who had the most to lose into the destruction of the public institutions on which they relied."

8   Connectivity does not necessarily mean subjective intentionality for being and staying connected. As I discuss in chapter 8, infrastructures of connectivity often work without consensus and intentional participation. Connectivity and disconnectivity are not necessary opposites but strategic positions within complex social relations. For an excellent discussion, see Staeheli (2012).

9   I borrow the idea of dis-connectivity and un-networking from the work of Staeheli (2012, 2013).

10 My understanding of experimentation is primarily derived from Deleuze and Guattari's (1987) *A Thousand Plateaus*. For Deleuze and Guattari, experimentation is the answer to and the way out of a series of dualisms such as stability/structure vs. change/flow, assemblage vs. elements, closed/fixed vs. indeterminate/open. Experimentation is about the abolition of dualism (see also chapter 4) that confines practice and thought to predefined positions. "One never knows in advance," says Deleuze (1987, p. 47); cautious experimentation is the center of practice (Deleuze and Guattari, 1987, p. 161). Here, experimentation goes hand in hand with the notion of experience, creative transformation, and creative involution (which are discussed throughout this book, particularly in chapter 8). In the field of science and technology studies I rely on Barad (2007), Schillmeier (2015), and Fischer's (2009) work on experimentalism and experimental systems—developed in discussion with Rheinberger (1997).

11 More broadly, everyday life and especially the "conduct of everyday life" is the epicenter of sociomaterial change that I explore in this book. I draw on the

important work of Schraube and Højholt (2016) to understand the link between the conduct of everyday life and social transformation. In particular, I am interested in how the conduct of everyday life often escapes and defies other more standardized and regulated aspects of everyday life (such as consumerism). What is crucial for my project is not to map or capture a definite image of everyday life but to explore how the uncontrollable excess that is always part of our mundane practices is transforming everyday life to a space for experimentation (Stephenson and Papadopoulos, 2006). I am interested in the moments when, with Debord (1981), we could say that everyday life turns against itself; that is, when unexpected, experimental aspects of everyday life set ordinary ways of being in motion.

12 Specifically about visual materials, I use them as enactments of stories rather than as representations of the topics discussed in the book—see the work of Banks (2014a, 2014b). With few exceptions, rather than providing a direct analysis of the images I let them narrate stories that I hope are different from but complementary to the stories I am advancing in the main text.

13 On problems and problematics, see Deleuze (1994). See also D. W. Smith (2012) and Maniglier (2012).

14 Several ethnographic and social research texts have inspired this methodology: Choy's *Ecologies of Comparison* (2011), Fischer's *Emergent Forms of Life and the Anthropological Voice* (2003), Ford's *Savage Messiah* (2011), Haraway's *Primate Visions* (1989), and the debates on the politics of ethnography presented in the volume *Insurgent Encounters* edited by Juris and Khasnabish (2013). My guiding star has been always Clifford's *Routes* (1997) and in particular chapter 12, "Fort Ross Meditation." A second source of permanent inspiration has been geophilosophy and geopoetics, such as Deleuze and Guattari's *A Thousand Plateaus* (1987) and the works of Glissant (1997) and White (1992).

15 My approach to the baroque is inspired by de Sousa Santos (2001) and Deleuze (1993); see also Flanagan (2009).

16 There are some possible parallels here with magic realism—Jameson (1986), Moses (2001), Orr (2015), Selmon (1988), and Wenzel (2006) have influenced how magic realism is mobilized in this book.

17 For different takes on abundance, see, for example, Bresnihan (2016a), Holmgren (2002), Collard, Dempsey, and Sundberg (2014), and Hoeschele (2010).

CHAPTER 1. DECOLONIAL POLITICS OF MATTER

1 I will discuss technoscience in chapter 8. My starting point here is Haraway (1997). See also Ravetz (2006) on postnormal science.

2 The assumption of a tight link between material transformation and historical change is not new—the reduction of a thing to an object of contemplation and its separation from actual material activity is something that already Marx wanted to overcome in order to establish his materialist approach to history. However, in this conception of materiality, the manipulation of ontology was

conceived as a possibility and as an aid to achieve other sociopolitical targets; the most prominent of them is social liberation—an idea that is so characteristic of Marx and Engels's foregrounding of technology as a tool for changing history. But if technoscience is not a tool in the hands of an actor forging other political goals but is politics as such, then the question is what kind of politics it is.

3 When I talk about ontological politics, I refer to the actual practice of it rather than to its theoretical conceptualizations such as Law (1999), Marres (2009), and Mol (1999). Ontological politics is a very specific version of the broader debates on ontology/ontologies emerging in the social sciences that I discuss extensively in chapter 8. Here I refer only to science and technology studies; see, for example, the seminal work of Haraway (1989) and the works of Pickering (1995, 2008), Star (1995), Strathern (1991), C. Thompson (2005), and Cussins (1996). For some examples, see Mol (2002), Moser (2008), Moreira (2006), and Law and Singleton (2005).

4 Rather than a multicultural one; see Viveiros de Castro (2004).

5 For an analysis, see Savransky (2017).

6 Matter is a multifaceted and difficult concept. I use it throughout this book when I do not talk about specific formations of matter—the epigenome, a fiber-optic cable, the brain, PET plastic, academic theories, the embodiments of the HIV virus, and so on—to capture all that it exists as a creative self-organizing process. My main references here are Barad (2007) and, primarily, Deleuze and Guattari (1987), who provide such an approach to matter (see, for example, Propositions VII and VIII in chap. 12).

7 See Chung et al. (2016) and Cromby et al. (2016).

8 See various accounts on the colonization of America that describe this double movement of the frontier: Bailyn (1988), Todorov (1984), and F. J. Turner (1993).

9 McClintock (1995) demonstrates how imperial political power, imperialist expansion, industrialism, and, finally, intimacy and the domestic space all get mixed and transformed in the process of the colonial contest.

10 See Quijano (2007) and also Mignolo and Escobar (2010).

11 See, for example, Vazquez (2011) and Maldonado-Torres (2007).

12 See Svirsky (2010) and also P. Patton (1996, 2000).

13 This maps on the two systems of power that Quijano sees emerging with the conquest and colonization of the Americas: the economic power of the capitalist system of production and labor and the social and political power structured around the idea of race (Quijano, 2000; Quijano and Wallerstein, 1992).

14 Moulier Boutang (1998) has analyzed this process in his work on the abolition of slavery, and Pratt (1992) has shown the connection between colonial expansion, travel, and systems of representation.

15 Representation as a process of constructing social groups that later can be included in the political institutions of contemporary liberal democracies; for further discussion, see Stephenson and Papadopoulos (2006) and Papadopoulos, Stephenson, and Tsianos (2008).

16 Ontological politics is not about setting up new "ontological choreographies" (Cussins, 1996) and "dances of agency" (Pickering and Guzik, 2008) but about embedding these ontological choreographies in the current system of production. For example, as Vora (2009) shows, not only does transnational surrogate motherhood assemble new "ontological choreographies" around it, but by doing so it facilitates the extraction of biocapital from "biological labor." Ontological politics is not an innocent description of the workings of contemporary technoscience, but it is the modus operandi of bioentrepreneurialism. This, of course, applies to all the examples I mentioned here—see, for example, Starosielski (2015) on the complex lives of subsea cables (which maintain transoceanic Internet traffic) and their entanglements with politics, economy, and the environment.

17 Innovation and the imaginary of productionism are tightly entangled. Both are considered the engine of social change—see Yusoff (2013b) and L. Suchman and Bishop (2000).

18 This form of politics is not concerned with representation and with the symmetrical inclusion of different entities in the political and productive arena of the constituted institutions, as these can exist only by erasing specific capacities of each actor in order to make them fit—see Stephenson and Papadopoulos (2006).

19 In Papadopoulos, Stephenson, and Tsianos (2008), we argue that this type of politics has primacy over the formation of control and drives historical transformation. This is the reason this politics is often called autonomous—not in the sense that it is independent but in the sense that it is not overdetermined by the existing system of power and control. I discuss autonomous politics in various places in this book, particularly in chapters 2, 5, and 8.

20 The work of the Ecological Humanities Group (http://www.ecologicalhumanities.org) has been crucial for developing these ideas. D. B. Rose's *Reports from a Wild Country* (2004) was an inspiration for thinking an ethics of decolonization in the context of the unfolding process of colonization in the frontier of matter. Van Dooren's work has been important for exploring relations to human and nonhuman others who have been made superfluous and for understanding "precisely *how* different communities (of humans and nonhumans) are entangled, and how these entanglements are implicated in the production of both extinctions and their accompanying patterns of amplified death" (van Dooren, 2011). See also D. B. Rose and van Dooren (2011).

21 And as with every desire it can unfold either as a manic and anxious chase of something that we don't have—Meillassoux's (2009) obsession to grasp the "in-itself" seems to be an expression of this form of desire in the politics of matter—or as a force to escape existing closures, experiment, and make novel connections and forms of life. Like every politics, politics of matter contains both. Here I refer to the latter while I silently presuppose the former as a repressed version of a politics of matter.

22 See Eglash (2016).

23 See Papadopoulos, Stephenson, and Tsianos (2008, pp. 71ff.). For an in-depth analysis, see W. Smith et al. (2015).

24 See here also the analysis of Schraube (2005).

25 See the historical work of P. H. Smith (2004), which demonstrates how artisans and artists, sculptors, locksmiths, and carpenters were key actors in the formation of modern science; see also H. Rose (1983).

26 On the centrality of scale for understanding global processes, see the work of A. Tsing (2000) and Glick-Schiller and Çaglar (2008).

27 See chapter 8 for a further discussion of the free-software movement.

CHAPTER 2. BIOFINANCIALIZATION AS TERRAFORMATION

1 Zalasiewicz et al. (2011). For a critical discussion of the implications of Anthropocene discourse for politics, see Beuret and Brown (2016).

2 Although this is not the focus of this chapter, the relation of social science/social scientists and the crisis is an important aspect of this topic. In a special-issue editorial on the 2008 financial crisis, Bryan, Martin, Montgomerie, and Williams discussed why extant knowledge in the social sciences failed to be attuned to the crisis. What if this "important failure" (Bryan et al., 2012, p. 302), is not due to inertia and lack of reflexivity but because academic research and practice has internalized the logic that sustains the socioeconomic mechanisms that contributed to the financial crisis, to the extent that failing to see the crisis was almost inevitable and the limited number of novel responses to the crisis even more so. Academics and knowledge producers, as a part of the professorial-managerial classes, are deeply embedded in a social and cultural environment that has been formed by the ascendance and consolidation of the financialization of economy and society—Beverungen, Dunne, and Hoedemaekers (2013) describe this situation persuasively.

3 For an analysis of such mobilizations, see chapter 8 and also Tsianos, Papadopoulos, and Stephenson (2012). See also Barthold, Dunne, and Harvie (2017).

4 I am working here with Oliver Ressler's piece of art *We Have a Situation Here* (2011) in order to rethink possibilities for the emergence of new alternative forms of subversion not only when established power collapses but also when those who oppose established power collapse and leave space for new initiatives (http://www.ressler.at/we_have_a_situation_here/).

5 For a discussion of autonomism, see my previous work (Papadopoulos et al., 2008) and the references included in the notes further down in this chapter.

6 I am thinking here of Harvie and Milburn's work on affective contagion in social movements in relation to this magical transformative moment in social movement action. As they say, activists need to be also "sorcerers" because they "are trying to conjure up something beyond themselves, something they cannot wholly know, something beyond the existing 'natural' limits of society; something 'supernatural'" (Harvie and Milburn, 2016, p. 12).

7 We encounter two types of labor struggles that put postwar social organization under pressure: organized industrial action and the refusal of work. The decline of industrial action and work refusal after the 1980s left a major gap in the contentious politics of this period but opened up space for cross-fertilization between labor struggles and other social movements. The effects of employers' responses to these struggles since the 1980s are well known: deindustrialization in the Global North, outsourcing of manufacturing, and the precarization of labor, which have become one of the targets of new labor social movements in the past decades (see Cosse, 2008; Hamm, 2011; Mattoni and Doerr, 2007; Murgia and Selmi, 2012; Papadopoulos et al., 2008, section V; and Tarì and Vanni, 2005). Simultaneously, we can observe a direct intensification of struggles in the Global South for better wages with endless mobilizations that constitute a direct pressure on capital's flight from organized labor in the North (Waterman et al., 2012). In addition, the working classes of the Global North flew into the credit system in order to increase their wages and social status. A wave of industrial action and a subsequent multiplication and hybridization of social struggles, the fight for higher wages in the South, and the access to higher loans in the North all characterize the various ways labor movements contested these transformations over the past decades.

8 The other strategy for lowering the cost of production apart from the exit from organized labor in the Global North is to lower the price of fixed capital and to intensify the appropriation of natural resources. The multiple ecological movements created a new consciousness about the limits of growth: environmental costs cannot be externalized to society, and current growth cannot be sustained (Wall, 1999). Of course, green development was proclaimed as one of the vehicles that could contribute to managing the 2008 crisis. More broadly, all these movements put an important issue on the agenda: They contested the strategy for social growth and welfare that has been privileged by moderate as well as radical supporters of the markets who in different configurations have defined the political orientation of Global North societies at least since the 1970s. These positions assert that the promotion and maintenance of social good is a positive externality of active markets and of a wealthy private sector. This has been questioned by calculating the balance sheet between positive and negative externalities of the markets and by challenging the idea that the promotion of social good could be a spillover effect of economic agents operating as freely as possible in the market.

9 See the comparison of profit rates between nonfinancial and financial corporations (Duménil and Lévy, 2005).

10 However, it is probably misleading to explain recent crises based solely on the assumption that underconsumption/overaccumulation and profitability are their primary cause (see, for example, Brenner, 2006; K. H. Roth, 2008; this position can be found also in Wolff, 2008). Georges Papanikolaou has made me aware that already Engels (1878/1987) in *Anti-Dühring* and Lenin (1899) in the

Development of Capitalism in Russia had vehemently criticized positions that hold underconsumption as the cause of capitalist crises.

11 Which happened as a series of crises; see, for example, Brenner (2000, 2004).

12 Among many others, two moments seem to be defining for the conditions of labor since the 1970s: higher levels of productivity on the one hand and the stagnation, often deterioration of the living and working conditions of the working classes on the other (Foster, Magdoff, and Magdoff, 2008; Mishel, Bernstein, and Allegretto, 2007; Yeldan, 2009).

13 For further discussion on financial-led accumulation, see Paulani (2009).

14 As Arrighi (2009, p. 82) puts it: "Incomes have been redistributed in favour of groups and classes that have high liquidity and speculative dispositions; so incomes don't go back into circulation in the form of effective demand, but they go into speculation, creating bubbles that burst regularly."

15 The background to this is a major shift in the strategy of investment in Global North societies and subsequently in the rest of the world. Duménil and Lévy (2004) argue that the potential benefits from the partial restoration of the profit rate after the mid-1980s were offset by the payment of dividends and interest. The increases of the rates of profit were redirected to debt sustained speculation (through financialization; see Blackburn, 2008) instead of being directed to investment.

16 See different approaches to these issues: Arrighi and Silver (1999), Barthold et al. (2017), Duménil and Lévy (2005), Orhangazi (2008).

17 See also Bryan and Rafferty (2006), Dowling and Harvie (2014), Langley (2008), and Pellandini-Simányi, Hammer, and Vargha (2015).

18 D. MacKenzie (2011) has discussed the role of evaluation cultures but mainly focused on variations in valuation of different financial instruments. I refer here primarily to sociological and anthropological approaches to valuation (see, for example, Barbier and Hawkins, 2012, and Beckert and Aspers, 2011).

19 What is the financial value of a novel compound, of an equation, of our academic work, of a scientific paper, of animal tissue, of a simulation of a neural network, of soil, of an oil spill, and what of the dying birds? As an example, see the processes, conflicts, and resistances entailed in the financialization of housing and urban space in Bresnihan and Byrne (2015), Byrne (2016a, 2016b), Colau and Alemany (2014), D'Avella (2014), Garcia-Lamarca and Kaika (2016), and Moore (2015).

20 This incommensurability and the process of imposing scales of values that are transferable to economic ranking and ratings has been studied in many fields of life, such as aesthetic valuation, wine markets, the valuation of knowledge and academic research, the valuation of death, and insurance to the valuation of the environment (for various examples, see Beckert and Aspers, 2011; Karpik, 2010; Moeran and Pedersen, 2011; Stark, 2009; Vargha, 2015; and Zelizer, 1979).

21 For different approaches to biofinancialization that have influenced the position presented in this chapter, see French and Kneale (2012), Fumagalli (2011), Marazzi (2010), R. Martin (2002), and Murphy (2013).

22 This exploitation of the future is intensified by the fact that we are entering a period where postcontractual employment is increasingly becoming common; see the note later in this chapter, and also Papadopoulos et al. (2008).

23 See, for example, the fascinating research of Kortright (2012) on transgenic rice and how the promise of a high-yielding crop shapes geopolitics, agrofood investments, research, and experimental labor (see also Cooper, 2008; Sunder Rajan, 2006). In earlier work (Papadopoulos et al., 2008, pp. 107ff.) we called this the "formation of emergent life, that is the attempt to develop means for the maximum control of life and to exploit life's emergent qualities in highly uncertain conditions."

24 See various approaches to the construction of instruments that allow the qualifications of different values in Callon, Millo, and Muniesa (2007), Busch (2011), and Karpik (2010).

25 Social sciences have been plagued by this pressure. The science-based-research and evidence-based-research movements came to dominate attempts to evaluate social scientific research outputs and to create tools and standards for measuring them (Howe, 2004; Morse, 2006; Ryan and Hood, 2004). More broadly the new culture of measuring and valuing outputs has been at the core of reorganizing and restructuring British higher education and universities worldwide (Beverungen et al., 2013; De Angelis and Harvie, 2009; Edu-Factory Collective, 2010; R. Martin, 2011; Newfield, 2011).

26 For different accounts on these practices, see C. W. Smith (1999) and Stark (2009).

27 The story of stagnating growth appears differently when viewed from the perspective of global labor: we can observe a considerable growth of the share in the production of the world GDP by some of the emerging countries. The emerging economies—in particular India and China—became powerful players in the past thirty-five years. The total share of Asia (excluding Japan, which experienced a similar slowdown to that of the other Global North Atlantic economies after 1970s) of the world GDP almost doubled from 16.4 percent in 1973 (twenty years earlier, in 1950, it was 15.4 percent) to 30.9 percent in 2001 (A. Maddison, 2003). Labor productivity was considerably higher for most of the emerging countries as well (Conference Board, 2009). The consequences of these transformations for global labor are severe. Most of the workforce in global manufacturing is unskilled, deskilled, or low-skilled and is exposed to intense marginalization, exploitation, and violation of basic workers' rights (Akyüz, 2003). The background to the rise of global productivity was the opening up of the "periphery" to the neoliberal policies of the 1970s, a process that was initiated in a moment where investments in Global North Atlantic societies started decreasing. Profit was instead streaming into the United States from the "periphery" (Duménil and Lévy, 2004). Cheap loans were offered to the "developing" world. This expansive movement of financialization achieved on the one hand the complete inclusion of the "periphery" into the new regime (you can call it globalization), and on the other hand it proletarianized

the working populations of these countries (Federici and Caffentzis, 2008; K. H. Roth, 2008). Global financialization created the technoeconomic infrastructure of a global economy and the new transnational neoliberal system. When toward the mid- to late 1980s interest rates went up and capital flows were redirected to the United States, a series of countries defaulted on their loans—a chain of regional crises spread throughout the world, and the IMF and World Bank imposed structural adjustment programs. Ironically the periphery become the center for the generation of capital profits (and simultaneously the center of exploitation).

28 Labor productivity in western Europe almost tripled during the "Golden Age of capitalism," from US$5.54 per hour in 1950 to $16.21 in 1973. By 1998 it had reached $28.53 per hour (A. Maddison, 2006). For the United States the rise is as follows: from $12.65 per hour in 1950 to $23.72 in 1973 to $34.55 in 1998.

29 And despite the crisis of the 1970s, this rate didn't seem to be much affected (Nordhaus, 2004); see also Resnick and Wolff (2006, chap. 17) and Glyn (2006, p. 151).

30 For a discussion of employees by sector in the United States see Foster et al. (2008). These trends are similar for other developed economies and are corroborated by the database for sectoral employment of the Groningen Growth and Development Centre; see Timmer and de Vries (2007).

31 At the core of precarious labor relations is the increase of precarious contracts—that is, nonstandard contract forms based on different configurations between the length and stability of the working contract (permanent contracts, fixed-term contracts, or informal or free/unpaid labor) and the working-time arrangements (full-time employment, part-time employment, or irregular working patterns). The less stable and regular the working contract is, the higher the degree of atypicality and the intensity of precarity. However, another important additional dimension characterizes precarious labor: In conditions of structural flexibilization of labor markets (Grimshaw et al., 2001), employment contracts become increasingly insecure and exploitation is maintained by breaking the bond of the contract, rather than through the contract itself. This results in an amplification of dependency: one is under increased pressure to ensure that one's future capacity to be "productive" will be compatible with the demands of the market (lifelong learning, continuous acquisition of skills, and innovation are keywords in this process). So the absence of permanent (or even long) contractual employment increases the "exploitation of the self" (Ehrenstein, 2006). Furthermore, one is not only exploited in the present, but also one's future is exploited. Exploitation of the self happens in the regime of precarious labor when someone tries to anticipate and explore the future through its dissemination into the present and to intensify their own efforts to ensure that they remain competitive. This postcontractual form of dependency is twofold: it is a dependency on the employer, who offers limited contracts, as well as a dependency on oneself to increase one's own capacity to get such contracts in the future.

32 Following a time where industrial capitalism (1800–1970) came after a long period of instability during the nineteenth century to replace merchant capitalism, which depended on the exploitation of slave labor and the territories of the colonies. Regarding the formation of third capitalism, see broader debates about post-Fordism in relation to work and value production (Bowring, 2002; Dyer-Witheford, 1999; Fleming and Spicer, 2003; Gorz, 1999; Hampson, Ewer, and Smith, 1994) as well as the various debates about cognitive capitalism (Azais, Corsani, and Dieuaide, 2001; Marazzi, 2007; Morini, 2007; Moulier Boutang, 2011; Vercellone, 2007). Both of them define the coordinates of what here is referred to as third, embodied value production.

33 See Heery and Salmon (2000) and Pollert and Charlwood (2009).

34 Paul Thompson, for example—arguing from the perspective of labor process theory (see Knights and Willmott, 1990)—emphasizes that the socialization of value production is empirically unsubstantiated and that the core site of value production remains the workplace, be it the manufacturing shop floor, the office, or the retail space (P. Thompson, 2005). But these claims are equally unsubstantiated; there is no extensive study, to my knowledge, that would support them either. The debates initiated by labor process theory on the socialization of value production seem to ignore the broader transformations of the social regime of accumulation—something that was part of its initial developments (see also Böhm and Land, 2012; for example, Burawoy, 1985; Jaros, 2005).

35 See, for example, Moulier Boutang (2011), Bollier (2003), C. Hess and Ostrom (2007), and de Peuter and Dyer-Witheford (2010).

36 That is, all these shared naturecultural creations that are neither public (maintained by the state) nor private but belong to all and in particular to everyone who contributes to maintaining a specific commons. There is a wealth of publications on the commons; see, for example, Helfrich and Stiftung (2012), Linebaugh (2008), Dolsak and Ostrom (2003), and Bollier and Helfrich (2012) and the discussion in chapter 8.

37 And very often these activities do not lie outside the workplace but also lie outside the direct sphere of capitalist organization as Gibson-Graham (2006) describe in their work. See also Morini (2007) and K. Weeks (2007).

38 See T. Nichols (1980).

39 See, for example, Hanlon (2012, 2014).

40 The mobilization of various aspects of one's own life in order to be able to work has been explored in many different settings; see, for example, A. Ross (2009), Brophy and de Peuter (2007), and Hesmondhalgh and Baker (2011).

41 In labor process theory the workplace is examined in relative isolation from the wider political economy (P. Thompson and Smith, 2001). This perspective can produce a wealth of insights into control and subordination in the workplace, but it misses broader social transformations that affect the workplace. For example, in the discussion of labor process in call centers, Warhurst, Thompson, and Nickson (2009, p. 101) clearly say that call center workers are not unskilled but need to have considerable social and interactive skills and

competencies. However, they refuse to discuss how these skills are acquired or learned, how they develop, and how they are nurtured. They thus refuse to see the social transformations that make workplace exploitation possible (see, for example, Burawoy et al., 2000; and Rowlinson and Hassard, 2001).

42 See, for example, De Angelis and Harvie (2009).

43 For further discussion see Boyle (2008), Bollier (2008), and, more broadly on the politics of cooperation, Ratner (2015).

44 For further discussion, see K. Weeks (2011) and Barbagallo and Federici (2012).

45 See, for example, Alberti (2011) and B. Anderson (2010),

46 Verticalization creates alliances between segments of different classes with segments of the state and the private sector beyond democratic regulation. The democratic deficit emerging from the verticalization of society can be described as postliberal polity (Papadopoulos et al., 2008; Tsianos et al., 2012). Global North states are steadily withdrawing their support and protection from their citizens, and the 2008 economic crisis seems to have been used as a vehicle to extend this project even further than it was thinkable beforehand. And although we see neoliberalism triumphant in economic terms (however, neoliberalism is a far broader and diverse project than expressed in its economic doctrines; see Papadopoulos, 2002; Peck and Tickell, 2002), we see also an abandonment of core liberal principles: for example, the retreat from the classic neoliberal doctrine of a minimal state to the support of a strong state that openly intervenes not to defend society, ecology, and democracy but to defend the positions of certain social classes and certain private actors—a public-private alliance maintained by the state. Moreover, there is a retreat from the ultimate liberal principle of state institutions as the guarantors of individual freedoms. State authorities are now prepared to legitimize illiberal practices in order to uphold this alliance. Consider, for example, the denationalization of citizens (Nyers, 2010), the undermining of personal data sovereignty, or the attacks on education as a public good (Newfield, 2008). This looks like an amplification of the neoliberal process minus liberal democracy, a move toward postliberal polity in which less state means less state for democracy, society, and the environment and more state for these vertical alliances. For an extensive discussion of verticalization in the context of postliberalism, see Papadopoulos, Stephenson, and Tsianos (2008, pp. 25–35). For further analyses of the concept, see the work of Stephenson (2010), Kippax and Stephenson (2010), Tsianos et al. (2012), and Papadopoulos et al. (2008).

47 See the note on postcontractual exploitation earlier in this chapter.

48 The mobilizations across many of the societies of the Global North between 2008 and 2012 show that this conflict traverses the whole of society and directly affects how politics is performed; see Tsianos and Papadopoulos (2012) and Tsianos et al. (2012).

49 Respectively, Savage (2002) and Newfield (2008).

50 See A. R. López and Weinstein (2012) and Li (2010).

51 See Hoffman, Postel-Vinay, and Rosenthal (2007) and Sullivan, Warren, and Westbrook (2000).

52 See Papadopoulos et al. (2008) and Tsianos et al. (2012).

53 Originally Tronti (2005); see also Bowring (2002), Cleaver (1992), Fleming (2012), Shukaitis (2014), and K. Weeks (2011, pp. 96ff.).

54 Such as the important movements described in Zibechi (2011).

55 The main problem with these responses is that they focus on a strategy of resistance. While resistance may be important in order to alleviate the immediate effects of current social conflicts, it cannot constitute a movement that can force power to reorganize itself in a deep transformative way. Historically, what we had is an aleatoric succession of events in which the subaltern classes attempted to escape their own conditions of existence and exploitation, and capital control was responding with always-new strategies for its survival. The way subaltern populations experience the development of capitalism is in the form of continuously novel forms of control. And at the same time subaltern populations acted in the ever-changing conditions of capital control by instigating new practices of escape and justice that didn't respond to these new configurations of control but created new forms of life and new conditions that were not easily visible from the viewpoint of power. It is like a Beckett play—the actors coexist on the stage and each actor's action is the precondition for the actions of the other, but they never respond to each other and never create a coherent dialogue; they simply act and change the other through the mere material effects of their doings. The new exodus won't be a response to the impasses of biofinancial accumulation. It'll be an exodus that will open something that operates on a terrain that is not fully organized by the command of embodied value production and the biofinancial regime. We have extensively discussed this thesis elsewhere (Papadopoulos et al., 2008; Tsianos et al., 2012); see also chapter 8.

56 For example, Hardt and Negri (2009) call for a subtraction of labor power from capital, and Holloway (2010b) suggests that our alternative doing can be outside and *against* abstract labor (labor that produces capitalist value). See also Bowring (2004).

57 See, for example, research on the valuation of complex financial instruments (D. MacKenzie, 2011, 2012), research on the performative capabilities of economics (D. MacKenzie and Millo, 2003; Millo and MacKenzie, 2009), or studies in material and semiotic arrangements used for the calculation of economic objects (Callon et al., 2007; Knorr-Cetina and Preda, 2006; Muniesa, 2007; Pryke, 2010).

58 The main driver behind this is the computerization of financial markets—that is, automatic trading as well as e-finance platforms.

59 This not only sounds like but also is a reproduction of neoclassical economics' core assumptions about the subjectivity and nature of the actors involved in markets (Papadopoulos, 2002, 2003).

60 As Rona-Tas and Hiss (2011, p. 226) write, "Measuring value as events yet to unfold in the future, rather than as costs or labor already expended in the past

or even as subjective needs or objective scarcity revealed in the present, introduces an element of fundamental uncertainty that poses a formidable challenge to valuation and price formation." See also Beckert (2009).

61 See also Lightfoot and Harvie (2016).

62 As, for example, D. MacKenzie (2009) implies.

63 And it is also important not to forget here that social studies of finance have willingly or unintentionally internalized some of the assumptions of neoclassical economic models in order to provide their analyses. See the critical interrogation of social studies of finance in Engelen et al. (2011).

64 A position that primarily was developed in regard to the role of working-class struggles in historical change: capital is not the driving force of change, but instead workers' refusal and insubordination force capital to reorganize itself (Cleaver, 1992; Dyer-Witheford, 1999; Negri, 1988). This perspective on autonomy is of course limited to the relation between capital and labor, but the question of autonomous politics exceeds this relation. In the wake of the new social movements that emerged from the Zapatista *encuentros* and the Seattle mobilizations in the middle to end of the 1990s, autonomy is explored in relation to technoscience, culture, feminist and queer politics, and the struggles for the commons (Berardi, 2009; Böhm, Dinerstein, and Spicer, 2010; G. Brown, 2007; Dinerstein, 2010; Papadopoulos et al., 2008; Shukaitis, Graeber, and Biddle, 2007).

65 On the idea of excess, see Free Association (2011) and Papadopoulos et al. (2008).

66 See http://www.hsbc.com/inthefuture. The series of adverts including the advertisements discussed here were issued by HSBC Holdings plc.

67 See the work of Waterton, Ellis, and Wynne (2013) on the ambiguities and intricacies of DNA bar-coding of species in order to prevent biodiversity loss. I wonder about the implications of bar-coding species for increasing market share.

68 On rent, see Fumagalli (2011), Marazzi (2010), Negri (2010), and Vercellone (2010).

69 In this sense social science fiction corresponds to Suvin's influential definition of science fiction as "a literary genre whose necessary and sufficient conditions are the presence and interaction of estrangement and cognition, and whose main formal device is an imaginative framework alternative to the author's empirical environment" (Suvin, 1979, pp. 7–8). Although this definition has been extended (see, for example, Freedman, 2000; Jameson, 2007) and challenged (for example, see several contributions included in Bould and Miéville, 2009), I am interested in Suvin's idea that this estrangement is capable of prefiguring "an *alternate reality*, one that possesses a *different historical time* corresponding to different human relationships and sociocultural norms actualized by the narration. This new reality overtly or tacitly presupposes the existence of the author's empirical reality, since it can be gauged and understood only as the empirical reality modified in such-and-such ways" (Suvin, 1979, p. 71; emphasis

in original). This potential for prefigurative politics (see, for example, Holloway, 2010a) is crucial for creating ontologies and material worlds alternative to our existing empirical realities, an idea that is central to the concept of autonomy that is discussed in several places in this book.

70 From Bergson via Deleuze, "the goal of fabulation is to break the continuities of received stories and deterministic histories, and at the same time to fashion images that are free of the entangling associations of conventional narratives and open to unspecified elaboration in the construction of a new mode of collective agency" (Bogue, 2007, p. 106). For a comprehensive discussion of the concept of fabulation as it is used here, see Barr (1992) and Bogue (2010).

71 Here I expand and modify the concept of "endocolonialism" coined by Virilio (1995, 1998) to describe not only how states colonize their own urban spaces and their citizens' bodies in postindustrial societies but also the process of an inward coloniality, the coloniality of matter itself in times where geocolonialism—that is, territorial colonialism—is exhausted (see also chapter 1).

72 Colonization from the Indies to Mars is the attempt to initially treat space— whether it is made up of people, other species, territories, or inanimate matter—as a resource that does not belong to anybody, *res nullius*, in order to appropriate it, enclose it in private or in some rare cases public spaces (that is, to transform it to *res privatae* and in some few cases to *res publicae*). But hardly ever to designate it as *res communes* (see also chapter 1).

73 Ghelfi (2015, p. 22) describes how a widespread vision of redemption of the commons is at the heart of current understandings of alternative politics and many social movements. This vision sees the commons as a force "potentially able to re-appropriate the whole (wealth produced and means of production). Here the commons are not just the condition of possibility for the development of the current regime of accumulation, but the commons become a singular name, a 'universal,' an object of desire that can be gained through the struggles of an antagonistic subject." Autonomy in these widespread beliefs refers to the potential of forming a historical subject that can resist its appropriation and instigate radical social change (see an example in Mezzadra, 2011b).

74 Terraforming appears as a science fiction theme already in the 1930s and 1940s in Jack Williamson's stories. This essay is inspired by Robinson's *Mars* trilogy (1992, 1993, 1996) and Butler's *Xenogenesis* trilogy (2000), which opened for me an alternative vision of and multiple perspectives on terraformation. I rely on the analyses of these works by B. Clarke (2008), Haraway (1991a), Jameson (2000), and Leane (2002).

75 See, for example, Beech (2009) and Fogg (1995).

76 See Bell and Parker (2009).

77 "Mars has what it takes. It's far enough away to free its colonists from intellectual, legal, or cultural domination by the old world, and rich enough in resources to give birth to a new civilisation," says Zubrin (1994), a former chairman of the National Space Society and a right-wing enthusiast of Mars colonization.

78 See Collis and Graham (2009) and also discussions in popular forums such as York (2002).

79 See Yusoff (2013a); on the public framing of geoengineering, see Bellamy and Lezaun (2015).

80 Here Hall's (1986; see also Slack, 1996) work is important; however, I want to think of articulation as a practice that pertains not only to the process of coding but also to material practices.

CHAPTER 3. ONTOLOGICAL ORGANIZING

1 When I refer to organizing here I mainly follow the work of alternative organization studies—see, for example, Parker et al. (2014), Parker, Fournier, and Reedy (2007), and Tadajewski et al. (2011).

2 The empirical materials that underpin the ideas presented in this chapter have been collected over a period of more than a decade in different political events and activism related to migration as well as in various migration transit sites across Greece between 2009 and 2011. This work is the outcome of my collaboration with Vassilis Tsianos, and many of the ideas and material developed here are based on our common work.

3 See, for example, Isin and Nielsen (2008), Isin and Nyers (2014), and Nyers and Rygiel (2012).

4 Julie Okmûn Rebouillat is a freelance photographer and member of the group Contre-Faits (http://www.contre-faits.org). Between 2008 and 2012 she worked on the issue of borders and migration and made a series of photographs exploring everyday life of illegalized migrants in Calais. The photographs included in this chapter are part of this series, titled "No Land's Men, the Struggle for Calais." She has also investigated the struggles of No Borders activists and the camps that have been organized at strategic points across Europe to demand the opening of borders to all (such as in Calais and Brussels). Julie is currently working on several other projects exploring life in squats, social movement mobilizations, and carnival parades. Her approach is based on immersion and engagement with the people that she portrays. This allows her to provide a direct and intimate representation of daily life, as opposed to the often sensational and distanced images circulating in the mass media on similar topics. In Calais, for example, she spent more than one month living in a migrant squat, and she participated in several No Border camps in order to share the activists' point of view on their practices and actions.

5 See Federici (2004) and Ignatieff (1978).

6 See von Oswald (2002).

7 See Moulier Boutang (1998) and Papadopoulos et al. (2008).

8 Demographics, aging, and boosting the working-age population are the most crucial arguments circulating in the Global North for maintaining mobility and accepting migrants.

9 See Kasparek (2016).

10 See our analysis of camps as technologies of temporality within global labor markets and as regulators of porous borders in Papadopoulos et al. (2008).

11 See Castles and Miller (2003) and Karakayali (2008).

12 See De Genova and Peutz (2010).

13 See Papadopoulos et al. (2008, chap. 11).

14 See Alberti (2011, 2014).

15 See T. Nichols (1980, p. 35).

16 See B. Anderson (2010) and Bosniak (2006).

17 See, for example, Brass and van der Linden (1997), Glenn (2004), Lowe (1996), Lucassen and Lucassen (1997), and Steinfeld (2001).

18 The idea of the double-R axiom is developed in Papadopoulos et al. (2008).

19 For an example, see Cuppini (2017).

20 See Bell and Binnie (2004), van Gunsteren (1998), Nyers (2009), and Sassen (2004).

21 See also Buckel, Fischer-Lescano, and Oberndorfer (2010).

22 See Karakayali and Tsianos (2005), Mezzadra (2011a), Papadopoulos et al. (2008), and Rodriguez (1996). See also Nyers (2015) and Martignoni and Papadopoulos (2014). For a summary of some of these debates see Casas-Cortes, Cobarrubias, and Pickles (2015).

23 As, for example, in Jessop and Sum (2006).

24 See Transit Migration Forschungsgruppe (2006).

25 See, for example, Schuster (2003) and De Genova (2010).

26 See Düvell (2006) and Sharma (2009).

27 See examples in Giordano (2014) and Martignoni (2015).

28 See O'Connell Davidson (2011).

29 For a typical example, see B. Jordan and Düvell (2002).

30 See the detailed work of Karakayali (2008).

31 See different approaches to this problem in Bishop (2011), De Genova (2015), King (2016), Lafazani (2012), Nyers and Rygiel (2012), Rigby and Schlembach (2013), and Rygiel (2011).

32 See Broeders and Engbersen (2007) and various examples in Papadopoulos et al. (2008, chaps. 11 and 12).

33 See chapters 6 and 12 in Papadopoulos et al. (2008).

34 For an extended critique of integrationism see Glick-Schiller and Çaglar (2008) and S. Hess, Binder, and Moser (2009).

35 The so-called European migrant and refugee crisis in 2015 shows how the movement of people has invalidated the existing European regime for the control of mobility—for a comment on this situation, see Tsianos (2015a, 2015b). For a broader take on this issue, see De Genova (2016).

36 For an extended discussion, see Papadopoulos and Tsianos (2007b) and Tsianos and Papadopoulos (2012).

37 See examples in Bishop (2011), Grazioli (2017b), King (2016), and Lafazani (2013).

38 See Bojadzijev, Karakayali, and Tsianos (2004), Mezzadra (2001), and Papadopoulos and Tsianos (2007a).

39 As, for example, Standing (2011) sees it. For a critical discussion, see Papado-poulos (2017), Shukaitis (2013), and Alberti, Holgate, and Tapia (2013).

40 For further discussion, see Grazioli (2017b), Lafazani (2013), and Papadopou-los, Tsianos, and Tsomou (2015).

41 See Alberti (2010), Lafazani (2011), and the documentation of the actions from the *Welcome to Europe* network: http://w2eu.net/nobordertv/pagani-detention-center-2/pagani-detention-center/.

42 See Kasparek (2016), Kuster and Tsianos (2016), and Tsianos (2017).

43 For an extensive discussion, see the ethnographic research of Ibrahim (2010).

44 See Panagiotidis and Tsianos (2007).

45 See Trimikliniotis, Parsanoglou, and Tsianos (2015b), see also Kuster (2011), Panagiotidis and Tsianos (2007), and Kuster (2016).

46 The idea of World 2 is developed in Papadopoulos (2006).

47 As discussed by B. Anderson et al. (2009).

48 For a discussion see Bishop (2011).

49 See also King (2016) and Trimikliniotis, Parsanoglou, and Tsianos (2015a, 2015b).

50 See, for example, Ticktin (2016).

51 See Puig de la Bellacasa (2012, 2017).

52 See Walsh (2010, 2013).

CHAPTER 4. ACTIVIST MATERIALISM

1 For a critical approach to the concept of the species from different perspec-tives, see Hartigan (2015b), Margulis and Sagan (2002), Stamos (2003), Kirksey (2015), Mora et al. (2011), and Waterton et al. (2013).

2 See Dyer-Witheford (2006).

3 This is echoed in object-oriented ontology and other recent debates that claim that the real exists even if we don't know what it is, as, for example, in Laruelle (2013), Harman (2011), and others. Of course, this proximity between the two positions quickly disappears when Marx turns to activism, which is inherently unacceptable for object-oriented ontology in its move to create a new grand ontology.

4 See, for example, Bennett (2010) or Dolphijn and van der Tuin (2015).

5 This tight relation between ontology and activism, materiality, and practice is still—more than 170 years later—present in humans' struggle to grasp their species being. Hartigan (http://www.aesopsanthropology.com/blog/?p=348) asks: "When self-reflexivity became an expectation of ethnography in the 1990s, the focus was on social diacritics, principally race, gender, and class—the positions that informed and biased perspectives, that needed to be ac-counted for in achieving analytical clarity. Today perhaps a second-wave of the reflexive turn is upon us, when the diacritics are components of species-being. What is it in my species-being that makes it so difficult for me to interact with these plants before me? My skin and proclivity for motion, for starters. Can I

calibrate these in such a way that I can learn from a plant before my attention wanes or my body aches to move?" See also the inspiring work of Myers (2014, 2015) on similar questions.

6 For an extended analysis of these issues, see Z. A. Jordan (1967).

7 A position that is already well developed in R. Williams (1977).

8 See J. Gilbert (2008), and Papadopoulos et al. (2008).

9 See R. Wilson and Connery (2005).

10 See Rorty (1967) and Gadamer (1989), respectively.

11 See examples in Lefebvre (1991), Debord (1981), and D. Harvey (1990).

12 See, for example, Csordas (1994b), Harré (1996), and Overton (1998).

13 See, for example, Bourdieu (1987), Braidotti (2002), and Fausto-Sterling (2000).

14 For a discussion see Blackman et al. (2008). See also Papadopoulos (2002, 2003).

15 See Hall (1990) and Papadopoulos (2006).

16 As discussed in Clifford (2000).

17 See Clifford (2001), Hall (1986), and Slack (1996).

18 See Grossberg (1986) and Laclau and Mouffe (1985).

19 For different approaches see J. Butler, Laclau, and Žižek (2000), Clifford (1986), Hall and Jefferson (1976), Mouffe (2000), Sedgwick (1990), Spivak (1999), and Warner (1999).

20 See Stephenson and Papadopoulos (2006) and the important work of Pickering (1994, 1995) for a detailed discussion and critique of representationalism.

21 Chesters and Welsh (2006), Papadopoulos et al. (2008).

22 See Holland (2005).

23 See, for example, De Landa (1997).

24 Deleuze and Guattari (1987, p. 367): "What we have . . . are two formally different conceptions of science, and, ontologically, a single field of interaction in which royal science continually appropriates the contents of vague or nomad science while nomad science continually cuts the contents of royal science loose." For an extended discussion of the relation of Deleuze's, and occasionally Guattari's, philosophy to science and technology, see Bonta and Protevi (2004, pp. 12–31), Jensen and Rødje (2010), Gaffney (2010), and Pickering (2008).

25 See also Klein (2012).

26 See Haraway (1997), Shapin and Schaffer (1985).

CHAPTER 5. INSURGENT POSTHUMANISM

1 Worlding resembles, in the words of Robbins (2011, pp. 1–2), an event that "has created its own unique local surround, a restricted time/space that replaces and cancels out any abstract planetary coordinates."

2 See R. Wilson and Connery (2005).

3 This chapter focuses solely on posthumanism as a problematization of the humanist split between society and ecology, mind and matter, as well as culture and nature. Transhumanism (the discourse of overcoming the human condition and human body in order to preserve the "human") touches only marginally (or

not at all) on these critiques of humanism and therefore is not considered here. See Pickering (1995, 2011) and earlier notes with specific references to the type of posthumanism that I use in this book.

4 These questions are inspired by critical organizational studies, for example, Beuret (2016), De Cock, Fleming, and Rehn (2007), Parker (2002), and Parker et al. (2007).

5 And this is even the case for events that marked social movements as targeting social power—see, for example, the note on the Paris commune in chapter 8.

6 The demise of posthumanist movements is not only the outcome of the rise of the wage labor system, as I discuss later in this chapter, but also of the broader emergence of the humanist subject that takes mainly two forms: The first form is the self-assertive secular subject; see, for example, Taylor (2007, p. 18). The second form is the subject endowed with the capacity to autonomously negotiate his or her individual morality and interiority. This is the "cult of inwardness" at the core of the liberal humanist subject; see Renaut (1997) and Taylor (1989). However, liberal secular humanism, although it is concomitant with the liberal assembled state, is also in continuous tension with it, especially because of the inherent illiberal tendencies of the capitalist state to delimit and restrict the autonomy and rights of the humanist subject.

7 See the discussion on verticalization of society and the state in chapter 2 and the relevant notes in the same chapter.

8 See Gramsci (1918), quoted in Pozzolini (1970, p. 76).

9 See Engels (1990b, chap. 9).

10 This ideological appearance of the state is believed to be grounded in what Lukács (1971) described as the process of reification where social relations attain a ghostly objectivity as if they were untouchable and existed always.

11 In the words of Kojève (1969, p. 67): "And this means that the transcendent Universal (God), who recognizes the Particular, must be replaced by a Universal that is immanent in the World. And for Hegel this immanent Universal can only be the State. What is supposed to be realized by God in the Kingdom of Heaven must be realized in and by State, in the earthly kingdom." See also Avineri (1972, p. 177).

12 By opposing Hegel's conception of the state as the sublation of pervasive social dichotomies, Marx and Engels see the state as a tool for governing antagonisms that is controlled by the interests of the ruling classes, the "Master-State" in the words of Kojève (1969, p. 57).

13 The name of this path is revolution—which seizes the state in order to organize the move to a stateless society (see Lenin, 1917, chap. 3). The revolutionary dictatorship of the proletariat was Marx's answer to the political dilemmas between anarchosyndicalism, on the one hand, and reformist democratic politics, on the other. These dilemmas became widespread after 1848. After the defeat of the Paris Commune, Marx attempted to overcome the focus on spontaneous revolt and to form a new kind of organization, the Marxist Parties, with the Social Democratic Party of Germany as one of the first (Marx, 1976, chap. 4).

Lenin, faithful to this militant action program, gradually replaced the dictatorship of the proletariat as such with the leading role of the party in this stage of transition.

14 The important work of Carroll (2012) on science, technology, and the state and of Mora-Gámez (2016) on rights, reparations, and statehood in Colombia has been guiding me on these issues. See also Carroll (2006), Passoth and Rowland (2010), Rowland and Passoth (2014), and the historical work of Guldi (2012).

15 For an extensive discussion, see Papadopoulos (2008).

16 The Marx of the 1844 *Manuscripts* and *The German Ideology* (unlike the later Marx of *The Gotha Programme*) captures some of these tight interdependences between people's action and the creativity of the material world (of "nature" in Marx's words) and investigates how alienation from the capacity for self-organized development is imposed on them by separating people into classes (and genders and races) and alienating them from "nature," as discussed in the previous chapter.

17 See, for example, Hindle (2003) and Shoemaker (2007).

18 See Merchant (1990) and Starhawk (1982). This process continues today with the multiplication of enclosures in the eco-commons—see, for example, Bollier (2003) and Shiva (2005).

19 For a further discussion of these sources, see Papadopoulos et al. (2008, section II) and De Angelis (2007).

20 The position described here is similar to Dyer-Witheford's take on the commons (see http://commonism.wordpress.com/ and de Peuter and Dyer-Witheford, 2010). It also corresponds to Harney's approach to the history of living labor and its capitalist capture—see Harney (2006, 2008) and also De Angelis (2007).

21 "So common rights differ from human rights. First, common rights are embedded in a particular ecology with its local husbandry. For commoners, the expression 'law of the land' from chapter 39 [of the Magna Carta] does not refer to the will of the sovereign. Commoners think first not of title deeds, but of human deeds: how will this land be tilled? Does it require manuring? What grows there? They begin to explore. You might call it a natural attitude. Second, commoning is embedded in a labor process; it inheres in a particular praxis of field, upland, forest, marsh, coast. Common rights are entered into by labor. Third, commoning is collective. Fourth, being independent of the state, commoning is independent also of the temporality of the law and state. Magna Carta does not list rights, it grants perpetuities. It goes deep into human history" (Linebaugh, 2008, pp. 44–45). See also De Angelis (2017).

22 See, for example, A. O. Thompson (2006), Fouchard (1981), Sayers (2014b), and C. J. Robinson (2000).

23 See Sayers (2014a), Baram (2012), and more broadly Orser and Funari (2001), Price (1979), and Weik (1997). I am grateful to Patrick Bresnihan for bringing some of these literatures to my attention and for our inspiring discussions.

24 See, for example, Bilby (2005).

25 This is an argument that Linebaugh and Rediker (2000) make in their extraordinary book *The Many-Headed Hydra: Sailors, Slaves, Commoners, and the Hidden History of the Revolutionary Atlantic*. See also A. Greer (2012) and B. Maddison (2010).

26 See J. Nichols (1800, pp. 125ff.).

27 See Hay (1975).

28 See J. Nichols (1800, p. 131).

29 E. P. Thompson (1991, p. 105); see also Porter and Tiusanen (2006), Palmer (1989; 1979, pp. 59ff.).

30 See some very different takes on the entanglement of ecology and the commons in Ashbrook and Hodgson (2013), Bresnihan (2016a), Cronon (1983), Linebaugh (2008), Newfont (2012), Olwig (2013), Reid and Taylor (2010), and Wall (2014).

31 For various approaches to commons that consider the dimension of local environments and practices of commoning in England, Wales, and Europe, see Neeson (1993), de Moor, Shaw-Taylor, and Warde (2002), and Rodgers et al. (2011). In the Americas, Southeast Asia, and Australia, see Chakravarty-Kaul (1996), Cronon (1983), Newfont (2012), Strang (1997), and Wade (1988).

32 For a critical analysis, see Carter and Jackson (2005) and N. Jackson and Carter (1998).

33 See the note earlier in this chapter for a brief discussion of some of these factors.

34 For an extended discussion of these issues, see O'Connell Davidson (2015), Papadopoulos et al. (2008), and Steinfeld (2001).

35 Furthermore, free and coerced labor coincide: they have coexisted in the same geographical spaces and historical periods throughout the whole development of the productive process up until today.

36 This position, of course, originates in Marx (1857/1973, chapter on Capital—Notebook II).

37 See Connery (2007).

38 I am here less interested in the mass mobilizations of the 1960s and more in the emergence of a lived alternative culture—see, for example, J. Savage (2015) and Russell (2010).

39 See Connery (2006, 2007).

40 For an extended discussion, see Papadopoulos et al. (2008, chaps. 2 and 3) and Stephenson and Papadopoulos (2006).

41 Some traces of this understanding of the state can be already found in the work of Poulantzas (1978), who highlights how the modern state evolves as a permanent but unstable equilibrium of compromises between different social subjectivities and classes. The state does not have the *resolution* of social conflicts—by absorbing or terminating them—as its ultimate aim.

42 For example, see Latour (2005b).

43 The logic of the network not only implies a specific way of ordering and making society but also reorganizes the very concept of the "subject." Latour is not

alone here. Nikolas Rose and other theorists of governmentality, for example, attempt to grasp the production of subjectivities in neoliberal networks of power; see, for example, Barry, Osborne, and Rose (1996).

44 See also Hallward (2008, pp. 120–121).

45 See different approaches: Tilly (2004), Touraine (1981), Della Porta and Diani (2006), and Welsh (2000).

46 See Deleuze and Guattari (1994, p. 159).

47 And with Spinoza (1994, 228) we know: the mob inspires fear when it acts, but it acts only when it is unafraid; therefore, it has to be tamed by the state and religion. From an activist perspective, see also the discussion of sadness in Colectivo Situaciones (2007).

48 Derrida (1992) attacked this logic of choice by assuming that undecidability is a permanent ingredient of any decision; the final undecidability of any process of making and actualization should not be the ground for "sad passions" but the necessity of practice.

49 See, for example, Starhawk (2002).

50 See Bakhtin (1984, p. 285; see also pp. 94ff.).

51 See Bakhtin (1984, p. 318; see also pp. 335ff.).

52 See also Kromidas (2014).

53 Experience in the sense developed in Stephenson and Papadopoulos (2006).

54 See Papadopoulos and Tsianos (2007a) and Kuster (2016).

55 Hypocritically, as many have argued, among them Wallerstein (1995).

56 For a discussion of the ambiguity of the term *Gewalt*, see Balibar (2009). In the quotations from the English translation of *Critique of Violence* I will keep the term *violence* as this is the term used by the translator; see Benjamin (1996a).

57 Benjamin (1996a, p. 242).

58 This focus on the ordinariness of divine *Gewalt* seems to run counter to many of the usual interpretations of Benjamin's short essay. These interpretations can be divided into two main groups. The most prominent example of the first interpretation is Agamben (2005). Agamben is again, as in so many of his other writings, fascinated with the idea of catastrophe and disaster. His starting point is the moment of exception. Not least because of this, he reads Benjamin in connection to Schmitt's state of exception. The legal vacuum that divine violence creates is, according to Agamben, a form of exception that opposes the legally imposed state of exception—that is, the law-created anomie. Agamben thus ends up misinterpreting Benjamin and reentering his idea of divine *Gewalt* into the dialectic of constituent and constituted power that is between a totalizing state and revolution, this time as a dialectic between their negations: as a law-created anomie against the anomie of the rebellious. Agamben's (2005, p. 62) spectacle of power is the gigantomachy "between state of exception and revolutionary violence," in which their relation is "so tight that the two players facing each other across the chessboard of history seem always to be moving a single pawn." We can find a good example of the second common interpretation of Benjamin's essay in Žižek (2008). Here Žižek popularizes and moralizes

divine *Gewalt* in an attempt to see it as a sign of an injustice emanating as a life force—typical Žižek you might say: sounds critical, perhaps even radical, and there is a lot of drama too as it shows how good reasons, benign sentiments, and cruel deeds can all coexist in one. Benjamin's concept of violence is a pure "drive of life" (Žižek, 2008, p. 198) that signifies a world that is "ethically "out of joint" (Žižek, 2008, p. 200). Žižek, who is so keen to make clear that Benjamin's divine *Gewalt* is not to be confused with either "terrorist acts" committed by religious fundamentalists or organized revolutionary terror (Žižek, 2008, p. 185), sees this violence as an event: "There are no 'objective' criteria enabling us to identify an act of violence as divine; the same act that, to an external observer, is merely an outburst of violence can be divine for those engaged in it . . . ; the risk of reading and assuming it as divine is fully the subject's own" (Žižek, 2008, p. 200). Žižek delivers the other side of Agamben: for the latter, divine *Gewalt* is always linked to the violence of state power, while for Žižek, divine violence is "when those outside the structured social field strike 'blindly'" (Žižek, 2008, p. 202). Zizek again reinforces the never-ending dialectic of constituent and constituted power in the form of exclusion and inclusion. Those who are outside the social field have the right to institute justice through divine violence. Here Žižek misinterprets Benjamin, whose main task was to break this vicious circle of defining violence by judging it in terms of the justice of the ends. Agamben's and Žižek's otherworldly readings of Benjamin's concept of divine *Gewalt* as something that "strikes out of nowhere" (Žižek, 2008, p. 202) are based on their understanding of divine *Gewalt* as an exceptional phenomenon. The main argument developed in this chapter points in the opposite direction: divine *Gewalt* is a radical everyday practice; the more radical it is, the less violent it is.

59 Benjamin (1996a, p. 239).

60 The political applicability and meaning of *general strike* is perhaps limited today, but one could think of the mutation of the general strike into a *metropolitan strike*; instead of withdrawing from their position as workers, in the metropolitan strike people withdraw from their positions as citizens of a particular polity. They use urban space in ways that are not compatible with their position as citizens of an existing polity—see Papadopoulos et al. (2015) and Tsianos et al. (2012). By doing this they create a vacuum that exposes the oppressive structure of political power and challenge its legitimacy. This exit from the position of the citizen as the fixed subject of rights and responsibilities into a position that forces the reorganization of the political itself is the main form of the metropolitan strike seen in the cycle of struggles and social movements mobilization between 2006 and 2012 (for a discussion, see the introduction and chapter 8). Metropolitan strikes turn the materiality of metropolitan space into something that can no longer serve the existing function of polity. This is what happened in most of these mobilizations: a new temporality of justice emerged when people started reclaiming and experimenting with urban space.

61 For an extended discussion of the meaning of the proletarian general strike in Benjamin, see Tomba (2009, pp. 139ff.).

62 Munro (2004).

63 Tomba (2009, p. 140).

64 Benjamin (1996a, p. 247).

65 There is a proliferation of these figures of mediation in current theory; for example, see Collins and Evans (2007), Latour (2004a), and Stengers (2005).

66 See Puig de la Bellacasa (2010).

67 "Just-ice takes places in the time of the stop," as Munro (2004, p. 64) says.

68 See, for example, Latour (1987), Law and Hassard (1999), and Stengers (2010).

69 The postanthropocentric dimension of posthuman politics of movements is neither about developing an ecological egalitarianism that considers the value of all nonhuman beings as equal—for a critique of this type of inclusive egalitarianism, see Rancière (1998); see also Stephenson and Papadopoulos (2006, chap. 6)—nor about creating the grounds for the articulation of constantly novel connections and concerns between "us" and "them"—see Puig de la Bellacasa (2010, 2011).

70 See Latour (1993).

71 As discussed in chapter 1; see also Quijano (2007) and Mignolo (2011).

CHAPTER 6. BRAIN MATTER

1 See Wittgenstein (1958).

2 See also Shenk (2010).

3 See, for example, S. Watson (1998).

4 This discussion relies on various approaches to the cultural history and politics of the body in the original works of Bordo (1990), Frank (1991), S. F. Gilbert (1997), Haraway (1991b), and E. Martin (1992).

5 See also Leder (1990).

6 See Fodor (1983) and Scholl and Leslie (1999).

7 See Breckenridge and Vogler (2001) and Schillmeier (2010).

8 For a discussion, see De Boever and Neidich (2013) and Neidich (2014).

9 See E. Martin (1990) and the detailed discussion of this in chapter 7 in this book.

10 See, for example, Crimp (1988) and J. Weeks (1995).

11 For a discussion see Irigaray (1985) and Roberts et al. (1996).

12 As Dennett (1995) proposes.

13 See S. F. Gilbert and Epel (2009) and Gottlieb (1997).

14 For an extended discussion of emergent architectures, see Cooper (2008) and Papadopoulos et al. (2008).

15 As critically discussed for example in S. J. Gould (1977).

16 See, for example, the work of Woodworth (1921, 1938).

17 See J. B. Watson (1913).

18 For example, see Baldwin (1897) and Dewey (1999).

19 See Gardner (1987, p. 28).

20 The work of McCulloch and Pitts (1943) on neural networks and the work of Goldstine and von Neumann (1963) on flow diagrams and cellular automata during the 1940s and the 1950s offered the visual images of neural processes and calculations that came to later to dominate the cognitivist modelling of thinking. See also Morris and Gotel (2006).

21 As, for example, in Churchland (1986).

22 See different approaches in Beaulieu (2003), Dumit (2004), and Joyce (2005).

23 See Star (1989).

24 See Karmiloff-Smith (1992) and Littlefield (2009).

25 See different approaches in S. D. Brown and Stenner (2009), Cromby (2007), Franks (2010), and J. W. Scott (2001).

26 See examples in Liben (1999), Thelen and Smith (1994), and E. A. Wilson (1998).

27 Changeux (1997), Edelman (1989), and Edelman and Tononi (2000).

28 Heidegger (1993) and Merleau-Ponty (1966).

29 See Cromby (2004); see also Middleton and Brown (2005), Stam (1996), and current debates on the relation of social science and biology in Meloni, Williams, and Martin (2016).

30 See, for example, Cromby (2015), Bayer and Malone (1996), Teo (2015), and Pulido Martinez (2013).

31 Vygotsky (1987), Wygotski (1987); see also Papadopoulos (2010).

32 See, for example, Colombetti (2014), Damasio (2004), Lakoff and Johnson (1999), Pickering (2010), and Varela, Thompson, and Rosch (1991)

33 See, for example, A. Clark (1997), Edelman (1992), Lewontin (2000), and S. Rose (1998).

34 See, for example, Csordas (1994b), Harré (1996), Overton (1998), and Sampson (1996).

35 See, for example, Bourdieu (1987), Braidotti (2002), Fausto-Sterling (2000), and Foucault (1978).

36 See Colombetti (2014) and Ziemke (2001).

37 As opposed to machines that act in virtual space or in protected, experimental environments; see Brooks (2001) and Chrisley and Ziemke (2002).

38 See Damasio (1999) and LeDoux (2002).

39 See, for example, cybernetics, developmental systems theory, and Gottlieb (1992), Lewontin (2000), Oyama (2000), and Pickering (2010).

40 See also Pickering (2010) and E. Thompson and Varela (2001).

41 See Blanche, Bhavnani, and Hook (1999).

42 See, for example, Alaimo and Hekman (2008), A. E. Clarke and Olesen (1999), and De Lauretis (1987).

43 See, for example, W. Bauchspies and Puig de la Bellacasa (2009), A. E. Clarke et al. (2010), Haraway (1991b), Latimer and Schillmeier (2009), Myers (2008), and Rapp (2000).

44 See, for example, Hayles (1999) and Lilley, Lightfoot, and Amaral (2004).

45 As in B. Turner (1984, p. 247), for example.

46 See Duden (2002) and Scarry (1985).

47 See, for example, Papadopoulos (2003).

48 Schull and Zaloom (2011).

49 See Pitts-Taylor (2010) and the editorial in Cromby, Newton, and Williams (2011).

50 See E. Martin (2010).

51 See E. Martin (2002) and Papadopoulos (2008).

52 Especially through the critique of cognitivist ideas within these fields; see Balsamo (1995) and Hayles (1999).

53 Already in the 1990s it became obvious that this perspective was untenable in the field of robotics, especially in relation to humanoid robots; see Brooks (1991), Hayles (1999), Pickering (2010, 2011), and Varela et al. (1991).

54 See, for example, Brooks (2002).

55 For a discussion of these developments, see A. Adam (1998), A. Clark (1998), Kember (2003), Puig de la Bellacasa (2009), Steels and Brooks (1995), and L. A. Suchman (2007).

56 Embodied approaches to robotics use semiopen connectionist nets to link different brain subsystems.

57 Sonigo (2005).

58 Self-reproduction has its origins in John von Neumann's cellular automata, mentioned earlier in this chapter. However, this recombinant plasticity refers not to abstract and universal criteria for self-reproducing machines but to real-life machines in open autopoietic systems, as in Varela et al. (1991). Cybernetics offers a similar position; see Pickering (2010).

59 See Pitts-Taylor (2010).

60 See Gottlieb, Wahlsten, and Lickliter (1998).

61 See Margulis (1998).

62 See S. F. Gilbert and Epel (2009).

63 See Robert (2004).

64 As described in Gottlieb (1992).

65 For example, see G. B. Müller (2007) and Sultan (2007).

66 See also S. F. Gilbert and Epel (2009), Huttenlocher (2002), and Jablonka and Lamb (2005).

67 See different overviews in Gottlieb (2007), Robert (2004), and van Speybroeck, van de Vijver, and de Waele (2002).

68 See discussion in Calvanese et al. (2009), S. F. Gilbert and Epel (2009), and Robert (2004).

69 See Gottlieb et al. (1998).

70 As for example in Lamb (1994) and Masterpasqua (2009).

71 See, for example, Kiefer (2007).

72 See discussion in Landecker (2011a).

73 See, for example, Bandyopadhyay and Medrano (2003).

74 See, for example, Champagne (2010).

75 For a discussion of current developments in epigenetics, see Landecker and Panofsky (2013); about broader cultural and social implications, see Landecker (2011a) and Niewohner (2011). See also S. F. Gilbert (2002) and Mitchell et al. (1996).

76 For an in-depth discussion of these issues, see Chung et al. (2016) and Cromby et al. (2016); see also Kenney and Müller (2016).

77 I borrow the concept of ecomorphs from Wainwright and Reilly (1994).

78 On standardization, see Bowker and Star (1999) and Busch (2011).

79 Maps of ecomorphs are the product of epigenetic sequencing in the same way a map of genes in the human genome database was the product of DNA sequencing. However, the number of ecomorphs will probably be far more than the approximately twenty-five thousand human genes. Ecomorphs will materialize the vision of truly learning how to create life and how to efficiently remake the brain.

80 The idea of autocreativity is inspired by Metzger's autodestructive/autocreative art and his 1960 and 1961 manifestos—see Metzger and Breitwieser (2005), and for an overview, see A. Wilson (2008).

81 See also Pickersgill, Martin, and Cunningham-Burley (2015) on how the promissory discourse of neuroplasticity is perceived by different groups of publics, and G. M. Thomas and Latimer (2015) on how values around bodies propelled by current biomedicine seep into the everyday life of the clinic and mundane health care environments.

82 See, for example, Abi-Rached and Rose (2010), Fitzgerald, Rose, and Singh (2016), N. Rose (2001), and E. A. Wilson (2004). For an in-depth discussion of such politics, see Pickersgill (2013).

83 For a broader critical discussion of governmentality, see Stephenson and Papadopoulos (2006).

84 Pitts-Taylor (2010, 2016) and Thornton (2011), for example, have argued that neural subjectification diligently reproduces and ultimately consolidates widespread neoliberal social relations.

85 See the work of A. E. Clarke et al. (2010), Kippax and Stephenson (2010, 2016), and Latimer (2013).

86 See, for example, Malabou (2008) and S. Watson (1998).

CHAPTER 7. COMPOSITIONAL TECHNOSCIENCE

1 See Eilperin (2013); see also Cohen (2001).

2 In a press conference on September 17, 1985.

3 It was on February 5, 1986, that President Reagan, during his visit to the Department of Health and Human Services (HHS) asked the surgeon general to prepare a report on AIDS, which was published in October 1986.

4 See, for example, E. Martin (1994), Crimp (1987), C. Patton (1990), and B. D. Adam and Sears (1996).

5 See Epstein (1996).

6 See Epstein (1997).

7 The are many reasons for this; for an extensive discussion, see the remarkable work of D. B. Gould (2009).

8 For a discussion of these issues, see B. D. Adam (1995), C. Patton (1990), and D. B. Gould (2009).

9 See Epstein (1996, 2007).

10 See Pickering (1992).

11 The traditional way in the philosophy of science to think of the relation between knowledge and politics was epistemology, where the content of technoscience and the associated politics were subordinated to ways of knowing and defining truth (Alcoff, 1998). Social studies of science emerging in the 1970s contested this view, but their attempts to question traditional epistemology devolved into a new cycle of epistemological disputes between proponents of rationality and sociality (Longino, 2002). The covert politics of epistemology culminated in the science wars of the 1990s.

12 My focus here is on debates that pertain directly to the question of politics. For a wider analysis of different intellectual genealogies in science and technology studies see the work of Fischer (2009), in particular chapter 2, which broadly discusses some of the concerns articulated in this current chapter.

13 See Collins and Evans (2007) and Lynch and Cole (2005).

14 For examples, see Edwards and Sheptycki (2009), Jenkins (2007), and Weinel (2007).

15 See Collins and Evans (2007, pp. 8–9) and Collins, Weinel, and Evans (2010).

16 See Collins and Evans (2007, chap. 1).

17 The broader theoretical framework of this approach is contractualism—see Rawls (1971).

18 The provenance of this approach is in Habermas (1993); see also Scanlon (1998).

19 See, for example, S. P. Turner (2003).

20 This is what Collins (2009) calls "elective modernism."

21 See, for example, Habermas (1984).

22 See critical approaches to communication in Taylor (1986) and Wellmer (1977).

23 For an extensive discussion of the main tenets of this position, see Wimmer (1980).

24 See Lengwiler (2008).

25 See different approaches to this in Irwin and Michael (2003), Leach, Scoones, and Wynne (2005), and Wynne (2005).

26 See various positions in Felt et al. (2008), Tutton (2007), Callon and Rabeharisoa (2008), Elliott, Harrop, and Williams (2010), Epstein (1996), and de Saille (2014).

27 See, for example, Hubbard et al. (1979).

28 See, for example, A. E. Clarke and Olesen (1999) and Rapp (2000).

29 See, for example, P. Brown (2007).

30 See, for example, Epstein (1995) and Rosengarten (2004).

31 Wynne (2003) has developed this position persuasively.

32 See, for example, Chilvers (2008), Rowe and Frewer (2004), Webster (2007), and Jasanoff (2003b).

33 See, for example, Fortun (2001).

34 See, for example, Elliott and Williams (2008).

35 See, for example, Mayer (2003), Munro and Mouritsen (1996), and Neyland and Woolgar (2002).

36 See Jasanoff (2003a).

37 See Nowotny, Scott, and Gibbons (2001).

38 See, for example, Liberatore and Funtowicz (2003).

39 See, for example, G. Davies (2006) and Hamlett (2003).

40 For a renewed interest in radical transformative approaches to institutions within current social movements, see the debates in the journal and online forum of the important theoretical virtual space "eipcp. European Institute for Progressive Cultural Policies," such as Salvini (2008), Raunig (2006, 2007), Universidad Nómada (2008), and Virno (2007).

41 See, for example, Epstein (2007).

42 For a background debate, see MacIntyre (1981) and Taylor (1989, 1991).

43 See Leach et al. (2005).

44 See, typically, Latour (1993, 2000).

45 See, for example, Callon (1987), Law (2004), and Mol (2002).

46 See also Stengers (2005).

47 See, for example, Callon and Rabeharisoa (2004).

48 See Barry et al. (1996).

49 See Burchell, Gordon, and Miller (1991).

50 See, for example, Mol and Law (2004), Law and Mol (2008), Law and Urry (2011), and Moser (2008). For a critical interrogation of the proliferation of network metaphors and visual representations of networks in popular and scientific cultures, see Munster (2013).

51 Governance has become one of the dominant tropes of political thought and practice; see, for example, Castells (1997), Rhodes (1997), and Rosenau and Czempiel (1992).

52 See different examples in Bevir and Rhodes (2003), European Commission (2001), and Edwards and Hughes (2005).

53 For a critical approach see Harney (2006).

54 For example, Callon, Lascoumes, and Barthe (2009) propose that the posthuman condition can be dealt with by entering into hybrid forums of negotiation and exchange. These forums can be described as representational arenas that foster communication between, and transformation of, the human and nonhuman actors involved.

55 Various positions question the effects of governance and the underlying logic of productionism; see Haraway (1992), Harney (2008), Negri (2005b), and Papadopoulos et al. (2008).

56 See Kippax and Stephenson (2010, 2012); see also Rosengarten and Michael (2009), Ribes and Polk (2015), and Barnett and Whiteside (2006).

57 See, for example, Biehl (2009), Kippax and Stephenson (2016), and Rosengarten (2009).
58 See also Latour (2003, p. 28).
59 As in Haraway (1988).
60 This line of knowledge politics started initially with feminist standpoint theory. It engages with different marginalized standpoints in a context of structural relations of power; see Harding (1991), Hartsock (1983), Hill Collins (1991), H. Rose (1994), and D. E. Smith (1987). For an in-depth discussion, see Puig de la Bellacasa (2013, 2014c).
61 See, for example, L. A. Suchman (2007).
62 See Feenberg (1991) and Winner (1993).
63 See Figueroa and Harding (2003) and Shiva (2005).
64 See Nowotny and Rose (1979), Ravetz (1990), and H. Rose and Rose (1969).
65 See different approaches in D. J. Hess (2007a), da Costa and Philip (2008), Earl and Kimport (2011), de Saille (2014), McCormick (2007), and Woodhouse et al. (2002).
66 See different examples in Feenberg (1995), Illich (1979), and Marcuse (1991).
67 See, for example, Kleinman (2000) and Wynne (2002).
68 See, for example, D. J. Hess (2004, 2007b), Welsh (2007), and Welsh, Plows, and Evans (2007).
69 Although situated approaches are historically grounded in classic Marxist positions, they also point toward a post-Marxist reading of social transformation that refuses to reduce all contradictions in a region of objectivity to the totalizing social dichotomy and underlying dialectical antagonism between capital and labor. Here there is no monolithic totalizing structure that determines the existence of a marginalized position (Marcuse, 1991); it is, rather, a decentered and multifaceted structure of injustice that produces a range of different subaltern positions (Althusser, 2001; Jameson, 1981). This understanding of marginalized positions is supported by the rejection of a unified reading of the composistion of oppositional experience (Sandoval, 1991). Instead, there is a focus on how different autonomous "subaltern" social groups in certain historical moments and places developed divergent self-organized experiences that do not simply reflect the main social antagonisms but diffract them into new forms of social existence.
70 Which is assumed to be a more authentic form of objectivity—see Harding (1986).
71 For an in-depth discussion, see Papadopoulos and Stephenson (2007).
72 See Clifford (2000), Hall (1990), and Papadopoulos (2006).
73 See also chapter 4 in this book and Papadopoulos (2008).
74 See the classic position of J. W. Scott (1991).
75 See Stephenson (2003); see also Papadopoulos and Stephenson (2007).
76 Following Whitehead (1979) we developed this idea in Stephenson and Papadopoulos (2006) and also in Papadopoulos et al. (2008).

77 I discuss the notion of involution and involvement in the next chapter; see also Hustak and Myers (2013).

78 See different takes on these issues in Bracke and Puig de la Bellacasa (2008), Haraway (1988), G. Rose (1997), Stephenson and Papadopoulos (2006), and Puig de la Bellacasa (2013, 2014c).

79 As discussed in M. R. Smith and Marx (1994).

80 Specifically about HIV, any parts of the history of the movement that do not fit into this teleological history are often seen as mistakes of the actors of the time or at least as failures. The work of D. B. Gould (2009) is a good example of this ambivalence, as she finds it difficult to avoid judging negatively the AIDS communities that struggled to form a coherent subject of action and enter into direct action in the early stages of the epidemic.

81 See also Crimp (2002).

82 See, for example, Altman (1986) for a rich account of the gay community's mobilizations, campaigns, and efforts during the beginning of the 1980s.

83 See also Latimer and Puig de la Bellacasa (2013) and Schillmeier and Domenech (2010).

84 See Treichler (1999, chap. 9).

85 See Hodge (2000), Stockdill (2003).

86 See, for example, M. Thompson (1994).

87 See Berkowitz, Callen, and Sonnabend (1983); see also Crimp (1988) and C. Patton and Kelly (1987).

88 See Shilts (1987).

89 See Treichler (1999).

90 See Kwitny (1992) and Chambré (2006).

91 On the emotional aspects of the AIDS movement, see D. B. Gould (2009).

92 See Watney (1997).

93 Especially in the first years the epidemic had a profound impact on the community's sense of its own existence because as C. Patton (1985, p. 6) says, even in the most politically radical lesbians and gay men there was this lurking fear that perhaps at the end "homosexuality is death."

94 See Bersani (1987) and Harrington (2008).

95 For an extended discussion of the article and its impact, see D. B. Gould (2009, pp. 92–104).

CHAPTER 8. CRAFTING ONTOLOGIES

1 See Derwent Valley Mills Partnership (2001).

2 The Derwent Valley Mills along the River Derwent in Derbyshire, UK, contains a series of eighteenth- and nineteenth-century silk and textile cotton mills. The mills are considered the birthplace of the modern factory system and were added to the UNESCO World Heritage List in 2001. See https://www.derwentvalleymills.org/.

3 See https://remakemuseum.tumblr.com/.

4 In 2015 the Re:Make project was nominated for a Museums and Heritage Award in the category Educational Initiative. Later in 2015 Derby Museums received substantial funding from the Heritage Lottery Fund for redeveloping the site and reopening the new Derby Silk Mill—Museum of Making in 2020. See https://www.derbymuseums.org/hlfsuccess/#.VbdN79Vviko.

5 See https://www.museum-id.com/idea-detail.asp?id=523 and https://themuseum ofthefuture.com/2014/07/21/the-convincing-transformation-process-of-the -derby-silk-mill/. Citizen participation and collaboration is probably the most crucial dimension of the Re:Make project; for a discussion see the work of N. Simon (2010) on the participatory museum (see also https://www.participatorymuseum .org/).

6 See Calladine (1993).

7 There are considerable debates about ontology and the so-called ontological turn in the social sciences. Although many of these debates form the background of this book, *ontology* as used here points toward the capacity of certain actors (such as a group of humans or members of an animal species or processes or certain objects, or some co-action between them) for changing the material configuration of their space of existence. The material configurations of ontologies change always in different ways and can be viewed along different scales depending on who and how one acts in them. They are never only local or only global; they connect different scales and processes, and what counts is which local/global aspects they mix. What is crucial for the use of the term *ontology* here is the idea that the material configuration of each ontology can change only in specific ways. I see this ordinary use of the term *ontology* in continuation with positions that promote a much more comprehensive understanding of the term (Blaser, 2012). And although these debates have far-reaching implications for many fields of study in the social sciences, such as anthropology (de la Cadena, 2010; Henare, Holbraad, and Wastell, 2007; Holbraad, Pedersen, and Viveiros de Castro, 2014), geography (Braun, 2008; N. Clark, 2011), science and technology studies (Fujimura, 1991; Lezaun, 2014; Pickering, 1995, 2008; Sismondo, 2015; Woolgar and Lezaun, 2013), political philosophy (Braun and Whatmore, 2010; Coole and Frost, 2010), literary theory (Boscagli, 2014), organization studies (Beyes and Steyaert, 2011; Carlile et al., 2013; Knox et al., 2008; Parker et al., 2014), and philosophy (Bryant, Srnicek, and Harman, 2011), the aim of the chapter is not to directly engage with them, nor to engage with broader discussions of ontology in social theory; see, for example, Agamben (1998), Badiou (2005a), Braidotti (2006), De Landa (2002), and Rabinow (1996). I will therefore try to limit the use of the term *ontology* to the definition I provided earlier, namely ontologies as spaces of existence in which matter is organized in a specific way and not another and can be changed in specific ways and not others. See also chapter 1 for a discussion of ontological politics.

8 See the public announcement "Derby Museums Secure Major Heritage Lottery Fund support," May 5, 2015, https://www.hlf.org.uk/about-us/media-centre /press-releases/derby-museums-secures-major-hlf-support.

9 See, for example, Savvides (2018), Delfanti and Söderberg (2012), and Söderberg and Delfanti (2015). More broadly on the relation between social movements, innovation, and science and technology, see, for example, D. J. Hess (2007a, 2007b, 2015) and Frickel and Gross (2005); on grassroots innovation, see Gupta (2016) and Maxigas and Troxler (2014).

10 See https://makerfaireafrica.com/maker-manifesto/.

11 The Maker Faire website says that "200,000 people annually attend the two flagship Maker Faires in the Bay Area and New York, with an average of 44% of attendees first timers at the Bay Area event, and 61% in New York. A family-friendly event, 50% attend the event with children. In 2017, over 190 independently-produced "Mini Maker Faires" plus over 30 larger-scale Featured Maker Faires will have taken place around the world, including Tokyo, Rome, Shenzhen, Taipei, Seoul, Paris, Berlin, Barcelona, Detroit, San Diego, Milwaukee, and Kansas City." See https://makerfaire.com/makerfairehistory/.

12 See Sivek (2011). Although similar to the maker culture, hacker culture developed in a very different context. For a genealogy of hackspaces and hacklabs. see Maxigas (2012) and Tabarés-Gutiérrez (2016); see also Stein (2017).

13 For an analysis, see Luckman (2016).

14 While the transnational and multiscalar dispersion of manufacturing process across many regions remains the characteristic of current global production networks (Coe et al., 2004; Henderson et al., 2002), increasing voices promote relocation and backshoring of advanced parts of production to the West. Already in 2012 Jeffrey R. Immelt, CEO of GE and head of President Obama's Council on Jobs and Competitiveness, wanted "business investment and export" to lead recovery after the 2008 economic crisis. With this investment "our workers will prove America's potential. People talk about the Darwinian nature of markets . . . but nothing is inevitable about the industrial decline of the United States" (Immelt, 2012). This became one of the key narratives of Donald Trump's presidential campaign in 2016.

15 See Chris Anderson (2012) and Marsh (2012). For a popularization of the third industrial revolution, see Rifkin (2011). See also popular coverage, from the conservative *The Economist* (2012) to the liberal technoentrepreneurial *Wired* (2011) featuring on its front cover open-hardware entrepreneur Limor Fried. On the entrepreneurial dimension of the maker and hacker culture, see Irani (2015) and Wolf and Troxler (2016).

16 In *Financial Times* we read, for example, that "Gadget Makers Drive US Manufacture Return" (March 27, 2013, https://www.ft.com/content/1445bab6-96b0-11e2-a77c-00144feabdc0).

17 See, for example, Banks (2010), Kozłowski et al. (2014), Terranova (2004), Lovink and Rossiter (2011), and Hanlon (2016).

18 This immersive engagement with the totality of the production process characterizes holistic technologies, according to Franklin (1999); see also Malafouris (2008).

19 See, for example, Gauntlett (2011).

20 See, for example, Bollier (2008), Kostakis and Bauwens (2014), Aigrain (2012), and Lezaun and Montgomery (2014). For a typology of collaborative and sharing platforms, see de Rivera et al. (2016).

21 See, for example, Flowers (2008) and Conway and Steward (2009).

22 See the activist research of the Precarious Workers Brigade and Carrot Workers Collective (2014). See also Beverungen, Böhm, and Land (2014), Terranova (2000), and Banks (2010).

23 From craftivism (Bratich, 2010; Bratich and Brush, 2011; Buszek and Robertson, 2011; B. Greer, 2014), craft entrepreneurialism (Luckman, 2016), and Lego art (Doyle, 2013) to arts and crafts (Adamson, 2013; Thurnell-Read, 2014), professional workmanship (Frayling, 2011; Sennett, 2008) and craft in science (Ingold, 2013; P. H. Smith, 2004), in fabrication labs (Walter-Herrmann and Büching, 2013) in relation to design (Marti et al., 2016) and in hackspaces (Maxigas, 2012; Maxigas and Troxler, 2014).

24 See Vaneigem (1996, p. 76). Some go even further and assign to creativity ontological status: creativity is the driving force for the self-making of the material world (McLean, 2009).

25 For example, see discussion on the capture of creativity in digital labor in Burston, Dyer-Witheford, and Hearn (2010) and Dyer-Witheford (2015).

26 See the work of Raqs Media Collective (2003).

27 A transformation that is connected to post-Fordism (Bowring, 2002; Dyer-Witheford, 1999; Fleming and Spicer, 2003; Gorz, 1999; Hampson et al., 1994) as well as to cognitive capitalism (Azais et al., 2001; Marazzi, 2007; Morini, 2007; Moulier Boutang, 2011; Vercellone, 2007).

28 Malafouris (2010) sees "material engagement" not as a historical mode of human practice pertaining to specific modes of human existence (such as artisanship, craft, and subsistence economies) but as an inherent transhistoric disposition of the human brain. In chapter 6, I problematize claims about a definite structure of the human mind and show that the way we understand our brains in a particular historical period also shapes the material functioning of the brain itself. Every epoch has its brain. Rather than "material engagement" being an essential quality of the human brain, it is a particular understanding of the brain that has material effects on its biomolecular and neural composition and the way our bodies operate. In this sense the return of material engagement to society and material life in the early twenty-first century shapes and is shaped by the way we conceptualize our brains and bodies.

29 See https://www.whitehouse.gov/nation-of-makers.

30 See Vaneigem and Obrist (2009).

31 As Hatch (2014, pp. 167ff.) claims.

32 See https://www.techshop.ws/.

33 More broadly on the relation between technoscience, culture, and everyday life, see Shaw (2008) and Michael (2006).

34 Masculinist not only in the sense of the gender composition of the maker culture (although one needs to be careful here because maker groups are very

different) but also in reproducing traditional and dominant masculinities. A critical view of the maker culture would also involve a discussion of its effects on social and urban environments (for example, the relation between maker culture and gentrification).

35 World 2 contains all these worlds that have been rendered invisible because of their inability or unwillingness to participate in Western modernity (Papadopoulos, 2006). In Chakrabarty's (2000) words this is History 2, the other to History 1, which represents the Eurocentric story of expansion of modern capital.

36 The accounts of Clifford (1989, 1997) and Pratt (1992) on traveling underpin my approach here. Traveling knowledges, experience, and technologies and the transnational and interspecies traffic in technoscience have been extensively discussed in various ways in STS; see, for example, Haraway (1997), W. Anderson (2002b), Turnbull (2002), Murphy (2012), and Latour (1987). For a discussion of specific examples, see W. K. Bauchspies (2014), Verran (2002), Morita (2013), Hayden (2007), Crane (2010), and Lezaun (2006).

37 See Mignolo (2002) and Grosfoguel (2007).

38 Kadri (2010). See also, for example, Lindtner (2015) and Kera (2012) on the maker culture in China and Asia, respectively.

39 See Clifford (2013). More specifically about indigenous movements, see Blaser et al. (2010).

40 See de la Cadena (2015); I discuss this work again later in the chapter. See also Cameron, de Leeuw, and Desbiens (2014).

41 See Simpson (2014), P. Wolfe (2006), and Hunt (2013).

42 See Coburn (2015) and Blaser et al. (2010).

43 See Tuck and Yang (2012), Wildcat et al. (2014), and Stewart-Harawira (2013).

44 My starting point here is Staeheli's (2013) work on disconnectivity.

45 See https://communityrejuvenation.blogspot.co.uk/2012/03/community-rejuvenation-projects-latest.html. Lavie Raven was one of the artists who created a monumental mural in East Oakland, California, with the message "Decolonize." It can be viewed here: https://crpbayarea.smugmug.com/Other/Decolonize-Mural/.

46 Instead of mental representations, material engagement presupposes an embedded and embodied mind; see Malafouris (2004) and Papadopoulos (2010).

47 See de la Cadena (2010, p. 345) and also Blaser (2012), Escobar (2011, 2012), and Mignolo (2013).

48 Stacking in permaculture refers mainly to the idea of layering/multilevel functions of elements and actions in each specific agricultural system—be it on a balcony, in an urban backyard, or on a farm or a large community agricultural system. This idea has been key in permaculture design from its beginning (Mollison, 1988). The term *stacking* first appears as a principle of permaculture in Mollison and Slay's *Introduction to Permaculture* (1991). *Stacking* here means that "all elements in the design should serve multiple functions, and all functions should serve multiple elements." The concept has developed in many directions as a fundamental principle of design by the permaculture com-

munity. See, for example, https://theurbanfarmer.ca/resources/permaculture
-design/, where stacking refers to three design strategies: "1) All elements in the
design should serve multiple functions. . . . 2) All functions in the system should
be served by multiple elements. This principle is essentially one of planning re-
dundancy into the system so there is less fragility and more resilience. . . . 3) Stack
elements in vertical and horizontal space as well as in time."

49 The term *stack* is also used in computer science and computer engineering,
where it invokes a rather static image of a vertical digital container operated
for storing data objects in a push-down list (every new object is placed on top
of all already existing objects). Against this background, cultural critics such
as Bratton (2014) and Terranova (2014) use the term *stack* to refer to a universal layered megastructure of software and hardware systems that englobes
all human activities and social life. As explicated here, the notion of *stack* that
informs this chapter comes from a very different context: permaculture and
related ecological design practices. Rather than as a metaphor for a planetary
system of domination (the Grand Stack), it is used to designate situated and
concrete practices for creating symbiotic environments. In this sense it is more
about stack*ing*—that is, grounded ecological making—rather than about *the*
stack, a dramatic universalist image of global domination.

50 See Holmgren (2002), Mollison (1988), and Mollison and Holmgren (1990).

51 See S. D. Brown (2012), Lury, Parisi, and Terranova (2012), and Parisi (2012).

52 Law (2011).

53 Deleuze and Guattari (1987, p. 20); see also chapter 4.

54 Deleuze (2001, p. 99).

55 Deleuze (2001, p. 95).

56 Deleuze (2005, p. 717). See also Deleuze and Guattari (1987). For further discussion, see also chapter 1.

57 See chapter 1 for a discussion of matter as used here. Ingold (2013) makes the
case that matter is always moving and can never be captured in fixed representations or conceived as containing preestablished entities; rather, matter
can only be followed and this is possible by developing close, intimate, and
intense relations to matter through the skill of craft—that is, through engaging
in a trajectory of movement that resonates with the trajectory of the specific
matter one is engaged with. Ghelfi (2015) describes this relation to matter as an
ecology of canals. Following Deleuze and Guattari (1987), he says that crafting
is about finding a "smooth space" inside matter. Matter exists in multiple modalities and becomes a constitutive element of (human) practice. The work of
McLean (2011) is illuminating this relation between more-than-human worlds
and human practice. See also Stubbe (2016).

58 See also Sundberg (2014).

59 On articulation, see Clifford (2001), Hall (1986), and Slack (1996).

60 This includes different practices of repurposing, reusing, creatively misusing,
renewing, and reappropriating existing things and materials that Malewitz
(2014) has called "rugged consumerism."

61 See another example in Boym (2002). On a theoretical level, see Bogost (2012) and Henry (2008) on approaching objects without reducing their openness.

62 See also Harman (2007).

63 See https://www.arduino.cc/. See also Ghelfi (2015).

64 See Rato and Ree (2012). See also Bratich (2010), Coleman (2013, pp. 93–122), Puig de la Bellacasa (2009), and Stubbe (2016) on craft and embodied work in digital technologies and software coding.

65 For a critical discussion of the politics of datafication, see Gray (2014), Liboiron and Pine (2015), and Vis (2013).

66 See Parisi (2013).

67 Raspberry Pi is a widely used compact computer motherboard that is a much more complex version of the Arduino microcontroller I mentioned in the preceding paragraphs.

68 For an insightful discussion of the political implications of this return, revival, and reinvention of craft, see Bratich (2010); specifically about different political traditions of hacking, see Maxigas (2012) and Karatzogianni (2015).

69 See https://www.dirtyelectronics.org/.

70 See also Stephenson and Papadopoulos (2006).

71 With Hartigan (2015b, p. 5) we need to ask: "If publics are decidedly human—self-reflexive readers, hailed by various nationally mediated cultural form—then how do we account for the presence of so many highlighted arrangements of multispecies life in their midst?" See also Kelly and Lezaun (2014).

72 The turn to a conceptualization of ecology as a "general" condition of existence—which underpins my understanding presented in this chapter—is discussed in Hoerl (2013).

73 See further discussion of ecological guilds in Kornan and Kropil (2014), Root (1967), Simberloff and Dayan (1991), and S. E. Williams and Hero (1998).

74 See Puig de la Bellacasa (2011). More broadly, "species thinking" (Chakrabarty, 2009, p. 213) deeply upsets any notion that invention power relies on humans. In fact, human activity and relationality more broadly cannot exist outside practices of interspecies engagement—be it interspecies care, labor, or even exploitation and destruction, as, for example, Hartigan (2015a), Pandian (2009), Schrader (2010), van Dooren (2014), and Kirksey and Helmreich (2010) show in their work.

75 See Haraway (1997), Ihde and Selinger (2003), and Weber (2010). The fusion of technology, science, and everyday life is not just another name for today's science; this fusion refers to something much wider, namely the acknowledgment that technology actively shapes basic research, which is increasingly concerned with impact on applications and the everyday. Translation and technological interoperability in technoscience is a constitutive moment of knowledge production directly linking technological innovation to basic research. But beyond that, what drives technoscience is that it continuously happens inside *and* outside formal research institutions (state-funded and private)—which I discuss in the second half of this chapter.

76 See Schraube (2009).

77 See McNeil (2013).

78 See also Papadopoulos (2014).

79 See Mollison (1988) and Mollison and Holmgren (1990).

80 For a discussion of some of these practices, see the work of Savvides (2018).

81 And this even if there are attempts to complicate the practice of translation; see, for example, Sakai (1997).

82 See S. D. Brown's (2013) reading of Serres's parasite.

83 See also Star (1995).

84 See the previous section in this chapter on involution, and, originally, Deleuze and Guattari (1987). See also Hustak and Myers (2013).

85 See Harman (2007).

86 See examples in Choy et al. (2009), Liberona and Myers (2016), and Rodríguez Giralt (2015).

87 See https://www.cac.lt/en/other/general/15/7719. For their other relevant works, see Urbonas and Urbonas (2008).

88 See some very different but excellent examples in Stamets (2005), A. L. Tsing (2015), and Choy et al. (2009).

89 See A. E. Clarke (2016), Puig de la Bellacasa (2014b), Star (1991), and Stewart (2013).

90 See also Kortright (2012).

91 See also Myers (2012).

92 See Schraube (2013). Experimental labor is an embodied, haptic process; see Myers and Dumit (2011) and Puig de la Bellacasa (2009).

93 On invention power, see, originally, Moulier Boutang (2011, p. 93) and Negri (2005a, p. 268).

94 In chapter 2 I described the underlying process that supports this multiplication of different types of precarious labor as the fusion of work and financialization. This fusion takes place across different levels: the financialization of everyday life (through debt, for example); the stagnation or decline of wages and compensation for this decline through lending; the increased valuation practices that define the quality of work outputs; the importance of "exploiting" one's own future as a working subject, something that we have called postcontractual exploitation; and, finally, the incorporation of nonwork processes to value production.

95 There are various approaches to precarious labor; in particular, see Armano and Murgia (2012a, 2012b), Berardi (2010), Edu-Factory Collective (2009), Morini and Fumagalli (2010), R. Müller and Kenney (2014), and Papadopoulos et al. (2008).

96 Many different perspectives on the commons and on social mobilizations gravitate around this idea (for further discussion, see Blomley, 2008; De Angelis, 2017; De Angelis and Harvie, 2013; Fournier, 2013; Linebaugh, 2008). Three different aspects pertain to the discussion about the commons: (1) On a *political* level, struggles for the commons include a myriad of campaigns and actions

around the naturecultural and the informational commons (Blomley, 2008; Bollier and Helfrich, 2012). (2) Since its (re-)introduction in the 1990s on a *conceptual* level, the notion of the commons is mainly anchored in critical/global political economy—either in a radical political perspective (Midnight Notes Collective, 1990; Ricoveri, 2013) or as an alternative institutional approach to the politics of scarcity and resource governance (Mehta, 2010; Ostrom, 1990). This reliance on political economy brings with it severe limitations: mainly a blunt humanism and anthropocentrism as discussed in chapter 5 (Ghelfi, 2015) and a too-easy reduction of the commons to the management of common-pool resources (Bresnihan, 2016a). (3) Finally, on a *practical* level, the notion of the commons can have very different implications for political practice. Central to the notion of the commons is dispossession—that is, the private enclosure and capitalist appropriation of the livelihoods and environments, the means of production, and the products of work of people and their communities (Bollier, 2003; Ricoveri, 2013). Now, the practical responses to this process of appropriation can vary significantly: a movement for refusal of the appropriation of work, that is, struggles engaging with the commons as the central and underlying moment of contemporary capitalist production, an approach inspired by the tensions that arise in the increased socialization of production (Read, 2011; Vercellone, 2010) and the crisis of social reproduction (Barbagallo and Federici, 2012); a movement for the creation of new commons, that is, the subtraction from capitalist appropriation and the making of new extracapitalist commons, an approach mainly inspired by the multitude of social justice campaigns in the Global South (Caffentzis, 2010; McCarthy, 2005; Ricoveri, 2013); and finally, a movement for reappropriation, that is, the organization of a historical subject that can achieve the recovering of what has been enclosed by capital, an approach that is aligned with traditional workerist grand narratives of social liberation (Mezzadra, 2011b).

97 The difference between the commons and the publics is that the latter is always in some form or another linked to the state, either as civil society operating in the symbolic and territorial realm of the state or as social groups that are activated by certain governmental institutions or as pressure groups that articulate their demands toward the state. One could go as far as to say that publics are proactively constructed by state institutions. On the other hand, the commons exist and can sustain themselves without the direct intervention of state institutions.

98 I am thinking here with the work of D. López (2010) on issues related to the relation between security and technology.

99 See Fumagalli (2011), Marazzi (2010) and Negri (2010). See also the work of Birch (2013, 2017a).

100 See https://blog.arduino.cc/2015/04/30/microsoft-and-arduino-new-partnership/. See also the discussion of Arduino in Eglash (2016).

101 See Latour (1983).

102 See, for example, Reardon (2012).

103　See, for example, Thorpe and Gregory (2010).

104　See chapter 3 for a further discussion.

105　See Bayat (2010).

106　See Dinerstein (2014), Sitrin (2012), Zibechi (2011), and Zibechi and Ryan (2010). See also the work of Mora-Gámez (2016).

107　The principal figure of postliberalism is not the state or private interests or the individual subject (which dominated neoliberal ideology); rather, it is an organic clumping of certain segments of the state together with some private interests, certain subjectivities, parts of social classes, or segments of the public. In previous work we have discussed the figure of the postliberal vertical aggregate to describe this (Papadopoulos et al., 2008). Postliberal aggregates do not cohere around a shared ideology. They entail an intermingling of various actors into large formations that coalesce along an imagined commonality of social domination. Postliberal modes of control condense economic, technoscientific, political, and cultural power and control decision-making processes. They reassemble parts of the fragmented society that was the outcome of forty years of neoliberal policies into vertical aggregates. The verticality here refers to segments of state, public, or private actors that act together and form global players that effectively go beyond the identity of each player involved. Against this postliberal turn of control, the movements that I describe here attempt to respond and to create open and shared alternative spaces of existence from below (Arditi, 2008).

108　Some of these ideas feature in recent debates about the prefigurative politics of social movements, for example, Pickerill and Chatterton (2006). Prefigurative politics are forms of alterontological practice that attempt to create other workable and livable forms of life such as ecological alternatives, independent media ecologies, feminist collectives, and autonomous learning environments (Chatterton and Pickerill, 2010; Meyerhoff, Johnson, and Braun, 2011).

109　See, for example, the work of Mora-Gámez (2016) on the system of rights restitution and reparations in "post-conflict" Colombia and how armed conflict "victims" create hybrid alternative spaces of existence within the formal infrastructures of reparation and compensation. See also Salvini (2013), an inspiring description of the remaking of urban space in Barcelona since the early 1980s through the everyday social and political struggles over the right to the city— and also studies on Athens, London, and Rome by Chatzidakis, Maclaran, and Bradshaw (2012), Ford (2011), and Grazioli (2017a), respectively. See also K. Ross's (2015) study of the Paris Commune and how this radical experiment in self-organization was not just structured by its revolutionary political aims (Merriman, 2014) but mainly through the artisanal remaking of the everyday conditions of life in the commune: the decentralization and diffusion into the ordinary of questions on self-subsistence, ecology, art, social relations, and work. This integration of experimentation in everyday life created the explosive political potential of the commune that continues long after its horrific suppression. Finally, on the imperceptible politics of commons movements, see Kanngieser and Beuret (2017).

110 See also De Smet (2016) and van de Sande (2013).

111 See, for example, Mander and Tauli-Corpuz (2006), Escárcega (2013), and N. Thomas (2010).

112 See Clifford (2013) and specific studies such as Clare Anderson (2012), N. Thomas (2010), Linebaugh and Rediker (2000), and Frykman et al. (2013).

113 And also enable the formation of "microworlds," as Verran (2002) calls the making of specific mundane material arrangements of collective life.

114 Provincializing formal experimental science and examining its colonial involvement and postcolonial futures has been discussed widely in science and technology studies; see, for example, W. Anderson and Adams (2007), Verran (2001), W. Anderson (2002a), Adams (2002), A. Tsing (2005), and Turnbull (2000).

115 See Olivera and Lewis (2004). It is also important to see how this mobilization evolved as the fold of the commons and the state described earlier in the chapter changed the nature and targets of the water movements. More recently, parts of the water movement in Bolivia (Bresnihan, 2016b) and in Uruguay (Taks, 2008) are gradually reintegrated into state governance of water by including self-managed water committees (Zibechi, 2009) in public-private administration of water resources. For a broader discussion, see Dupuits and García (2016).

116 See also Fitting (2006) and Mann (2011).

117 See also North and Grinspun (2016).

118 Something that raises questions about food sovereignty within the whole region; see Altieri and Toledo (2011) and McKay, Nehring, and Walsh-Dilley (2014).

119 Similar studies in postcolonial technoscience and indigenous knowledge discuss bioprospecting (N. Harvey, 2001; Hayden, 2003), genetic contamination (van Dooren, 2010), and the trafficking of genetic materials (Hayden, 2007; Kowal, Radin, and Reardon, 2013).

120 See Barlow (2008; 2012, p. 20). More broadly I take inspiration from the work of Kohn (2013), C. B. Jensen (2015), Gow (2001), and Raffles (2002). All of them describe different cases in which humans, other animal species, plants, and their riverine environments and riparian rain forests create each other and all contribute to the making of these ontologies.

121 See also Hope (2008).

122 On traditional resistance and protest movements, see Caygill (2013), Tyler (2013), and Douzinas (2013). When we commonly think of resistance, as in these works, it is as the other to power. Here I propose to reverse this understanding. When we talk about resistance as confrontation to power, then I propose to talk about protest, opposition, or revolt, as, for example, in Tyler's work. In this configuration, power and "resistance" (as protest, opposition, or revolt) are in tight and infrangible connection, one opposing the other. In fact, one could say that "resistance" as opposition becomes part of power and enters the nexus of existing power relations—Checchi (2015) provides an insight-

ful analysis of these different readings and meanings of resistance. Following Checci's work and previous research (Papadopoulos et al., 2008) I discuss here an alternative reading of resistance. Resistance (that is, resistance as resistance, not as protest, opposition, or revolt) unfolds as a movement of exit and escape from the prevalent power relation in a certain field. Resistance is not primarily about direct confrontation but about changing the conditions in which power operates. This is possible only by changing the ontological conditions of life in a certain field. In chapter 3 I talked about "organizational ontology" as the practice of resisting power by organizing alternative ontologies. Organizational ontology involves all these mundane practices that change the materiality of life in ways that escape power (and by doing this, they force power to change). J. C. Scott (2009, 2012) talks in a similar way about "infrapolitics," the deliberate attempt to remain outside (state) power (see also Chatterjee, 2004; Feigenbaum, Frenzel, and McCurdy, 2013; Salvini, 2013; and Sitrin, 2012).

123 For a discussion, see Della Porta and Diani (2006). This is why most of social movements theory and research has focused on the identities of social movements (Melucci, 1996; Touraine, 1981). However, this has been criticized from many different perspectives and various authors (Chesters and Welsh, 2005; Della Porta and Rucht, 2013; Rucht, 1991; Sitrin, 2012).

124 See Chesters and Welsh (2006) and Papadopoulos et al. (2008).

125 See also A. E. Clarke and Olesen (1999), Rapp (2000), Hubbard et al. (1979), P. Brown et al. (2006), and A. E. Clarke et al. (2010).

126 See Dinerstein (2010), Kokkinidis (2014), Ozarow and Croucher (2014), Ratner (2013), and Vieta (2009).

127 See the important work of Kokkinidis (2014). See also Sitrin's book *Horizontalism: Voices of Popular Power in Argentina* (2006), which inspired me while writing this section.

128 See Epstein (2007).

129 See Wynne (2003).

130 See, for example, Nowotny and Rose (1979), H. Rose (1983), H. Rose and Rose (1969, 1976a, 1976b), and Ravetz (1971).

131 As discussed in the section on participation in the previous chapter, there are two main traditions in participatory politics. The first one focuses on public engagement, inclusion of social actors in science policy, and making scientific institutions more accountable. The second one starts from a radical critique of technological, scientific, and medical rationalities and is grounded in the new social movements after the 1990s. It attempts to develop a practice of radical democratic participation and science activism. This second one is very close to the idea of more-than-social movements presented in the following sections.

132 See, for example, Welsh and Wynne (2013) for discussion of antinuclear movements, anti-GMO mobilizations, and climate change campaigns. See also McNeil and Haran (2013), Haran (2013), and Reynolds (2013). These debates privilege a form of politics that attempts to create channels of communication and exchange between science and publics by facilitating deliberative

and participatory inclusion of publics in science policy. But when publics and science implode into each other, as described in this chapter, rather than one clear-cut conflict between science and publics there is a multiplicity of exchanges and conflicts that emerge as the separation between publics and science disappears. This collapse opens different political possibilities beyond the modernist/humanist ideal of a potential enlightened public against authoritarian science and its realignment with a possible democratic science.

133 This is the case not only for movements that directly target technoscientific issues but also for many traditional social movements that engage with technoscientific knowledge in their organizing practices, as, for example, the 15M movement in Spain in 2011; see an analysis in Calleja López (2017) and Ghelfi (2015).

134 See Nancy (2000, p. 24); see also N. Clark (2011).

135 See Troutt (2006) and Nancy (1991).

136 See Claeson et al. (1996). See also Laugksch (2000).

137 See, for example, Savvides (2017) and W.-M. Roth and Barton (2004). See also Barad (2001), Bencze and Alsop (2014), and Dreessen, Schepers, and Leen (2016).

138 See https://excotwincities.org/.

139 See also Haworth and Elmore (2018). More broadly, on the question of participation as a pathway for democratizing literacy, see Noorani, Blencowe, and Brigstocke (2013).

140 Savvides (2018), for example, has done extensive research on different modes of learning and education within maker and hackspaces. See also Maxigas and Troxler (2014).

141 See different examples in Latimer and Puig de la Bellacasa (2013), Liboiron (2016), Pérez-Bustos (2016), Sánchez Criado, Rodríguez Giralt, and Mencaroni (2016), and Puig de la Bellacasa (2015).

142 The idea of problematics originates in Deleuze (1994) and Deleuze and Guattari (1987). See also D. W. Smith (2006).

143 See also Star (2010); for a broader discussion of similar issues, see J.-H. Passoth, Peuker, and Schillmeier (2012).

144 See Staeheli (2012).

145 See the work of Rappert (2014) and Rappert and Bauchspies (2014) on secrecy.

146 Deleuze and Guattari (1987, p. 238): "Finally, becoming is not an evolution, at least not an evolution by descent and filiation. . . . If evolution includes any veritable becomings, it is in the domain of symbioses that bring into play beings of totally different scales and kingdoms, with no possible filiation. . . . There is a block of becoming between young roots and certain microorganisms, the alliance between which is effected by the materials synthesized in the leaves (rhizosphere). . . . The term we would prefer for this form of evolution between heterogeneous terms is 'involution,' on the condition that involution is in no way confused with regression. Becoming is involutionary, involution is creative."

147 See Deleuze and Guattari (1987) and earlier note; see also Ansell-Pearson (1999), Hansen (2000), and Sterling (1977).

148 Margulis's (1998) theory of symbiogenesis refers to the appearance of new forms of life through the stable association of two organisms of different species. See also S. F. Gilbert and Epel (2009).

149 Virilio (1995, 1998). See also Krautwurst (2007).

150 See also S. D. Brown (2013).

151 See Beuret (2018).

152 See Figueroa Sarriera and Gray (2016).

153 See Star (2010); more generally, on the question of consensus in recent social mobilizations, see Nunes (2012, 2014); on the question of infrastructures and protest, see Feigenbaum et al. (2013).

154 See also S. J. Jackson et al. (2007) and Larkin (2013).

155 On the temporality of infrastructures, see Hetherington (2014, 2016) and Withers (2013). See an example in the work of Withers (2015).

156 Mora-Gámez (2016) calls them "entrelazando infrastructures," or *weaving* infrastructures, and Withers (2016) talks about "the politics of transmission." See also Sims (2007).

157 See Easterling (2005, 2014), Guldi (2012), and Truscello and Gordon (2013).

158 The designations "open" and "free" software account for a small difference but of crucial importance. Free and open software are not very different in terms of how they are made and their intrinsic qualities, but free software is made explicitly as an attempt to promote the value of nonproprietary software—that is, to promote justice by challenging copyright—while open software is promoting the software itself as an infrastructural tool for facilitating open information access (Coleman, 2013; Coleman and Golub, 2008; Stallman, 2013). We already know today that this difference, although so crucial for the development of free/open software, has been almost lost. As Kelty (2013, p. 3) puts it, "There is no free software. And the problem it solved is yet with us." In this sense there is no longer open-source software as fully separate from proprietary software, since both feed into each other in order to exist, and there is no free software as distinct from open and proprietary software because it is simply contributing to the making of the same infrastructures of codes despite the political differences and values that motivate it. See Carlson (2010), Delfanti (2011), and Hope (2008) for similar discussions in the field of open biology.

159 My guidance on these topics comes from the work of Pizio (in preparation). See also Infoaut (2013) and Dafermos and Söderberg (2009). The political conditions for the emergence of these new movements is discussed in DeNardis (2012, 2014). Here DeNardis shows how existing Internet infrastructures and their technological aspects have been gradually appropriated by private and state actors for their own economic and political interests. See also Musiani (2013, 2015).

160 The practice of making technoscience, whether it is mundane or formal, has created an alternative vision of technoscience as open. *Open technoscience* is a contested terrain, not a given reality or a definite program. Depending on the specific subfield and topic, the quest for openness addresses different issues and

different levels on which technoscience is operating: (1) open research agendas, (2) open standards, (3) open hardware, (4) open data repositories, and (5) open access to research outputs. Not all of these take place necessarily and simultaneously in every subfield, but technoscience is challenged from inside by combinations of these alternative practices.

161 For example, see an open-source and commons-based technoscientific project initiated by the Ecuadorian government: https://floksociety.org/.

162 Blaser (2012), for example, has developed a similar position that leads, however, to a different conclusion that attempts to change our epistemic perception and to epistemologically accommodate the existence of the world as multiple: "The political sensibility can be described as a commitment to the pluriverse—the partially connected unfolding of worlds—in the face of the impoverishment implied by universalism and potentially by the project of a common world. I say partially connected because the idea here is that these worlds are not sealed off from each other, with clear boundaries—they are certainly connected, yet there is no overarching principle that can be deduced from these connections and that would make this multiplicity a universe" (Blaser, 2012, p. 55).

163 I am thankful to Sebastian Dieterich for sharing with me his work on composting and fermentation. See also B. Turner (2014) and Abrahamsson and Bertoni (2014).

164 This could be considered a general feature of the development of technological systems. See, for example, Misa (2011).

REFERENCES

Abi-Rached, J. M., and Rose, N. (2010). The birth of the neuromolecular gaze. *History of the Human Sciences, 23*(1), 11–36. doi:10.1177/0952695109352407.

Abrahamsson, S., and Bertoni, F. (2014). Compost politics: Experimenting with togetherness in vermicomposting. *Environmental Humanities, 4*, 125–148.

Adam, A. (1998). *Artificial knowing: Gender and the thinking machine.* New York: Routledge.

Adam, B. D. (1995). *The rise of a gay and lesbian movement.* New York: Twayne.

Adam, B. D., and Sears, A. (1996). *Experiencing HIV: Personal, family, and work relationships.* New York: Columbia University Press.

Adams, V. (2002). Randomized controlled crime: Postcolonial sciences in alternative medicine research. *Social Studies of Science, 32*(5–6), 659–690. doi:10.1177/030631270203200503.

Adamson, G. (2013). *The invention of craft.* London: Bloomsbury Academic.

Agamben, G. (1998). *Homo sacer: Sovereign power and bare life.* Stanford, CA: Stanford University Press.

Agamben, G. (2005). *State of exception.* Chicago: University of Chicago Press.

Aigrain, P. (2012). *Sharing: Culture and the economy in the Internet age.* Amsterdam: Amsterdam University Press.

Akyüz, Y. (Ed.) (2003). *Developing countries and world trade: Performance and prospects.* Geneva: United Nations Conference on Trade and Development; London: Zed Books.

Alaimo, S., and Hekman, S. J. (Eds.). (2008). *Material feminisms.* Bloomington: Indiana University Press.

Alberti, G. (2010). Across the borders of Lesvos: The gendering of migrants' detention in the Aegean. *Feminist Review, 94*, 138–147.

Alberti, G. (2011). Transient working lives: Migrant women's everyday politics in London's hospitality industry. PhD diss., Cardiff University.

Alberti, G. (2014). Mobility strategies, "mobility differentials" and "transnational exit": The experiences of precarious migrants in London's hospitality jobs. *Work, Employment and Society, 28*(6), 865–881. doi:10.1177/0950017014528403.

Alberti, G., Holgate, J., and Tapia, M. (2013). Organising migrants as workers or as migrant workers? Intersectionality, trade unions and precarious work. *International Journal of Human Resource Management, 24*(22), 4132–4148. doi:10.108 0/09585192.2013.845429.

Alcoff, L. (Ed.) (1998). *Epistemology: The big questions*. Oxford: Blackwell.

Althusser, L. (2001). Ideology and ideological state apparatuses: Notes towards an investigation. In L. Althusser (Ed.), *Lenin and philosophy, and other essays* (pp. 85–126). London: Monthly Review Press.

Althusser, L. (2006). *Philosophy of the encounter: Later writings, 1978–1987*. London: Verso.

Altieri, M. A., and Toledo, V. M. (2011). The agroecological revolution in Latin America: Rescuing nature, ensuring food sovereignty and empowering peasants. *Journal of Peasant Studies, 38*(3), 587–612. doi:10.1080/03066150.2011.582947.

Altman, D. (1986). *AIDS in the mind of America*. Garden City, NY: Anchor Press/ Doubleday.

Anders, G. (2002). *Die Antiquiertheit des Menschen. Bd.1: Über die Seele im Zeitalter der zweiten industriellen Revolution*. Munich: Beck.

Anderson, B. (2010). Migration, immigration controls and the fashioning of precarious workers. *Work, Employment and Society, 24*(2), 300–317. doi:10.1177/0950017010362141.

Anderson, B., Sharma, N., and Wright, C. (2009). Editorial: Why no borders? *Refuge, 26*(2), 5–18.

Anderson, Chris. (2012). *Makers: The new industrial revolution*. London: Random House Business.

Anderson, Clare. (2012). *Subaltern lives: Biographies of colonialism in the Indian Ocean world, 1790–1920*. Cambridge: Cambridge University Press.

Anderson, W. (2002a). Introduction: Postcolonial technoscience. *Social Studies of Science, 32*(5–6), 643–658. doi:10.1177/030631270203200502.

Anderson, W. (2002b). Postcolonial technoscience. *Social Studies of Science, 32*(5–6), 643–658.

Anderson, W., and Adams, V. (2007). Pramoedya's chickens: Postcolonial studies of technoscience. In E. Hackett, O. Amsterdamska, M. Lynch, and J. Wajcman (Eds.), *The handbook of science and technology studies* (3rd ed., pp. 181–204). Cambridge, MA: MIT Press.

Ansell-Pearson, K. (1999). *Germinal life: The difference and repetition of Deleuze*. London: Routledge.

Anzaldúa, G. (1987). *Borderlands/la frontera: The new mestiza*. San Francisco: Aunt Lute.

Arditi, B. (2008). Arguments about the left turns in Latin America: A post-liberal politics? *Latin American Research Review, 43*(3), 59–81.

Armano, E., and Murgia, A. (2012a). *Mappe della precarietà, vol. 1: Spazi, rappresentazioni, esperienze e critica delle politiche del lavoro che cambia*. Bologna: Emil di Odoya.

Armano, E., and Murgia, A. (2012b). *Mappe della precarietà*, vol. 2: *Knowledge workers, creatività, saperi e dispositivi di soggettivazione*. Bologna: Emil di Odoya.

Arrighi, G. (2009). In retrospect. *New Left Review, 56*, 61–96.

Arrighi, G., and Silver, B. J. (1999). *Chaos and governance in the modern world system*. Minneapolis: University of Minnesota Press.

Ashbrook, K., and Hodgson, N. (2013). *Finding common ground*. Henley-on-Thames, UK: Open Spaces Society.

Avineri, S. (1972). *Hegel's theory of the modern state*. Cambridge: Cambridge University Press.

Azais, C., Corsani, A., and Dieuaide, P. (Eds.). (2001). *Vers un capitalisme cognitif: Entre mutations du travail et territoires*. Paris: Harmattan.

Badiou, A. (2005a). *Being and event*. London: Continuum.

Badiou, A. (2005b). *Metapolitics*. London: Verso.

Badmington, N. (2004). *Alien chic: Posthumanism and the other within*. London: Routledge.

Bailyn, B. (1988). *The peopling of British North America: An introduction*. New York: Vintage.

Baker, C., and American Association for the Advancement of Science. (1997). *Your genes, your choices: Exploring the issues raised by genetic research*. Washington, DC: American Association for the Advancement of Science.

Bakhtin, M. M. (1984). *Rabelais and his world*. Bloomington: Indiana University Press.

Baldwin, J. M. (1897). *Social and ethical interpretations in mental development: A study in social psychology*. New York: Macmillan.

Balibar, E. (2009). Violence. *Historical Materialism, 17*(1), 99–125.

Balsamo, A. (1995). Forms of technological embodiment: Reading the body in contemporary culture. In M. Featherstone and R. Burrows (Eds.), *Cyberspace/cyberbodies/cyberpunk: Cultures of technological embodiment* (pp. 215–237). London: Sage.

Bandyopadhyay, D., and Medrano, E. E. (2003). The emerging role of epigenetics in cellular and organismal aging. *Experimental Gerontology, 38*(11–12), 1299–1307.

Banks, M. (2010). Craft labour and creative industries. *International Journal of Cultural Policy, 16*(3), 305–321. doi:10.1080/10286630903055885.

Banks, M. (2014a). Analysing images. In U. Flick (Ed.), *The Sage handbook of qualitative data analysis* (pp. 394–408). London: Sage.

Banks, M. (2014b). Slow research: Exploring one's own visual archive. *Cadernos de Arte e Antropologia, 3*(2), 57–67.

Barad, K. (2001). Scientific literacy—>agential literacy=(learning+doing) science responsibly. In M. Mayberry, B. Subramaniam, and L. Weasel (Eds.), *Feminist science studies: A new generation* (pp. 226–246). New York: Routledge.

Barad, K. (2007). *Meeting the universe halfway: Quantum physics and the entanglement of matter and meaning*. Durham, NC: Duke University Press.

Baram, U. (2012). Cosmopolitan meanings of old Spanish fields: Historical archaeology of a maroon community in southwest Florida. *Historical Archaeology, 46*(1), 108–122.

Barbagallo, C., and Federici, S. (2012). "Care work" and the commons [Special issue]. *Commoner, 15,* 1–431.

Barbier, J.-C., and Hawkins, P. (Eds.). (2012). *Evaluation cultures: Sense-making in complex times.* New Brunswick, NJ: Transaction.

Barlow, M. (2008). *Blue covenant: The global water crisis and the coming battle for the right to water.* New York: New Press.

Barlow, M. (2012). *Our water commons: Toward a new freshwater narrative.* Ottawa: Council of Canadians.

Barnett, T., and Whiteside, A. (2006). *AIDS in the twenty-first century: Disease and globalization* (2nd ed.). Basingstoke, UK: Palgrave Macmillan.

Barr, M. S. (1992). *Feminist fabulation: Space/postmodern fiction.* Iowa City: University of Iowa Press.

Barry, A., Osborne, T., and Rose, N. (1996). *Foucault and political reason: Liberalism, neo-liberalism, and rationalities of government.* Chicago: University of Chicago Press.

Barthold, C., Dunne, S., and Harvie, D. (2017). Resisting financialisation with Deleuze and Guattari: The case of Occupy Wall Street. *Critical Perspectives on Accounting.* doi:10.1016/j.cpa.2017.03.010.

Bauchspies, W., and Puig de la Bellacasa, M. (2009). Re-tooling subjectivities, exploring the possible with feminist science and technology studies [Special issue]. *Subjectivity, 28*(1), 227–361.

Bauchspies, W. K. (2014). Presence from absence: Looking within the triad of science, technology and development. *Social Epistemology, 28*(1), 56–69. doi:10.1080/02691728.2013.862877.

Bayat, A. (2010). *Life as politics. How ordinary people change the Middle East.* Amsterdam: Amsterdam University Press.

Bayer, B. M., and Malone, K. R. (1996). Feminism, psychology and matters of the body. *Theory and Psychology, 6*(4), 667–692. doi:10.1177/0959354396064007.

Beaulieu, A. (2003). Brains, maps and the new territory of psychology. *Theory and Psychology, 13*(4), 561–568. doi:10.1177/09593543030134006.

Beckert, J. (2009). The social order of markets. *Theory and Society, 38*(3), 245–269. doi:10.2307/40587527.

Beckert, J., and Aspers, P. (Eds.). (2011). *The worth of goods: Valuation and pricing in the economy.* Oxford: Oxford University Press.

Beech, M. (2009). *Terraforming: The creating of habitable worlds.* New York: Springer.

Beier, A. L. (1985). *Masterless men: The vagrancy problem in England 1560–1640.* London: Methuen.

Bell, D., and Binnie, J. (2004). Authenticating queer space: Citizenship, urbanism and governance. *Urban Studies, 41,* 1807–1820.

Bell, D., and Parker, M. (Eds.). (2009). *Space travel and culture: From Apollo to space tourism.* Malden, MA: Wiley-Blackwell.

Bellamy, R., and Lezaun, J. (2015). Crafting a public for geoengineering. *Public Understanding of Science, 26*(4), doi:10.1177/0963662515600965.

Bencze, L., and Alsop, S. (Eds.). (2014). *Activist science and technology education.* New York: Springer.

Benjamin, W. (1996a). Critique of violence. In W. Benjamin (Ed.), *Selected writings,* vol. 1: *1913–1926* (pp. 236–252). Cambridge, MA: Harvard University Press.

Benjamin, W. (1996b). On Language as such and on the language of man. In W. Benjamin (Ed.), *Selected writings,* vol. 1: *1913–1926* (pp. 62–74). Cambridge, MA: Harvard University Press.

Bennett, J. (2010). *Vibrant matter: A political ecology of things.* Durham, NC: Duke University Press.

Berardi, F. B. (2009). *Precarious rhapsody: Semiocapitalism and the pathologies of post-alpha generation.* London: Minor Compositions.

Berardi, F. B. (2010). Precariousness, catastrophe and challenging the blackmail of the imagination. *Affinities: A Journal of Radical Theory, Culture, and Action, 4*(2), 1–4.

Berardi, F. B. (2011). Cognitarian subjectivation. In J. Aranda, B. K. Wood, and A. Vidokle (Eds.), *Are you working too much? Post-Fordism, precarity, and the labor of art* (pp. 134–146). Berlin: Sternberg.

Berkowitz, R., Callen, M., and Sonnabend, J. (1983). *How to have sex in an epidemic: One approach.* New York: News from the Front.

Berman, E. P. (2013). Not just neoliberalism: Economization in US science and technology policy. *Science, Technology, and Human Values, 39*(3), 397–431. doi:10.1177/0162243913509123.

Bersani, L. (1987). Is the rectum a grave? *October, 43,* 197–222.

Beuret, N. (2016). Organizing against the end of the world: The praxis of ecological catastrophe. PhD diss., University of Leicester.

Beuret, N. (2018). Counting carbon: Calculative activism and slippery infrastructure. *Antipode, 49*(5).

Beuret, N., and Brown, G. (2016). *The Walking Dead:* The Anthropocene as a ruined Earth. *Science as Culture,* 1–25. doi:10.1080/09505431.2016.1257600.

Beverungen, A., Böhm, S., and Land, C. (2014). Free labour, social media, management: Challenging Marxist organization studies. *Organization Studies, 36*(4), 473–489. doi:10.1177/0170840614561568.

Beverungen, A., Dunne, S., and Hoedemaekers, C. (2013). The financialisation of business ethics. *Business Ethics: A European Review, 22*(1), 102–117. doi:10.1111/beer.12011.

Bevir, M., and Rhodes, R. A. W. (2003). *Interpreting British governance.* London: Routledge.

Beyes, T., and Steyaert, C. (2011). Spacing organization: Non-representational theory and performing organizational space. *Organization, 19*(1), 45–61. doi:10.1177/1350508411401946.

Biehl, J. (2009). *Will to live: AIDS therapies and the politics of survival.* Princeton, NJ: Princeton University Press.

Bilby, K. M. (2005). *True-born maroons*. Gainesville: University Press of Florida.

Birch, K. (2013). The political economy of technoscience: An emerging research agenda. *Spontaneous Generations: A Journal for the History and Philosophy of Science, 7*(1). doi:10.4245/sponge.v7i1.19556.

Birch, K. (2017a). Rethinking value in the bio-economy: Finance, assetization, and the management of value. *Science, Technology, and Human Values, 42*(3), 460–490. doi:10.1177/0162243916661633.

Birch, K. (2017b). Technoscience rent: Towards a theory of rentiership. Unpublished manuscript.

Bishop, H. (2011). The politics of care and transnational mobility. PhD diss., Cardiff University.

Blackburn, R. (2008). The subprime crisis. *New Left Review, 50*, 63–108.

Blackman, L., Cromby, J., Hook, D., Papadopoulos, D., and Walkerdine, V. (2008). Creating subjectivities. *Subjectivity, 22*(1), 1–27.

Blanche, M. T., Bhavnani, K.-K., and Hook, D. (Eds.). (1999). *Body politics: Power, knowledge and the body in social sciences*. Johannesburg: Histories of the Present Press.

Blaser, M. (2012). Ontology and indigeneity: On the political ontology of heterogeneous assemblages. *Cultural Geographies, 21*(1), 49–58. doi:10.1177/1474474012462534.

Blaser, M., de Costa, R., McGregor, D., and Coleman, W. D. (Eds.). (2010). *Indigenous peoples and autonomy: Insights for a global age*. Vancouver: UBC Press.

Blomley, N. (2008). Enclosure, common right and the property of the poor. *Social and Legal Studies, 17*(3), 311–331. doi:10.1177/0964663908093966.

Bogost, I. (2012). *Alien phenomenology, or, What it's like to be a thing*. Minneapolis: University of Minnesota Press.

Bogue, R. (2007). *Deleuze's way: Essays in transverse ethics and aesthetics*. Aldershot, UK: Ashgate.

Bogue, R. (2010). *Deleuzian fabulation and the scars of history*. Edinburgh: Edinburgh University Press.

Böhm, S., Dinerstein, A. C., and Spicer, A. (2010). (Im)possibilities of autonomy: Social movements in and beyond capital, the state and development. *Social Movement Studies, 9*(1), 17–32. doi:10.1080/14742830903442485.

Böhm, S., and Land, C. (2012). The new "hidden abode": Reflections on value and labour in the new economy. *Sociological Review, 60*(2), 217–240.

Bojadzijev, M., Karakayali, S., and Tsianos, V. (2004). Le mystère de l'arrivée: Des camps et des spectres. *Multitudes, 19*, 41–52.

Bollier, D. (2003). *Silent theft: The private plunder of our common wealth*. London: Routledge.

Bollier, D. (2008). *Viral spiral: How the commoners built a digital republic of their own*. New York: New Press.

Bollier, D., and Helfrich, S. (Eds.). (2012). *The wealth of the commons: A world beyond market and state*. Amherst, MA: Levellers.

Bonneuil, C., and Levidow, L. (2012). How does the World Trade Organization know? The mobilization and staging of scientific expertise in the GMO trade dispute. *Social Studies of Science, 42*(1), 75–100. doi:10.1177/0306312711430151.

Bonta, M., and Protevi, J. (2004). *Deleuze and geophilosophy: A guide and glossary.* Edinburgh: Edinburgh University Press.

Bordo, S. (1990). Reading the slender body. In M. Jacobus, E. F. Keller, and S. Shuttleworth (Eds.), *Body/politics: Women and the discourses of science* (pp. 83–112). New York: Routledge.

Boscagli, M. (2014). *Stuff theory: Everyday objects, radical materialism.* London: Bloomsbury.

Bosniak, L. (2006). *The citizen and the alien: Dilemmas of contemporary membership.* Princeton, NJ: Princeton University Press.

Bould, M., and Miéville, C. (Eds.). (2009). *Red planets: Marxism and science fiction.* London: Pluto.

Bourdieu, P. (1987). *Sozialer Sinn: Kritik der theoretischen Vernunft.* Frankfurt am Main: Suhrkamp.

Bowker, G. C., and Star, S. L. (1999). *Sorting things out: Classification and its consequences.* Cambridge, MA: MIT Press.

Bowring, F. (2002). Post-Fordism and the end of work. *Futures, 34*(2), 159–172.

Bowring, F. (2004). From the mass worker to the multitude: A theoretical contextualisation of Hardt and Negri's "Empire." *Capital and Class, 28*(2), 101–132.

Boyle, J. (2008). *The public domain: Enclosing the commons of the mind.* New Haven, CT: Yale University Press.

Boym, C. (2002). *Curious Boym: Design works.* New York: Princeton Architectural Press.

Bracke, S., and Puig de la Bellacasa, M. (2008). Sister, will you join my act? (Women's) silenced voices and (feminist) standpoints. In R. Buikema and I. van der Tuin (Eds.), *Gender studies and culture.* London: Zed Books.

Braidotti, R. (2002). *Metamorphoses: Towards a materialist theory of becoming.* Cambridge: Polity.

Braidotti, R. (2006). *Transpositions: On nomadic ethics.* Cambridge: Polity.

Braidotti, R. (2013). *The posthuman.* Cambridge: Polity.

Brandt, M. (2014). Zapatista corn: A case study in biocultural innovation. *Social Studies of Science, 44*(6), 874–900.

Brass, T., and van der Linden, M. (Eds.). (1997). *Free and unfree labour: The debate continues.* Bern: Peter Lang.

Bratich, J. (2010). The digital touch: Craft-work as immaterial labour and ontological accumulation. *Ephemera: Theory and Politics in Organization, 10*(3/4), 303–318.

Bratich, J., and Brush, H. (2011). Fabricating activism: Craft-work, popular culture, gender. *Utopian Studies, 22*(2), 233–260.

Bratton, B. (2014). The black stack. *e-flux, 53.* http://www.e-flux.com/journal/53/59883/the-black-stack/.

Braun, B. (2008). Environmental issues: Inventive life. *Progress in Human Geography, 32*(5), 667–679. doi:10.1177/0309132507088030.

Braun, B., and Whatmore, S. (Eds.). (2010). *Political matter: Technoscience, democracy, and public life*. Minneapolis: University of Minnesota Press.

Breckenridge, C. A., and Vogler, C. (2001). The critical limits of embodiment: Disability's criticism. *Public Culture, 13*(3), 349–357.

Brenner, R. (2000). The boom and the bubble. *New Left Review, 6*, 5–43.

Brenner, R. (2004). New boom or new bubble? The trajectory of the US economy. *New Left Review, 25*, 57–100.

Brenner, R. (2006). *The economics of global turbulence: The advanced capitalist economies from long boom to long downturn, 1945–2005*. London: Verso.

Bresnihan, P. (2013). John Clare and the manifold commons. *Environmental Humanities, 3*, 71–91.

Bresnihan, P. (2016a). *Transforming the fisheries: Neoliberalism, nature, and the commons*. Lincoln: University of Nebraska Press.

Bresnihan, P. (2016b). What does it mean to win? Everyday water politics and the struggle for alternatives in Cochabamba, Bolivia. In L. Cortesi and K. L. Joy (Eds.), *Spilt waters: Examining conflicts related to water and their narration*. India: Forum for Policy Dialogue on Water Conflicts.

Bresnihan, P., and Byrne, M. (2015). Escape into the city: Everyday practices of commoning and the production of urban space in Dublin. *Antipode, 47*(1), 36–54. doi:10.1111/anti.12105.

Broeders, D., and Engbersen, G. (2007). The fight against illegal migration: Identification policies and immigrants' counterstrategies. *American Behavioral Scientist, 50*(12), 1592–1609.

Brooks, R. A. (1991). Intelligence without representation. *Artificial Intelligence, 47*(1–3), 139–159.

Brooks, R. A. (2001). Künstliche Intelligenz und Roboter-Entwicklung. In A. Münkel and Sonderausstellung Computer.Gehirn (Eds.), *Computer.Gehirn: Was kann der Mensch? Was können die Computer? Begleitpublikation zur Sonderausstellung im Heinz-Nixdorf-MuseumsForum* (pp. 14–37). Paderborn, Germany: Schöningh.

Brooks, R. A. (2002). *Flesh and machines: How robots will change us*. New York: Pantheon.

Brophy, E., and de Peuter, G. (2007). Immaterial labour, precarity, and recomposition. In C. McKercher and V. Mosco (Eds.), *Knowledge workers in the information society* (pp. 177–191). Lanham, MD: Lexington Books.

Brown, G. (2007). Mutinous eruptions: Autonomous spaces of radical queer activism. *Environment and Planning A, 39*(11), 2685–2698.

Brown, P. (2007). *Toxic exposures: Contested illnesses and the environmental health movement*. New York: Columbia University Press.

Brown, P., McCormick, S., Mayer, B., Zavestoski, S., Morello-Frosch, R., Altman, R. G., and Senier, L. (2006). "A lab of our own": Environmental causation of breast cancer and challenges to the dominant epidemiological paradigm. *Science, Technology, and Human Values, 31*(5), 499–536.

Brown, S. D. (2012). Memory and mathesis: For a topological approach to psychology. *Theory, Culture and Society, 29*(4–5), 137–164. doi:10.1177 /0263276412448830.

Brown, S. D. (2013). In praise of the parasite: The dark organizational theory of Michel Serres. *Informarica na educação: Teoria and prática, 16*(1), 83–100.

Brown, S. D., and Stenner, P. (2009). *Psychology without foundations: History, philosophy and psychosocial theory.* London: Sage.

Bruner, J. (1967). Preface. *Soviet Psychology, 5*(3), 3–5.

Bruner, J., Goodnow, J., and Austin, G. A. (1956). *A study of thinking.* New York: Wiley.

Bryan, D., Martin, R., Montgomerie, J., and Williams, K. (2012). An important failure: Knowledge limits and the financial crisis. *Economy and Society, 41*(3), 299–315. doi:10.1080/03085147.2012.661632.

Bryan, D., and Rafferty, M. (2006). *Capitalism with derivatives: A political economy of financial derivatives, capital and class.* New York: Palgrave Macmillan.

Bryant, L., Srnicek, N., and Harman, G. (Eds.). (2011). *The speculative turn: Continental materialism and realism.* Melbourne: re.press.

Buckel, S., Fischer-Lescano, A., and Oberndorfer, L. (2010). Postneoliberale Rechtsordnung? Suchprozesse in der Krise: Einleitung in den Schwerpunkt. *Kritische Justiz, 43*(4), 375–383.

Burawoy, M. (1985). *The politics of production: Factory regimes under capitalism and socialism.* London: Verso.

Burawoy, M., Blum, J. A., George, S., Gille, Z., Gowan, T., Haney, L., et al. (Eds.). (2000). *Global ethnography: Forces, connections, and imaginations in a postmodern world.* Berkeley: University of California Press.

Burchell, G., Gordon, C., and Miller, P. (Eds.). (1991). *The Foucault effect: Studies in governmentality.* Chicago: University of Chicago Press.

Burston, J., Dyer-Witheford, N., and Hearn, A. (2010). Digital labour: Workers, authors, citizens. [Special issue]. *Ephemera: Theory and Politics in Organization, 10*(3/4), 214–221.

Busch, L. (2011). *Standards: Recipes for reality.* Cambridge, MA: MIT Press.

Busch, L. (2017). *Knowledge for sale: The neoliberal takeover of higher education.* Cambridge, MA: MIT Press.

Buszek, M. E., and Robertson, K. (2011). Introduction (on craftivism). *Utopian Studies, 22*(2), 197–200.

Butler, J. (1997). Against proper objects. In E. Weed and N. Schor (Eds.), *Feminism meets queer theory* (pp. 1–30). Bloomington: Indiana University Press.

Butler, J., Laclau, E., and Žižek, S. (2000). *Contingency, hegemony, universality: Contemporary dialogues on the left.* London: Verso.

Butler, O. E. (1987). *Xenogenesis.* London: V. Gollancz.

Byrne, M. (2016a). "Asset price urbanism" and financialization after the crisis: Ireland's National Asset Management Agency. *International Journal of Urban and Regional Research, 40*(1), 31–45. doi:10.1111/1468-2427.12331.

Byrne, M. (2016b). Entrepreneurial urbanism after the crisis: Ireland's "bad bank" and the redevelopment of Dublin's docklands. *Antipode, 48*(4), 899–918. doi:10.1111/anti.12231.

Caffentzis, G. (2010). The future of "the commons": Neoliberalism's "Plan B" or the original disaccumulation of capital? *New Formations, 69*(1), 23–41. doi:10.3898/newf.69.01.2010.

Calladine, A. (1993). Lombe's Mill: An exercise in reconstruction. *Industrial Archaeology Review, 16*(1), 82–99.

Calleja López, A. (2017). Since 15M: The technopolitical reassembling of democracy in Spain. PhD diss., University of Exeter.

Callon, M. (1987). Society in the making: The study of technology as a tool for sociological analysis. In W. E. Bijker, T. P. Hughes, and T. J. Pinch (Eds.), *The social construction of technological systems: New directions in the sociology and history of technology* (pp. 83–103). Cambridge, MA: MIT Press.

Callon, M. (Ed.) (1998). *The laws of the markets.* Oxford: Blackwell.

Callon, M., Lascoumes, P., and Barthe, Y. (2009). *Acting in an uncertain world: An essay on technical democracy.* Cambridge, MA: MIT Press.

Callon, M., Millo, Y., and Muniesa, F. (Eds.). (2007). *Market devices.* Oxford: Blackwell.

Callon, M., and Rabeharisoa, V. (2004). Gino's lesson on humanity: Genetics, mutual entanglements and the sociologist's role. *Economy and Society, 33*(1), 1–27.

Callon, M., and Rabeharisoa, V. (2008). The growing engagement of emergent concerned groups in political and economic life: Lessons from the French association of neuromuscular disease patients. *Science, Technology, and Human Values, 33*(2), 230–261.

Calvanese, V., Lara, E., Kahn, A., and Fraga, M. F. (2009). The role of epigenetics in aging and age-related diseases. *Ageing Research Reviews, 8*(4), 268–276.

Cameron, E., de Leeuw, S., and Desbiens, C. (2014). Indigeneity and ontology [Special issue]. *Cultural Geographies, 21*(1), 19–168. doi:10.1177/1474474013500229.

Carlile, P. R., Nicolini, D., Langley, A., and Tsoukas, H. (Eds.). (2013). *How matter matters: Objects, artifacts, and materiality in organization studies.* Oxford: Oxford University Press.

Carlson, R. H. (2010). *Biology is technology: The promise, peril, and new business of engineering life.* Cambridge, MA: Harvard University Press.

Carroll, P. (2006). *Science, culture, and modern state formation.* Berkeley: University of California Press.

Carroll, P. (2012). Water and technoscientific state formation in California. *Social Studies of Science, 42*(4), 489–516. doi:10.1177/0306312712437977.

Carse, A. (2012). Nature as infrastructure: Making and managing the Panama Canal watershed. *Social Studies of Science, 42*(4), 539–563. doi:10.1177/0306312712440166.

Carter, P., and Jackson, N. (2005). Laziness. In C. Jones and D. O'Doherty (Eds.), *Manifestos for the business school of tomorrow* (pp. 87–93). Åbo, Finland: Dvalin Books.

Casas-Cortes, M., Cobarrubias, S., and Pickles, J. (2015). Riding routes and itinerant borders: Autonomy of migration and border externalization. *Antipode, 47*(4), 894–914. doi:10.1111/anti.12148.

Castellano, K. (in press). Moles, molehills, and common right in John Clare's poetry. *Studies in Romanticism.*

Castells, M. (1997). *The power of identity.* Malden, MA: Blackwell.

Castles, S., and Miller, M. J. (2003). *The age of migration: International population movements in the modern world.* Basingstoke, UK: Palgrave Macmillan.

Caygill, H. (2013). *On resistance: A philosophy of defiance.* London: Bloomsbury.

Chakrabarty, D. (2000). *Provincializing Europe: Postcolonial thought and historical difference.* Princeton, NJ: Princeton University Press.

Chakrabarty, D. (2009). The climate of history: Four theses. *Critical Inquiry, 35*(2), 197–222.

Chakravarty-Kaul, M. (1996). *Common lands and customary law: Institutional change in north India over the past two centuries.* Delhi: Oxford University Press.

Chambré, S. M. (2006). *Fighting for our lives: New York's AIDS community and the politics of disease.* New Brunswick, NJ: Rutgers University Press.

Champagne, F. A. (2010). Epigenetic influence of social experiences across the lifespan. *Developmental Psychobiology, 52*(4), 299–311.

Changeux, J.-P. (1997). *Neuronal man: The biology of mind.* Princeton, NJ: Princeton University Press.

Chatterjee, P. (2004). *The politics of the governed: Reflections on popular politics in most of the world.* New York: Columbia University Press.

Chatterton, P., and Pickerill, J. (2010). Everyday activism and transitions towards post-capitalist worlds. *Transactions of the Institute of British Geographers, 35*(4), 475–490. doi:10.1111/j.1475-5661.2010.00396.x.

Chatzidakis, A., Maclaran, P., and Bradshaw, A. (2012). Heterotopian space and the utopics of ethical and green consumption. *Journal of Marketing Management, 28*(3–4), 494–515. doi:10.1080/0267257x.2012.668922.

Checchi, M. (2015). The primacy of resistance: A historical and conceptual exploration. PhD diss., University of Leicester.

Chesters, G., and Welsh, I. (2005). Complexity and social movement(s): Process and emergence in planetary action systems. *Theory, Culture and Society, 22*(5), 187–211. doi:10.1177/0263276405057047.

Chesters, G., and Welsh, I. (2006). *Complexity and social movements: Multitudes at the edge of chaos.* London: Routledge.

Chilvers, J. (2008). Deliberating competence: Theoretical and practitioner perspectives on effective participatory appraisal practice. *Science, Technology, and Human Values, 33*(2), 155–185.

Choy, T. K. (2011). *Ecologies of comparison: An ethnography of endangerment in Hong Kong.* Durham, NC: Duke University Press.

Choy, T. K., Faier, L., Hathaway, M. J., Inoue, M., Satsuka, S., and Tsing, A. (2009). A new form of collaboration in cultural anthropology: Matsutake worlds. *American Ethnologist, 36*(2), 380–403. doi:10.1111/j.1548-1425.2009.01141.x.

Chrisley, R., and Ziemke, T. (2002). Embodiment. In L. Nadel (Ed.), *Encyclopedia of cognitive science* (pp. 1102–1108). London: Macmillan.

Chung, E., Cromby, J., Papadopoulos, D., and Tufarelli, C. (2016). Social epigenetics: A science of social science? *Sociological Review Monographs, 64*(1), 168–185. doi:10.1111/2059-7932.12019.

Churchland, P. S. (1986). *Neurophilosophy: Toward a unified science of the mind-brain.* Cambridge, MA: MIT Press.

Claeson, B., Martin, E., Richardson, W., Schoch-Spana, M., and Taussig, K.-S. (1996). Scientific literacy: What it is, why it is important, and why scientists think we don't have it: The case of immunology and the immune system. In L. Nader (Ed.), *Naked science: Anthropological inquiry into boundaries, power, and knowledge* (pp. 101–116). New York: Routledge.

Clark, A. (1997). *Being there: Putting brain, body, and world together again.* Cambridge, MA: MIT Press.

Clark, A. (1998). Where brain, body and world collide. *Daedalus, 127*(2), 257–280.

Clark, N. (2011). *Inhuman nature: Sociable life on a dynamic planet.* London: Sage.

Clarke, A. E. (2016). Anticipation work: Abduction, simplification, hope. In G. C. Bowker, S. Timmermans, A. E. Clarke, and E. Balka (Eds.), *Boundary objects and beyond: Working with Leigh Star.* Cambridge, MA: MIT Press.

Clarke, A. E., Mamo, L., Fosket, J. R., Fishman, J. R., and Shim, J. K. (Eds.). (2010). *Biomedicalization: Technoscience, health, and illness in the U.S.* Durham, NC: Duke University Press.

Clarke, A. E., and Olesen, V. L. (Eds.). (1999). *Revisioning women, health, and healing: Feminist, cultural, and technoscience perspectives.* New York: Routledge.

Clarke, B. (2008). *Posthuman metamorphosis: Narrative and systems.* New York: Fordham University Press.

Cleaver, H. (1992). The inversion of class perspective in Marxian theory: From valorisation to self-valorisation. In W. Bonefeld, R. Gunn, and K. Psychopedis (Eds.), *Open Marxism,* vol. 2: *Theory and practice* (pp. 106–144). London: Pluto Press.

Clifford, J. (1986). Partial truths. In J. Clifford and G. E. Marcus (Eds.), *Writing culture: The poetics and politics of ethnography* (pp. 1–27). Berkeley: University of California Press.

Clifford, J. (1989). Notes on theory and travel. In J. Clifford and V. Dhareshwa (Eds.), *Inscriptions,* vol. 5: *Traveling theories: Traveling theorists* (pp. 177–185). Santa Cruz, CA: Center for Cultural Studies.

Clifford, J. (1997). *Routes: Travel and translation in the late twentieth century.* Cambridge, MA: Harvard University Press.

Clifford, J. (2000). Taking identity politics seriously: "The contradictory, stony ground. . . ." In P. Gilroy, L. Grossberg, and A. McRobbie (Eds.), *Without guarantees: In honour of Stuart Hall* (pp. 94–112). London: Verso.

Clifford, J. (2001). Indigenous articulations. *Contemporary Pacific, 13*(2), 468–490.

Clifford, J. (2004). Looking several ways: Anthropology and native heritage in Alaska. *Current Anthropology, 45*(1), 5–30.

Clifford, J. (2009). Hau'ofa's Hope: Association for Social Anthropology in Oceania 2009 Distinguished Lecture. *Oceania, 79*(3), 238–249.

Clifford, J. (2013). *Returns: Becoming indigenous in the twenty-first century.* Cambridge, MA: Harvard University Press.

Clifford, J., and Marcus, G. E. (Eds.). (1986). *Writing culture: The poetics and politics of ethnography.* Berkeley: University of California Press.

Coburn, E. (Ed.) (2015). *More will sing their way to freedom: Indigenous resistance and resurgence.* Halifax: Fernwood.

Coe, N. M., Hess, M., Yeung, H. W.-C., Dicken, P., and Henderson, J. (2004). "Globalizing" regional development: A global production networks perspective. *Transactions of the Institute of British Geographers, 29*(4), 468–484.

Cohen, J. (2001). *Shots in the dark: The wayward search for an AIDS vaccine.* New York: Norton.

Colau, A., and Alemany, A. (2014). *Mortgaged lives: From the housing bubble to the right to housing.* Los Angeles: Journal of Aesthetics and Protest Press.

Colectivo Situaciones. (2007). Politicising sadness. *Turbulence, 1.* http://turbulence .org.uk/turbulence-1/politicising-sadness/

Coleman, E. G. (2013). *Coding freedom: The ethics and aesthetics of hacking.* Princeton, NJ: Princeton University Press.

Coleman, E. G., and Golub, A. (2008). Hacker practice: Moral genres and the cultural articulation of liberalism. *Anthropological Theory, 8*(3), 255–277. doi:10.1177/1463499608093814.

Collard, R.-C., Dempsey, J., and Sundberg, J. (2014). A manifesto for abundant futures. *Annals of the Association of American Geographers, 105*(2), 322–330. doi:10.1080/00045608.2014.973007.

Collins, H. M. (2009). We cannot live by scepticism alone. *Nature, 458*(7234), 30–31.

Collins, H. M., and Evans, R. (2007). *Rethinking expertise.* Chicago: University of Chicago Press.

Collins, H. M., Weinel, M., and Evans, R. (2010). The politics of the third wave and elective modernism. *Critical Policy Studies, 4*(2), 185–201.

Collis, C., and Graham, P. (2009). Political geographies of Mars: A history of Martian management. *Management and Organizational History, 4*(3), 247–261. doi:10.1177/1744935909337750.

Colombetti, G. (2014). *The feeling body: Affective science meets the enactive mind.* Cambridge, MA: MIT Press.

Conference Board. (2009). Performance 2009. Productivity, employment, and growth in the world's economies. https://www.conference-board.org/pdf_free /Productivity2009.pdf.

Connery, C. (2006). The Asian sixties: An unfinished project. *Inter-Asia Cultural Studies, 7*(4), 545–553. doi:10.1080/14649370600982826.

Connery, C. (2007). The world sixties. In R. Wilson and C. Connery (Eds.), *The Worlding Project: Doing cultural studies in the era of globalization* (pp. 77–108). Santa Cruz, CA: New Pacific.

Conway, S., and Steward, F. (2009). *Managing and shaping innovation*. Oxford: Oxford University Press.

Coole, D. H., and Frost, S. (Eds.). (2010). *New materialisms: Ontology, agency, and politics*. Durham, NC: Duke University Press.

Cooper, M. (2008). *Life as surplus: Biotechnology and capitalism in the neoliberal era*. Seattle: University of Washington Press.

Cosse, E. (2008). The precarious go marching. *In the Middle of a Whirlwind*. https://inthemiddleofthewhirlwind.wordpress.com/2008/10/03/the-precarious-go-marching/.

Crane, J. (2010). Adverse events and placebo effects: African scientists, HIV, and ethics in the "global health sciences." *Social Studies of Science, 40*(6), 843–870. doi:10.1177/0306312710371145.

Crimp, D. (1987). AIDS: Cultural analysis/cultural activism. *October, 43*, 3–16.

Crimp, D. (1988). How to have promiscuity in an epidemic. In D. Crimp and L. Bersani (Eds.), *AIDS: Cultural analysis, cultural activism* (pp. 237–271). Cambridge, MA: MIT Press.

Crimp, D. (2002). *Melancholia and moralism: Essays on AIDS and queer politics*. Cambridge, MA: MIT Press.

Cromby, J. (2004). Between constructionism and neuroscience: The societal co-constitution of embodied subjectivity. *Theory and Psychology, 14*(6), 797–821.

Cromby, J. (2007). Integrating social science with neuroscience: Potentials and problems. *BioSocieties, 2*(2), 149–169.

Cromby, J. (2015). *Feeling bodies: Embodying psychology*. London: Palgrave Macmillan.

Cromby, J., Chung, E., Papadopoulos, D., and Talbot, C. (2016). Reviewing the epigenetics of schizophrenia. *Journal of Mental Health*, 1–9. doi:10.1080/0963 8237.2016.1207229.

Cromby, J., Newton, T., and Williams, S. (2011). Neuroscience and subjectivity [Special issue]. *Subjectivity, 4*(3), 215–226.

Cronon, W. (1983). *Changes in the land: Indians, colonists, and the ecology of New England*. New York: Hill and Wang.

Csordas, T. J. (1994a). Introduction: The body as representation and being-in-the-world. In T. J. Csordas (Ed.), *Embodiment and experience: The existential ground of culture and self* (pp. 1–24). Cambridge, UK: Cambridge University Press.

Csordas, T. J. (Ed.) (1994b). *Embodiment and experience: The existential ground of culture and self*. Cambridge: Cambridge University Press.

Cuppini, N. (2017). Dissolving Bologna: Tensions between citizenship and the logistics city. *Citizenship Studies*, 1–13. doi:10.1080/13621025.2017.1307608.

Cussins, C. (1996). Ontological choreography: Agency through objectification in infertility clinics. *Social Studies of Science, 26*(3), 575–610.

D'Avella, N. (2014). Ecologies of investment: Crisis histories and brick futures in Argentina. *Cultural Anthropology, 29*(1), 173–199. doi:10.14506/ca29.1.10.

da Costa, B., and Philip, K. (Eds.). (2008). *Tactical biopolitics: Art, activism, and technoscience*. Cambridge, MA: MIT Press.

Dafermos, G., and Söderberg, J. (2009). The hacker movement as a continuation of labour struggle. *Capital and Class, 33*(1), 53–73.

Damasio, A. R. (1999). *The feeling of what happens: Body and emotion in the making of consciousness.* New York: Harcourt Brace.

Damasio, A. R. (2004). *Looking for Spinoza: Joy, sorrow and the feeling brain.* London: Vintage.

Davies, G. (2006). Mapping deliberation: Calculation, articulation and intervention in the politics of organ transplantation. *Economy and Society, 35*(2), 232–258. doi:10.1080/03085140600635722.

De Angelis, M. (2007). *The beginning of history: Value struggles and global capital.* London: Pluto.

De Angelis, M. (2017). Omnia sunt communia: *On the commons and the transformation to postcapitalism.* London: Zed.

De Angelis, M., and Harvie, D. (2009). "Cognitive capitalism" and the rat-race: How capital measures immaterial labour in British universities. *Historical Materialism, 17*(3), 3–30.

De Angelis, M., and Harvie, D. (2013). The commons. In M. Parker, G. Cheney, V. Fournier, and C. Land (Eds.), *The Routledge companion to alternative organisation* (pp. 280–294). London: Routledge.

De Boever, A., and Neidich, W. (Eds.). (2013). *The psychopathologies of cognitive capitalism, part one.* Berlin: Archive Books.

De Cock, C., Fleming, P., and Rehn, A. (2007). Organizing revolution? *Management and Organizational History, 2*(2), 107–114.

De Genova, N. (2010). The deportation regime: Sovereignty, space and the freedom of movement. In N. De Genova and N. M. Peutz (Eds.), *The deportation regime: Sovereignty, space, and the freedom of movement* (pp. 33–65). Durham, NC: Duke University Press.

De Genova, N. (2015). Border struggles in the migrant metropolis. *Nordic Journal of Migration Research, 5*(1), 3–10. doi:10.2478/njmr-2015-0005.

De Genova, N. (2016). The European question. *Social Text, 34*(3 [128]), 75–102. doi:10.1215/01642472-3607588.

De Genova, N., and Peutz, N. M. (Eds.). (2010). *The deportation regime: Sovereignity, space, and the freedom of movement.* Durham, NC: Duke University Press.

de la Cadena, M. (2010). Indigenous cosmopolitics in the Andes: Conceptual reflections beyond "politics." *Cultural Anthropology, 25*(2), 334–370. doi:10.1111/j.1548-1360.2010.01061.x.

de la Cadena, M. (2015). *Earth beings: Ecologies of practice across Andean worlds.* Durham, NC: Duke University Press.

De Landa, M. (1997). *A thousand years of nonlinear history.* New York: Zone Books.

De Landa, M. (2002). *Intensive science and virtual philosophy.* New York: Continuum.

De Lauretis, T. (1987). *Technologies of gender: Essays on theory, film, and fiction.* Bloomington: Indiana University Press.

de Moor, M., Shaw-Taylor, L., and Warde, P. (2002). *The management of common land in North West Europe, c. 1500–1850.* Turnhout, Belgium: Brepols.

de Peuter, G., and Dyer-Witheford, N. (2010). Commons and cooperatives. *Affinities: A Journal of Radical Theory, Culture, and Action, 4*(1), 30–56.

de Rivera, J., Gordo, Á., Cassidy, P., and Apesteguía, A. (2016). A netnographic study of P2P collaborative consumption platforms' user interface and design. *Environmental Innovation and Societal Transitions.* doi:10.1016/j.eist.2016.09.003.

de Saille, S. (2014). Fighting science with social science: Activist scholarship in an international resistance project. *Sociological Research Online, 19*(3). doi:10.5153/sro.3331.

De Smet, B. (2014). *A dialectical pedagogy of revolt: Gramsci, Vygotsky, and the Egyptian revolution.* Boston: Brill.

De Smet, B. (2016). *Gramsci on Tahrir: Revolution and counter-revolution in Egypt.* London: Pluto.

de Sousa Santos, B. (2001). Nuestra America: Reinventing a subaltern paradigm of recognition and redistribution. *Theory, Culture and Society, 18*(2–3), 185–217.

de Sousa Santos, B. (2004). A critique of lazy reason: Against the waste of experience. In I. M. Wallerstein (Ed.), *The modern world-system in the longue durée* (pp. 157–197). Boulder, CO: Paradigm.

Debord, G. (1981). Perspectives for conscious alterations in everyday life. In K. Knabb (Ed.), *Situationist International anthology* (pp. 68–75). Berkeley, CA: Bureau of Public Secrets.

Deleuze, G. (1987). *Dialogues with Claire Parnet.* New York: Columbia University Press.

Deleuze, G. (1993). *The Fold: Leibniz and the Baroque.* London: Athlone.

Deleuze, G. (1994). *Difference and repetition.* London: Athlone.

Deleuze, G. (2001). Dualism, monism and multiplicities (desire-pleasure-jouissance). *Contretemps, 2,* 92–108. http://www.usyd.edu.au/contretemps/contretemps2.html.

Deleuze, G. (2005). Pericles and Verdi: The philosophy of François Châtelet. *Opera Quarterly, 21,* 716–725.

Deleuze, G., and Guattari, F. (1987). *A thousand plateaus: Capitalism and schizophrenia.* Minneapolis: University of Minnesota Press.

Deleuze, G., and Guattari, F. (1994). *What is philosophy?* London: Verso.

Delfanti, A. (2011). Hacking genomes: The ethics of open and rebel biology. *International Review of Information Ethics, 15,* 52–57.

Delfanti, A., and Söderberg, J. (2012). Bio/hardware hacking. *Journal of Peer Production, 12.*

Della Porta, D., and Diani, M. (2006). *Social movements: An introduction* (2nd ed.). Malden, MA: Blackwell.

Della Porta, D., and Rucht, D. (2013). *Meeting democracy: Power and deliberation in global justice movements.* Cambridge: Cambridge University Press.

DeNardis, L. (2012). Hidden levers of Internet control. *Information, Communication and Society, 15*(5), 720–738. doi:10.1080/1369118x.2012.659199.

DeNardis, L. (2014). *The global war for Internet governance.* New Haven, CT: Yale University Press.

Dennett, D. C. (1995). *Darwin's dangerous idea: Evolution and the meanings of life.* New York: Simon and Schuster.

Derrida, J. (1992). Force of law: The mystical foundation of authority. In D. Cornell, M. Rosenfeld, and D. G. Carlson (Eds.), *Deconstruction and the possibility of justice* (pp. 3–67). New York: Routledge.

Derwent Valley Mills Partnership. (2001). *The Derwent Valley Mills and their communities.* Derbyshire, UK: County Hall.

Dewey, J. (1999). *Democracy and education: An introduction to the philosophy of education.* New York: Free Press.

Dinerstein, A. C. (2010). Autonomy in Latin America: Between resistance and integration: Echoes from the Piqueteros experience. *Community Development Journal, 45*(3), 356–366. doi:10.1093/cdj/bsq029.

Dinerstein, A. C. (2014). *The politics of autonomy in Latin America: The art of organising hope.* Basingstoke, UK: Palgrave.

Doidge, N. (2008). *The brain that changes itself: Stories of personal triumph from the frontiers of brain science.* London: Penguin.

Dolphijn, R., and van der Tuin, I. (2015). *New materialism: Interviews and cartographies.* Ann Arbor, MI: Open Humanities Press.

Dolsak, N., and Ostrom, E. (2003). *The commons in the new millennium: Challenges and adaptation.* Cambridge, MA: MIT Press.

Douzinas, C. (2013). *Philosophy and resistance in the crisis: Greece and the future of Europe.* Cambridge: Polity.

Dowling, E., and Harvie, D. (2014). Harnessing the social: State, crisis and (big) society. *Sociology, 48*(5), 869–886. doi:10.1177/0038038514539060.

Doyle, M. (2013). *Beautiful LEGO.* San Francisco: No Starch Press.

Dreessen, K., Schepers, S., and Leen, D. (2016). From hacking things to making things: Rethinking making by supporting non-expert users in a FabLab. *Interaction Design and Architecture(s) Journal (IxD&A), 30,* 47–64.

Duden, B. (2002). *Die Gene im Kopf—der Fötus im Bauch: Historisches zum Frauenkörper.* Hanover: Offizin.

Duménil, G., and Lévy, D. (2004). Neo-liberal dynamics: Toward a new phase? In L. Assassi, D. Wigan, and K. van der Pijl (Eds.), *Global regulation: Managing crises after the imperial turn.* New York: Palgrave Macmillan.

Duménil, G., and Lévy, D. (2005). Costs and benefits of neoliberalism: A class analysis. In G. A. Epstein (Ed.), *Financialization and the world economy* (pp. 17–45). Cheltenham, UK: Edward Elgar.

Dumit, J. (2004). *Picturing personhood: Brain scans and biomedical identity.* Princeton, NJ: Princeton University Press.

Dumit, J. (2012). *Drugs for life: How pharmaceutical companies define our health.* Durham, NC: Duke University Press.

Dupuits, E., and García, M. M. (2016). Water and (neo)extractivism in Latin America [Special issue]. *Alternautas, 3*(2), 3–124.

Düvell, F. (2006). *Europäische und internationale Migration: Einführung in historische, soziologische und politische Analysen.* Hamburg: Lit.

Dyer-Witheford, N. (1999). *Cyber-Marx: Cycles and circuits of struggle in high-technology capitalism.* Urbana: University of Illinois Press.

Dyer-Witheford, N. (2006). Species-being and the new commonism: Notes on an interrupted cycle of struggles. *Commoner, 11,* 15–32.

Dyer-Witheford, N. (2015). *Cyber-proletariat: Global labour in the digital vortex.* London: Pluto.

Dyke, E., and Meyerhoff, E. (2017). Toward an anti- and alter-university: Thriving in the mess of studying, organizing, and relating with Ex Co of the Twin Cities. In R. H. Haworth and J. M. Elmore (Eds.), *Out of the ruins: The emergence of radical informal learning spaces* (pp. 174–194). Oakland, CA: PM Press.

Earl, J., and Kimport, K. (2011). *Digitally enabled social change: Activism in the Internet age.* Cambridge, MA: MIT Press.

Easterling, K. (2005). *Enduring innocence: Global architecture and its political masquerades.* Cambridge, MA: MIT Press.

Easterling, K. (2014). *Extrastatecraft: The power of infrastructure space.* London: Verso.

Eco, U. (1989). *The open work.* Cambridge, MA: Harvard University Press.

Economist. (2012). The third industrial revolution. April 21. http://www.economist.com/node/21553017.

Edelman, G. M. (1989). *Neural Darwinism: The theory of neuronal group selection.* Oxford: Oxford University Press.

Edelman, G. M. (1992). *Bright air, brilliant fire: On the matter of the mind.* New York: Basic Books.

Edelman, G. M., and Tononi, G. (2000). *A universe of consciousness: How matter becomes imagination.* New York: Basic Books.

Edu-Factory Collective. (2010). The double crisis: Living on the borders. *EduFactory WebJournal,* 0 issue (January), 4–9. https://destructural.files.wordpress.com/2010/06/edufactory.pdf.

Edu-Factory Collective (Ed.). (2009). *Toward a global autonomous university.* New York: Autonomedia.

Edwards, A., and Hughes, G. (2005). Comparing the governance of safety in Europe: A geo-historical approach. *Theoretical Criminology, 9,* 345–363.

Edwards, A., and Sheptycki, J. (2009). Third wave criminology: Guns, crime and social order. *Criminology and Criminal Justice, 9*(3), 379–397.

Eglash, R. (2016). An introduction to generative justice. *Teknokultura: Journal of Digital Culture and Social Movements, 13*(2), 369–404. doi:10.5209/rev_TEKN.2016.v13.n2.52847.

Ehrenreich, B. (1989). *Fear of falling: The inner life of the middle class.* New York: Pantheon.

Ehrenstein, A. (2006). Social relationality and affective experience in precarious labour conditions: A study of young immaterial workers in the arts industries in Cardiff. PhD diss., Cardiff University.

Eilperin, J. (2013). How attitudes toward AIDS have changed, in the White House and beyond. *Washington Post*, December 4, 2013.

Elliott, E., Harrop, E., and Williams, G. (2010). Contesting the science: Public health knowledge and action in controversial land-use developments. In P. Bennett and K. Calman (Eds.), *Risk communication and public health* (pp. 181–196). Oxford: Oxford University Press.

Elliott, E., and Williams, G. (2008). Developing public sociology through health impact assessment. *Sociology of Health and Illness*, 30(7), 1101–1116.

Elman, J. L., Bates, E., Johnson, M., Karmiloff-Smith, A., Parisi, D., and Plunkett, K. (1996). *Rethinking innateness: A connectionist perspective on development.* Cambridge, MA: MIT Press.

Engelen, E., Ertuk, I., Froud, J., Johal, S., Leaver, A., Moran, M., et al. (2011). *After the great complacence: Financial crisis and the politics of reform.* Oxford: Oxford University Press.

Engels, F. (1987). Anti-Dühring: Herr Eugen Dühring's revolution in science. In K. Marx and F. Engels (Eds.), *Collected works*, vol. 25. London: Lawrence and Wishart. (Original work published 1878.)

Engels, F. (1990a). Ludwig Feuerbach and the end of classical German philosophy. In K. Marx and F. Engels (Eds.), *Collected works*, vol. 26. London: Lawrence and Wishart. (Original work published 1886.)

Engels, F. (1990b). The origin of the family, private property and the state. In K. Marx and F. Engels (Eds.), *Collected works*, vol. 26. London: Lawrence and Wishart. (Original work published 1886.)

Epstein, S. (1995). The construction of lay expertise: AIDS activism and the forging of credibility in the reform of clinical trials. *Science, Technology, and Human Values*, 20(4), 408–437.

Epstein, S. (1996). *Impure science: AIDS, activism, and the politics of knowledge.* Berkeley: University of California Press.

Epstein, S. (1997). Activism, drug regulation, and the politics of therapeutic evaluation in the AIDS era: A case study of ddC and the "surrogate markers" debate. *Social Studies of Science*, 27(5), 691–726.

Epstein, S. (2007). *Inclusion: The politics of difference in medical research.* Chicago: University of Chicago Press.

Escárcega, S. (2013). The global indigenous movement and paradigm wars: International activism, network building, and transformative politics. In J. S. Juris and A. Khasnabish (Eds.), *Insurgent encounters: Transnational activism, ethnography, and the political* (pp. 129–150). Durham, NC: Duke University Press.

Escobar, A. (2011). Sustainability: Design for the pluriverse. *Development*, 54(2), 137–140. doi:10.1057/dev.2011.28.

Escobar, A. (2012). Notes on the ontology of design: Parts I, II, III. http://sawyerseminar.ucdavis.edu/files/2012/12/ESCOBAR_Notes-on-the-Ontology-of-Design-Parts-I-II-%5f-III.pdf.

Esteva, G., and Marielle, C. (Eds.). (2003). *Sin maiz, no hay pais*. Mexico City: Consejo Nacional para la Cultura y las Artes.

European Commission. (2001). *European governance: A white paper*. http://europa .eu/rapid/press-release_DOC-01-10_en.htm.

Fausto-Sterling, A. (2000). *Sexing the body: Gender politics and the construction of sexuality*. New York: Basic Books.

Federici, S. (2004). *Caliban and the witch*. New York: Autonomedia.

Federici, S., and Caffentzis, G. (2008). Must the molecules fear as the engine dies? Notes on the Wall Street "meltdown." *Radical Perspectives on the Crisis*. http://sites.google.com/site/radicalperspectivesonthecrisis/finance-crisis/on -the-origins-of-the-crisis-beyond-finance/midnight-notes-on-crisis.

Feenberg, A. (1991). *Critical theory of technology*. New York: Oxford University Press.

Feenberg, A. (1995). *Alternative modernity: The technical turn in philosophy and social theory*. Berkeley: University of California Press.

Feigenbaum, A., Frenzel, F., and McCurdy, P. (2013). *Protest camps*. London: Zed.

Felt, U., Fochler, M., Mager, A., and Winkler, P. (2008). Visions and versions of governing biomedicine: Narratives on power structures, decision-making and public participation in the field of biomedical technology in the Austrian context. *Social Studies of Science, 38*, 233–258.

Figueroa, R., and Harding, S. (Eds.). (2003). *Science and other cultures: Diversity in the philosophy of science and technology*. New York: Routledge.

Figueroa Sarriera, H., and Gray, C. H. (2016). Generative justice [Special issue]. *Teknokultura: Journal of Digital Culture and Social Movements, 13*(2), 361–637. doi:10.5209/rev_TK.2016.v13.n1.52388.

Fischer, M. M. J. (2003). *Emergent forms of life and the anthropological voice*. Durham, NC: Duke University Press.

Fischer, M. M. J. (2009). *Anthropological futures*. Durham, NC: Duke University Press.

Fitting, E. (2006). Importing corn, exporting labor: The neoliberal corn regime, GMOs, and the erosion of Mexican biodiversity. *Agriculture and Human Values, 23*(1), 15–26. doi:10.1007/s10460-004-5862-y.

Fitzgerald, D., Rose, N., and Singh, I. (2016). Living well in the *Neuropolis*. *Sociological Review Monographs, 64*(1), 221–237. doi:10.1111/2059-7932.12022.

Flanagan, T. (2009). The thought of history in Benjamin and Deleuze. In J. A. Bell and C. Colebrook (Eds.), *Deleuze and history* (pp. 103–120). Edinburgh: Edinburgh University Press.

Fleming, P. (2012). The birth of biocracy and its discontents at work. *Research in the Sociology of Organizations, 35*, 177–199.

Fleming, P., and Spicer, A. (2003). Working at a cynical distance: Implications for power, subjectivity and resistance. *Organization, 10*(1), 157–179.

Flowers, S. (2008). Harnessing the hackers: The emergence and exploitation of Outlaw Innovation. *Research Policy, 37*(2), 177–193. doi:10.1016/j. respol.2007.10.006.

Fodor, J. A. (1983). *The modularity of mind: An essay on faculty psychology*. Cambridge, MA: MIT Press.

Fogg, M. J. (1995). *Terraforming: Engineering planetary environments.* Warrendale, PA: Society of Automotive Engineers.

Ford, L. O. (2011). *Savage messiah.* London: Verso.

Fortun, K. (2001). *Advocacy after Bhopal: Environmentalism, disaster, new global orders.* Chicago: University of Chicago Press.

Foster, J. B., Magdoff, H., and Magdoff, F. (2008). The new face of capitalism: Slow growth, excess capital, and a mountain of debt. *Monthly Review, 53*(11). http://monthlyreview.org/0402editr.htm.

Foucault, M. (1978). *The history of sexuality*, vol. 1: *An introduction.* London: Penguin.

Fouchard, J. (1981). *The Haitian maroons: Liberty or death.* New York: Blyden.

Fournier, V. (2013). Commoning: On the social organisation of the commons. *Management and Organizational History, 16*(4), 433–453.

Frank, A. (1991). For a sociology of the body: An analytical review. In M. Featherstone, M. Hepworth, and B. S. Turner (Eds.), *The body: Social process and cultural theory* (pp. 36–102). London: Sage.

Franklin, U. (1999). *The real world of technology* (Rev ed.). Toronto: Anansi.

Franks, D. D. (2010). *Neurosociology: The nexus between neuroscience and social psychology.* New York: Springer.

Frayling, C. (2011). *On craftsmanship: Towards a new Bauhaus.* London: Oberon Books.

Free Association. (2011). *Moments of excess: Movements, protest and everyday life.* Oakland, CA: PM Press.

Freedman, C. H. (2000). *Critical theory and science fiction.* Middletown, CT: Wesleyan University Press.

French, S., and Kneale, J. (2012). Speculating on careless lives. *Journal of Cultural Economy, 5*(4), 391–406. doi:10.1080/17530350.2012.703619.

Frickel, S., and Gross, N. (2005). A general theory of scientific/intellectual movements. *American Sociological Review, 70*(2), 204–232.

Frykman, N., Anderson, C., van Voss, L. H., and Rediker, M. (2013). *Mutiny and maritime radicalism in the age of revolution: A global survey.* Cambridge: Cambridge University Press.

Fujimura, J. H. (1991). On methods, ontologies and representation in the sociology of science: Where do we stand? In D. Maines (Ed.), *Social organization and social process: Essays in honor of Anselm L. Strauss* (pp. 207–248). Hawthorne, NY: De Gruyter.

Fumagalli, A. (2011). Twenty theses on cognitive capitalism (cognitive biocapitalism). *Angelaki, 16*(3), 7–17. doi:10.1080/0969725x.2011.626555.

Gadamer, H. G. (1989). *Truth and method* (2nd rev. ed.). New York: Crossroad.

Gaffney, P. (2010). *The force of the virtual: Deleuze, science, and philosophy.* Minneapolis: University of Minnesota Press.

Garcia-Lamarca, M., and Kaika, M. (2016). "Mortgaged lives": The biopolitics of debt and housing financialisation. *Transactions of the Institiute of British Geographers, 41*(3), 313–327. doi:10.1111/tran.12126.

Gardner, H. (1987). *The mind's new science: A history of cognitive revolution.* New York: Basic Books.

Gauntlett, D. (2011). *Making is connecting: The social meaning of creativity, from DIY and knitting to YouTube and Web 2.0.* Cambridge: Polity.

Geertz, C. (1993). *The interpretation of cultures: Selected essays.* London: Fontana.

Ghelfi, A. (2015). Worlding politics: Justice, commons and technoscience. PhD diss., University of Leicester.

Gibson-Graham, J. K. (2006). *A postcapitalist politics.* Minneapolis: University of Minnesota Press.

Gilbert, J. (2008). *Anticapitalism and culture: Radical theory and popular politics.* Oxford: Berg.

Gilbert, S. F. (1997). Bodies of knowledge: Biology and the intercultural university. In P. J. Taylor, S. E. Halfon, and P. N. Edwards (Eds.), *Changing life: Genomes, ecologies, bodies, commodities* (pp. 36–55). Minneapolis: University of Minnesota Press.

Gilbert, S. F. (2002). The genome in its ecological context: Philosophical perspectives on interspecies epigenesis. *Annals of the New York Academy of Sciences, 981,* 202–218.

Gilbert, S. F., and Epel, D. (2009). *Ecological developmental biology: Integrating epigenetics, medicine, and evolution.* Sunderland, MA: Sinauer Associates.

Gilroy, P. (2011). Paul Gilroy speaks on the riots, August 2011, Tottenham, North London. http://dreamofsafety.blogspot.com/2011/08/paul-gilroy-speaks-on-riots-august-2011.html.

Gilroy, P. (2013). 1981 and 2011: From social democratic to neoliberal rioting. *South Atlantic Quarterly, 112*(3), 550–558. doi:10.1215/00382876-2146467.

Giordano, C. (2014). *Migrants in translation: Caring and the logics of difference in contemporary Italy.* Berkeley: University of California Press.

Glenn, E. N. (2004). *Unequal freedom: How race and gender shaped American citizenship and labor.* Cambridge, MA: Harvard University Press.

Glick-Schiller, N., and Çaglar, A. (2008). *Migrant incorporation and city scale: Towards a theory of locality in migration studies.* Malmö: Malmö Institute for Studies of Migration, Diversity and Welfare and Department of International Migration and Ethnic Relations, Malmö University.

Glissant, E. (1997). *Poetics of relation* (B. Wing, Trans.). Ann Arbor: University of Michigan Press.

Glyn, A. (2006). *Capitalism unleashed: Finance globalization and welfare.* New York: Oxford University Press.

Goldstine, H. H., and von Neumann, J. (1963). Planning and coding of problems for an electronic computing instrument, part II, vol. 1, 1947. In A. H. Taub (Ed.), *John von Neumann, collected works,* vol. 5: *Design of computers, theory of automata and numerical analysis* (pp. 80–151). New York: Pergamon.

Gorz, A. (1999). *Reclaiming work: Beyond the wage-based society.* Cambridge: Polity.

Gottlieb, G. (1992). *Individual development and evolution: The genesis of novel behavior.* New York: Oxford University Press.

Gottlieb, G. (1997). *Synthesizing nature-nurture: Prenatal roots of instinctive behavior*. Mahwah, NJ: Lawrence Erlbaum.

Gottlieb, G. (2007). Probabilistic epigenesis. *Developmental Science, 10*(1), 1–11. doi:10.1111/j.1467-7687.2007.00556.x.

Gottlieb, G., Wahlsten, D., and Lickliter, R. (1998). The significance of biology for human development: A developmental psychobiological systems view. In W. Damon and R. M. Lerner (Eds.), *Handbook of child psychology*, vol. 1: *Theoretical models of human development* (pp. 233–273). New York: Wiley.

Gould, D. B. (2009). *Moving politics: Emotion and ACT UP's fight against AIDS*. Chicago: University of Chicago Press.

Gould, S. J. (1977). *Ontogeny and phylogeny*. Cambridge, MA: Belknap Press.

Gow, P. (2001). *An Amazonian myth and its history*. Oxford: Oxford University Press.

Gray, C. H. (2014). Big data, actionable information, scientific knowledge and the goal of control. *Teknokultura: Journal of Digital Culture and Social Movements, 11*(3), 529–554.

Grazioli, M. (2017a). From citizens to citadins? Rethinking right to the city inside housing squats in Rome, Italy. *Citizenship Studies, 21*(4), 1–16. doi:10.1080/136 21025.2017.1307607.

Grazioli, M. (2017b). The right to the city in the post-welfare metropolis: Everyday life, autonomous infrastructures and the mobile commons in Rome's self-organised housing. PhD diss., University of Leicester.

Greer, A. (2012). Commons and enclosure in the colonization of North America. *American Historical Review, 117*(2), 365–386.

Greer, B. (2014). *Craftivism: The art of craft and activism*. Vancouver: Arsenal Pulp Press.

Grimshaw, D., Ward, K. G., Rubery, J., and Beynon, H. (2001). Organisations and the transformation of the internal labour market. *Work, Employment and Society, 15*(1), 25–54. doi:10.1017/S0950017001000022.

Grosfoguel, R. (2007). The epistemic decolonial turn. *Cultural Studies, 21*(2), 211–223. doi:10.1080/09502380601162514.

Grossberg, L. (1986). On postmodernism and articulation: An interview with Stuart Hall. *Journal of Communication Inquiry, 10*(2), 45–60.

Guldi, J. (2012). *Roads to power: Britain invents the infrastructure state*. Cambridge, MA: Harvard University Press.

Gupta, A. K. (2016). *Grassroots innovation: Minds on the margin are not marginal minds*. Haryana, India: Random House.

Habermas, J. (1984). *The theory of communicative action*. Cambridge: Polity.

Habermas, J. (1993). *Justification and application: Remarks on discourse ethics*. Cambridge, MA: MIT Press.

Hackett, E. J. (2014). Academic capitalism. *Science, Technology, and Human Values, 39*(5), 635–638. doi:10.1177/0162243914540219.

Hall, S. (1986). Gramsci's relevance for the study of race and ethnicity. *Journal of Communication Inquiry, 10*(2), 5–27.

Hall, S. (1990). Cultural identity and diaspora. In J. Rutherford (Ed.), *Identity: Community, culture, difference* (pp. 222–237). London: Lawrence and Wishart.

Hall, S., and Jefferson, T. (1976). *Resistance through rituals: Youth subcultures in post-war Britain.* London: Hutchinson.

Hallward, P. (2008). Order and event: On Badiou's logics of worlds. *New Left Review, 53,* 97–122.

Hamlett, P. W. (2003). Technology theory and deliberative democracy. *Science, Technology, and Human Values, 28*(1), 112–140.

Hamm, M. (2011). Performing protest: Media practices in the trans-urban Euromayday movement of the precarious. PhD diss., University of Lucerne.

Hampson, I., Ewer, P., and Smith, M. (1994). Post-Fordism and workplace change: Towards a critical research agenda. *Journal of Industrial Relations, 36*(2), 231–257.

Hanlon, G. (2012). The "Google model" of production—the entrepreneurial function, the immaterial and the return to rent. Paper presented at the Center for Philosophy and Political Economy, University of Leicester, January 25, 2012.

Hanlon, G. (2014). The entrepreneurial function and the capture of value: Using Kirzner to understand contemporary capitalism. *Ephemera: Theory and Politics in Organization, 14*(2), 177–195.

Hanlon, G. (2016). *The dark side of management: A secret history of management theory.* London: Routledge.

Hansen, M. (2000). Becoming as creative involution? Contextualizing Deleuze and Guattari's biophilosophy. *Postmodern Culture, 11*(1).

Haran, J. (2013). The UK hybrid embryo controversy: Delegitimising counterpublics. *Science as Culture, 22*(4), 567–588. doi:10.1080/09505431.2013.768017.

Haraway, D. J. (1988). Situated knowledges: The science question in feminism and the privilege of partial perspective. *Feminist Studies, 14*(3), 575–599.

Haraway, D. J. (1989). *Primate visions: Gender, race, and nature in the world of modern science.* New York: Routledge.

Haraway, D. J. (1991a). A cyborg manifesto: Science, technology, and socialist-feminism in the late twentieth century. In D. Haraway, *Simians, cyborgs and women: The reinvention of nature,* (pp. 149–181). New York: Routledge.

Haraway, D. J. (1991b). *Simians, cyborgs, and women: The reinvention of nature.* New York: Routledge.

Haraway, D. J. (1992). The promises of monsters: A regenerative politics for inappropriate/d others. In L. Grossberg, C. Nelson, and P. A. Treichler (Eds.), *Cultural studies* (pp. 295–337). London: Routledge.

Haraway, D. J. (1997). *Modest_Witness@Second_Millennium. FemaleMan©_Meets_OncoMouse™: Feminism and technoscience.* New York: Routledge.

Haraway, D. J. (2013). Sowing worlds: A seed bag for terraforming with Earth others. In M. Grebowicz and M. Helen (Eds.), *Beyond the cyborg: Adventures with Donna Haraway* (pp. 137–146). New York: Columbia University Press.

Harding, S. (1986). *The science question in feminism*. Ithaca, NY: Cornell University Press.

Harding, S. (1991). *Whose science? Whose knowledge? Thinking from women's lives*. Ithaca, NY: Cornell University Press.

Hardt, M., and Negri, A. (2009). *Commonwealth*. Cambridge, MA: Harvard University Press.

Harman, G. (2007). On vicarious causation. *Collapse, 2*, 171–205.

Harman, G. (2011). Realism without materialism. *SubStance, 40*(2), 52–72. doi:10.1353/sub.2011.0011.

Harney, S. (2006). Governance, state, and living labour. *Cultural Logic: An Electronic Journal of Marxist Theory and Practice*. http://clogic.eserver.org/2006/harney.html.

Harney, S. (2008). Governance and the undercommons. *InterActivist*. http://slash.autonomedia.org/node/10926.

Harré, R. (1996). The necessity of personhood as embodied being. *Theory and Psychology, 5*(3), 369–373.

Harrington, M. (2008). AIDS activists and people with AIDS: A movement to revolutionize research and for universal access to treatment. In B. da Costa and K. Philip (Eds.), *Tactical biopolitics: Art, activism, and technoscience* (pp. 323–340). Cambridge, MA: MIT Press.

Hartigan, J. (2015a). *Care of the species: Cultivating biodiversity in Spain and Mexico*. Minneapolis: University of Minnesota Press.

Hartigan, J. (2015b). Plant publics: Multispecies relating in Spanish botanical gardens. *Anthropological Quarterly, 88*(2), 481–507. doi:10.1353/anq.2015.0024.

Hartsock, N. (1983). The feminist standpoint: Developing the ground for a specifically feminist historical materialism. In S. Harding and M. Hintikka (Eds.), *Discovering reality: Feminist perspectives on epistemology, metaphysics, methodology and philosophy of science* (pp. 283–310). Dordrecht, Netherlands: Reidel.

Harvey, D. (1990). *The condition of postmodernity: An enquiry into the origins of cultural change*. Oxford: Blackwell.

Harvey, N. (2001). Globalisation and resistance in post–cold war Mexico: Difference, citizenship and biodiversity conflicts in Chiapas. *Third World Quarterly, 22*(6), 1045–1061.

Harvie, D., and Milburn, K. (2016). On the uses of fairy dust: Contagion, sorcery and the crafting of other worlds. *Culture and Organization*, 1–17. doi:10.1080/14759551.2015.1118636.

Hatch, M. (2014). *The maker movement manifesto: Rules for innovation in the new world of crafters, hackers, and tinkerers*. New York: McGraw-Hill Education.

Hau'ofa, E. (1993). Our sea of islands. In E. Waddell, V. Naidu, and E. Hau'ofa (Eds.), *A new Oceania: Rediscovering our sea of islands* (pp. 2–16). Suva, Fiji: University of the South Pacific.

Haworth, R. H., and Elmore, J. M. (Eds.). (2017). *Out of the ruins: The emergence of radical informal learning spaces*. Oakland, CA: PM Press.

Hay, D. (1975). Poaching and the game laws on Cannock Chase. In D. Hay, P. Linebaugh, J. G. Rule, E. P. Thompson, and C. Winslow (Eds.), *Albion's fatal tree: Crime and society in eighteenth-century England* (pp. 189–253). London: Allen Lane.

Hayden, C. (2003). *When nature goes public: The making and unmaking of bio-prospecting in Mexico.* Princeton, NJ: Princeton University Press.

Hayden, C. (2007). Taking as giving: Bioscience, exchange, and the politics of benefit-sharing. *Social Studies of Science, 37*(5), 729–758.

Hayden, C. (2010). The proper copy: The insides and outsides of domains made public. *Journal of Cultural Economy, 3*(1), 85–102. doi:10.1080 /17530351003617602.

Hayles, N. K. (1999). *How we became posthuman: Virtual bodies in cybernetics, literature, and informatics.* Chicago: University of Chicago Press.

Heery, E., and Salmon, J. (Eds.). (2000). *The insecure workforce.* London: Routledge.

Heidegger, M. (1993). *Sein und Zeit.* Tübingen, Germany: Niemeyer.

Helfrich, S., and Stiftung, H. B. (Eds.). (2012). *Commons: Für eine neue Politik jenseits von Markt und Staat.* Bielefeld, Germany: Transcript.

Henare, A. J. M., Holbraad, M., and Wastell, S. (Eds.). (2007). *Thinking through things: Theorising artefacts ethnographically.* London: Routledge.

Henderson, J., Dicken, P., Hess, M., Coe, N., and Yeung, H. W.-C. (2002). Global production networks and the analysis of economic development. *Review of International Political Economy, 9*(3), 436–464. doi:10.1080/09692290210 150842.

Henry, M. (2008). *Material phenomenology.* New York: Fordham University Press.

Herbrechter, S. (2013). *Posthumanism: A critical analysis.* London: Bloomsbury.

Hesmondhalgh, D., and Baker, S. (2011). *Creative labour: Media work in three cultural industries.* London: Routledge.

Hess, C., and Ostrom, E. (2007). *Understanding knowledge as a commons: From theory to practice.* Cambridge, MA: MIT Press.

Hess, D. J. (2004). Guest editorial: Health, the environment and social movements. *Science as Culture, 13*(4), 421–427.

Hess, D. J. (2007a). *Alternative pathways in science and industry: Activism, innovation, and the environment in an era of globalization.* Cambridge, MA: MIT Press.

Hess, D. J. (2007b). Crosscurrents: Social movements and the anthropology of science and technology. *American Anthropologist, 109*(3), 463–472.

Hess, D. J. (2015). Undone science and social movements: A review and typology. In M. Gross and L. McGoey (Eds.), *Routledge international handbook of ignorance studies* (pp. 141–154). London: Routledge.

Hess, S., Binder, J., and Moser, J. (Eds.). (2009). *No integration?! Kulturwissenschaftliche Beiträge zur Integrationsdebatte in Europa.* Bielefeld, Germany: Transcript.

Hetherington, K. (2014). Waiting for the surveyor: Development promises and the temporality of infrastructure. *Journal of Latin American and Caribbean Anthropology, 19*(2), 195–211. doi:10.1111/jlca.12100.

Hetherington, K. (2016). Surveying the future perfect: Anthropology, development and the promise of infrastructure. In P. Harvey, B. Jensen Casper, and A. Morita (Eds.), *Infrastructures and social complexity: A companion* (pp. 40–50). London: Routledge.

Hill Collins, P. (1991). *Black feminist thought: Knowledge, consciousness, and the politics of empowerment.* New York: Routledge.

Hindle, S. (2003). Crime and popular protest. In B. Coward (Ed.), *A companion to Stuart Britain* (pp. 130–147). Malden, MA: Blackwell.

Hodge, G. D. (2000). Retrenchment from a queer ideal: Class privilege and the failure of identity politics in AIDS activism. *Environment and Planning D: Society and Space, 18*(3), 355–376.

Hoerl, E. (2013). A thousand ecologies: The process of cyberneticization and general ecology. In D. Diederichsen and A. Franke (Eds.), *The whole Earth: California and the disappearance of the outside* (pp. 121–130). Berlin: Sternberg.

Hoeschele, W. (2010). *The economics of abundance: A political economy of freedom, equity, and sustainability.* Farnham, UK: Gower.

Hoffman, P. T., Postel-Vinay, G., and Rosenthal, J.-L. (2007). *Surviving large losses: Financial crises, the middle class, and the development of capital markets.* Cambridge, MA: Harvard University Press.

Holbraad, M., Pedersen, M. A., and Viveiros de Castro, E. (2014). The politics of ontology: Anthropological positions. Theorizing the contemporary, *Cultural Anthropology Online.* http://www.culanth.org/fieldsights/461-the-politics-of -ontology.

Holland, E. W. (2005). Desire. In C. J. Stivale (Ed.), *Gilles Deleuze: Key concepts.* (pp. 55–64). Durham, NC: Acumen.

Holloway, J. (2010a). *Crack capitalism.* London: Pluto.

Holloway, J. (2010b). Cracks and the crisis of abstract labour. *Antipode, 42*(4), 909–923. doi:10.1111/j.1467-8330.2010.00781.x.

Holmgren, D. (2002). *Permaculture: Principles and pathways beyond sustainability.* Hepburn, Australia: Holmgren Design Services.

Hope, J. (2008). *Biobazaar: The open source revolution and biotechnology.* Cambridge, MA: Harvard University Press.

Horstman, J. (2010). *The Scientific American brave new brain: How neuroscience, brain-machine interfaces, neuroimaging, psychopharmacology, epigenetics, the Internet, and our own minds are stimulating and enhancing the future of mental power.* San Francisco: Jossey-Bass.

Howe, K. R. (2004). A critique of experimentalism. *Qualitative Inquiry, 10*(1), 42–61. doi:10.1177/1077800403259491.

Hubbard, R., Henifin, M. S., Fried, B., Druss, V., and Star, S. L. (1979). *Women look at biology looking at women: A collection of feminist critiques.* Cambridge, MA: Schenkman.

Hunt, S. E. (2013). Witnessing the colonialscape: Lighting the intimate fires of indigenous legal pluralism. PhD diss., Simon Fraser University.

Hustak, C., and Myers, N. (2013). Involutionary momentum: Affective ecologies and the sciences of plant/insect encounters. *Differences, 23*(3), 74–118. doi:10.1215/10407391-1892907.

Huttenlocher, P. R. (2002). *Neural plasticity: The effects of environment on the development of the cerebral cortex.* Cambridge, MA: Harvard University Press.

Ibrahim, A. (2010). Eritreische Frauen auf der Flucht. In Connection e.V. Pro Asyl (Ed.), *Eritrea: Desertion, Flucht und Asyl* (pp. 36–48). Frankfurt: Pro Asyl.

Ignatieff, M. (1978). *A just measure of pain: The penitentiary in the industrial revolution, 1750–1850.* New York: Columbia University Press.

Ihde, D., and Selinger, E. (Eds.). (2003). *Chasing technoscience: Matrix for materiality.* Bloomington: Indiana University Press.

Illich, I. (1979). *Tools for conviviality.* London: Fontana.

Immelt, J. R. (2012, March). The CEO of General Electric on sparking an American manufacturing renewal. *Harvard Business Review.*

Infoaut. (2013). "Freedom and rights? You have to sweat blood for them! On the internet, too." Infoaut interviews Autistici/Inventati. *InfoFreeFlow.* http://infofreeflow.noblogs.org/post/2013/09/02/freedom-and-rights-you-have-to-sweat-blood-for-them-on-the-internet-too-infoaut-interviews-autisticiinventati/.

Ingold, T. (2013). *Making: Anthropology, archaeology, art and architecture.* London: Routledge.

Irani, L. (2015). Hackathons and the making of entrepreneurial citizenship. *Science, Technology, and Human Values, 40*(5), 799–824.

Irigaray, L. (1985). *This sex which is not one.* Ithaca, NY: Cornell University Press.

Irwin, A., and Michael, M. (2003). *Science, social theory and public knowledge.* Maidenhead, UK: Open University Press.

Isin, E., and Nielsen, G. M. (Eds.). (2008). *Acts of citizenship.* London: Zed Books.

Isin, E., and Nyers, P. (Eds.). (2014). *Routledge handbook of global citizenship studies.* Abingdon, UK: Routledge.

Jablonka, E., and Lamb, M. J. (2005). *Evolution in four dimensions: Genetic, epigenetic, behavioral, and symbolic variation in the history of life.* Cambridge, MA: MIT Press.

Jackson, N., and Carter, P. (1998). Labour as dressage. In A. McKinlay and K. Starkey (Eds.), *Foucault, management and organization theory: From panopticon to technologies of self* (pp. 49–64). London: Sage.

Jackson, S. J., Edwards, P. N., Bowker, G. C., and Knobel, C. P. (2007). Understanding infrastructure: History, heuristics and cyberinfrastructure policy. *First Monday, 12*(6).

Jameson, F. (1981). *The political unconscious: Narrative as a socially symbolic act.* London: Methuen.

Jameson, F. (1986). On magic realism in film. *Critical Inquiry, 12*(2), 301–325.

Jameson, F. (2000). "If I find one good city I will spare the man": Realism and utopia in Kim Stanley Robinson's Mars trilogy. In P. Parrinder (Ed.), *Learning from other worlds: Estrangement, cognition, and the politics of science fiction and utopia* (pp. 208–232). Liverpool, UK: Liverpool University Press.

Jameson, F. (2007). *Archaeologies of the future: The desire called utopia and other science fictions.* London: Verso.

Januškevičiūtė, V. (Ed.) (2015). *XII Baltic Triennial exhibition guide.* Vilnius, Lithuania: Contemporary Art Centre.

Jaros, S. J. (2005). Marxian critiques of Thompson's (1990) "core" labour process theory: An evaluation and extension. *Ephemera: Theory and Politics in Organization, 5*(1), 5–25.

Jasanoff, S. (2003a). Breaking the waves in science studies: Comment on H. M. Collins and Robert Evans, "The third wave of science studies." *Social Studies of Science, 33*(3), 389–400.

Jasanoff, S. (2003b). Technologies of humility: Citizen participation in governing science. *Minerva, 41*(3), 223–244.

Jenkins, L. D. (2007). Bycatch: Interactional expertise, dolphins and the US tuna fishery. *Studies in History and Philosophy of Science, 38*(4), 698–712.

Jensen, C. B. (2015). Mekong scales: Domains, test sites, and the micro-uncommons. Paper presented at the Sawyer seminar workshop Uncommons, University of California, Davis.

Jensen, C. B., and Rødje, K. (Eds.). (2010). *Deleuzian intersections: Science, technology, anthropology.* New York: Berghahn.

Jessop, B., and Sum, N.-L. (2006). *Beyond the regulation approach: Putting capitalist economies in their place.* Cheltenham, UK: Edward Elgar.

Johnson, M. L. (1999). Embodied reason. In G. Weiss and H. F. Haber (Eds.), *Perspectives on embodiment: The intersections of culture and nature* (pp. 81–102). London: Routledge.

Jordan, B., and Düvell, F. (2002). *Irregular migration: The dilemmas of transnational mobility.* Cheltenham, UK: Edward Elgar.

Jordan, Z. A. (1967). *The evolution of dialectical materialism: A philosophical and sociological analysis.* London: Macmillan.

Joyce, K. (2005). Appealing images: Magnetic resonance imaging and the production of authoritative knowledge. *Social Studies of Science, 35*(3), 437–462.

Juris, J. S., and Khasnabish, A. (2013). *Insurgent encounters: Transnational activism, ethnography, and the political.* Durham, NC: Duke University Press.

Kadri, M. (2010). Tinkers, hackers, farmers, crafters. http://designobserver.com/feature/tinkers-hackers-farmers-crafters/14908.

Kanngieser, A., and Beuret, N. (2017). Refusing the world: Silence, commoning, and the Anthropocene. *South Atlantic Quarterly, 116*(2), 363–380. doi:10.1215/00382876-3829456.

Karakayali, S. (2008). *Gespenster der Migration: Zur Genealogie illegaler Einwanderung in der Bundesrepublik Deutschland.* Bielefeld, Germany: Transcript.

Karakayali, S., and Tsianos, V. (2005). Mapping the order of new migration: Undokumentierte Arbeit und die Autonomie der Migration. *Peripherie, 97/98,* 35–64.

Karatzogianni, A. (2015). *Firebrand waves of digital activism 1994–2014: The rise and spread of hacktivism and cyberconflict.* Houndmills, UK: Palgrave Macmillan.

Karfakis, N. (2013). The biopolitics of chronic fatigue syndrome. PhD diss., University of Leicester.

Karmiloff-Smith, A. (1992). *Beyond modularity: A developmental perspective on cognitive science.* Cambridge, MA: MIT Press.

Karpik, L. (2010). *Valuing the unique: The economics of singularities.* Princeton, NJ: Princeton University Press.

Kasparek, B. (2016). Complementing Schengen: The Dublin system and the European border and migration regime. In H. Bauder and C. Matheis (Eds.), *Migration policy and practice: Interventions and solutions* (pp. 59–78). London: Palgrave Macmillan.

Kelly, A. H., and Lezaun, J. (2014). Urban mosquitoes, situational publics, and the pursuit of interspecies separation in Dar es Salaam. *American Ethnologist, 41*(2), 368–383. doi:10.1111/amet.12081.

Kelty, C. M. (2008). *Two bits: The cultural significance of free software.* Durham, NC: Duke University Press.

Kelty, C. M. (2013). There is no free software. *Journal of Peer Production, 3.*

Kember, S. (2003). *Cyberfeminism and artificial life.* London: Routledge.

Kenney, M., and Müller, R. (2016). Of rats and women: Narratives of motherhood in environmental epigenetics. *BioSocieties.* doi:10.1057/s41292-016-0002-7.

Kera, D. (2012). Hackerspaces and DIYbio in Asia: Connecting science and community with open data, kits and protocols. *Journal of Peer Production, 2.*

Kiefer, J. C. (2007). Epigenetics in development. *Developmental Dynamics, 236*(4), 1144–1156.

King, N. (2016). *No borders: The politics of immigration control and resistance.* London: Zed Books.

Kippax, S., and Stephenson, N. (2010). Infectious diseases and globalisation. In B. Turner (Ed.), *The Routledge international handbook of globalization studies* (pp. 205–226). Abingdon, UK: Routledge.

Kippax, S., and Stephenson, N. (2012). Beyond the distinction between biomedical and social dimensions of HIV prevention through the lens of a social public health. *American Journal of Public Health, 102*(5), 789–799. doi:10.2105/AJPH.2011.300594.

Kippax, S., and Stephenson, N. (2016). *Socialising the biomedical turn in HIV prevention.* London: Anthem.

Kirksey, S. E. (2015). Species: A praxiographic study. *Journal of the Royal Anthropological Institute, 21,* 758–780.

Kirksey, S. E., and Helmreich, S. (2010). The emergence of multispecies ethnography. *Cultural Anthropology, 25*(4), 545–576. doi:10.1111/j.1548-1360.2010.01069.x.

Klein, U. (2012). Artisanal-scientific experts in eighteenth-century France and Germany [Special issue]. *Annals of Science, 69*(3), 303–434.

Kleinman, D. L. (Ed.) (2000). *Science, technology, and democracy.* Albany: State University of New York Press.

Knights, D., and Willmott, H. (Eds.). (1990). *Labour process theory.* Basingstoke, UK: Macmillan.

Knorr-Cetina, K., and Preda, A. (Eds.). (2006). *The sociology of financial markets.* Oxford: Oxford University Press.

Knox, H., O'Doherty, D., Vurdubakis, T., and Westrup, C. (2008). Enacting airports: Space, movement and modes of ordering. *Organization, 15*(6), 869–888.

Kohn, E. (2013). *How forests think: Toward an anthropology beyond the human.* Berkeley: University of California Press.

Kojève, A. (1969). *Introduction to the reading of Hegel: Lectures on the phenomenology of spirit.* New York: Basic Books.

Kokkinidis, G. (2014). Spaces of possibilities: Workers' self-management in Greece. *Organization, 22*(6), 847–871. doi:10.1177/1350508414521098.

Korňan, M., and Kropil, R. (2014). What are ecological guilds? Dilemma of guild concepts. *Russian Journal of Ecology, 45*(5), 445–447. doi:10.1134/s1067413614050178.

Kortright, C. (2012). C4 rice and hoping the sun can end hunger: Tales of plants, evolution, transgenics and crisis. PhD diss., University of California Davis.

Kortright, C. (2013). On labor and creative transformations in the experimental fields of the Philippines. *East Asian Science, Technology and Society, 7*(4), 557–578. doi:10.1215/18752160-2392710.

Kostakis, V., and Bauwens, M. (2014). *Network society and future scenarios for a collaborative economy.* Basingstoke, UK: Palgrave Macmillan.

Kowal, E., Radin, J., and Reardon, J. (2013). Indigenous body parts, mutating temporalities, and the half-lives of postcolonial technoscience. *Social Studies of Science, 43*(4), 465–483. doi:10.1177/0306312713490843.

Kozłowski, M., Kurant, A., Sowa, J., Szadkowski, K., and Szreder, J. (Eds.). (2014). *Joy forever: The political economy of social creativity.* London: Mayfly.

Kramer, L. (1989). *Reports from the holocaust: The making of an AIDS activist.* New York: St. Martin's.

Krautwurst, U. (2007). Cyborg anthropology and/as endocolonisation. *Culture, Theory and Critique, 48*(2), 139–160. doi:10.1080/14735780701723090.

Kromidas, M. (2014). The "savage" child and the nature of race: Posthuman interventions from New York City. *Anthropological Theory, 14*(4), 422–441. doi:10.1177/1463499614552739.

Kuster, B. (2011). Camps und Heterotopien der Gegenwart. À props de Rien ne vaut que la vie, mais la vie même ne vaut rien. In M.-H. Gutberlet and S. Helff (Eds.), *Die Kunst der Migration: Aktuelle Positionen zum europäisch-afrikanischen Diskurs* (pp. 147–157). Bielefeld, Germany: Transcript.

Kuster, B. (2016). Engfügungen. Grenze. Film. Überquerung. PhD diss., Akademie der Bildenden Künste Wien.

Kuster, B., and Tsianos, V. (2016). How to liquefy a body on the move: Eurodac and the making of the European digital border. In R. Bossong and H. Carrapico (Eds.), *EU Borders and Shifting Internal Security* (pp. 45–63). New York: Springer.

Kwitny, J. (1992). *Acceptable risks.* New York: Poseidon.

Laclau, E., and Mouffe, C. (1985). *Hegemony and socialist strategy: Towards a radical democratic politics.* London: Verso.

LaDuke, W. (2010). Our home on Earth. In J. Walljasper (Ed.), *All that we share: A field guide to the commons* (pp. 81–88). New York: New Press.

Lafazani, O. (2011). The border from the border: Critical perspectives on EU borders migration policies and beyond. Paper presented at the Conference Mobile Borders, University of Geneva, September 6–9, 2011.

Lafazani, O. (2012). The border between theory and activism. *ACME: An International E-Journal for Critical Geographies, 11*(2), 189–193.

Lafazani, O. (2013). A border within a border: The migrants' squatter settlement in Patras as a heterotopia. *Journal of Borderlands Studies, 28*(1), 1–13. doi:10.1080/08865655.2012.751731.

Lakoff, G., and Johnson, M. L. (1999). *Philosophy in the flesh: The embodied mind and its challenge to Western thought.* New York: Basic Books.

Lamb, M. J. (1994). Epigenetic inheritance and aging. *Reviews in Clinical Gerontology, 4,* 97–105.

Landecker, H. (2011a). Food as exposure: Nutritional epigenetics and the new metabolism. *BioSocieties, 6*(2), 167–194.

Landecker, H. (2011b). Food is licking is plastic: The social as signal in environmental epigenetics. Paper presented at the Workshop Anticipation: Exploring Technoscience, Life, Affect, Temporality, San Francisco, March, 27–29, 2011.

Landecker, H., and Panofsky, A. (2013). From social structure to gene regulation, and back: A critical introduction to environmental epigenetics for sociology. *Annual Review of Sociology, 39*(1), 333–357. doi:10.1146/annurev-soc-071312-145707.

Langley, P. (2008). *The everyday life of global finance: Saving and borrowing in Anglo-America.* Oxford: Oxford University Press.

Larkin, B. (2013). The politics and poetics of infrastructure. *Annual Review of Anthropology, 42*(1), 327–343. doi:10.1146/annurev-anthro-092412-155522.

Laruelle, F. (2013). *Principles of non-philosophy.* New York: Bloomsbury.

Latimer, J. (2013). *The gene, the clinic, and the family: Diagnosing dysmorphology, reviving medical dominance.* London: Routledge.

Latimer, J., and Puig de la Bellacasa, M. (2013). Re-thinking the ethical: Everyday shifts of care in biogerontology. In N. Priaulx and A. Wrigley (Eds.), *Ethics, law and society* (pp. 153–176). Farnham, UK: Ashgate.

Latimer, J., and Schillmeier, M. (2009). *Un/knowing bodies.* Malden, MA: Blackwell.

Latour, B. (1983). Give me a laboratory and I will raise the world. In K. Knorr-Cetina and M. Mulkay (Eds.), *Science observed: Perspectives on the social study of science* (pp. 141–170). London: Sage.

Latour, B. (1987). *Science in action: How to follow scientists and engineers through society.* Cambridge, MA: Harvard University Press.

Latour, B. (1993). *We have never been modern.* Cambridge, MA: Harvard University Press.

Latour, B. (2000). When things strike back: A possible contribution of "science studies" to the social sciences. *British Journal of Sociology, 51*(1), 107–123.

Latour, B. (2003). The promises of constructivism. In D. Ihde and E. Selinger (Eds.), *Chasing technoscience: Matrix for materiality* (pp. 27–46). Bloomington: Indiana University Press.

Latour, B. (2004a). *Politics of nature: How to bring the sciences into democracy.* Cambridge, MA: Harvard University Press.

Latour, B. (2004b). Why has critique run out of steam? From matters of fact to matters of concern. *Critical Inquiry, 30*(2), 225–248.

Latour, B. (2005a). From realpolitik to dingpolitik or how to make things public. In B. Latour and P. Weibel (Eds.), *Making things public: Atmospheres of democracy* (pp. 4–31). Cambridge, MA: MIT Press.

Latour, B. (2005b). *Reassembling the social: An introduction to actor-network-theory.* Oxford: Oxford University Press.

Latour, B. (2007). Turning around politics: A note on Gerard de Vries' paper. *Social Studies of Science, 37*, 811–820.

Laugksch, R. C. (2000). Scientific literacy: A conceptual overview. *Science Education, 84*(1), 71–94.

Law, J. (1999). After ANT: Complexity, naming and topology. In J. Law and J. Hassard (Eds.), *Actor network theory and after* (pp. 1–14). Oxford: Blackwell.

Law, J. (2004). *After method: Mess in social science research.* London: Routledge.

Law, J. (2011). What's wrong with a one-world world. http://www.heterogeneities .net/publications/Law2011WhatsWrongWithAOneWorldWorld.pdf.

Law, J., and Hassard, J. (Eds.). (1999). *Actor network theory and after.* Oxford: Blackwell.

Law, J., and Mol, A. (2008). The actor-enacted: Cumbrian sheep in 2001. In C. Knappett and L. Malafouris (Eds.), *Material agency* (pp. 55–77). New York: Springer.

Law, J., and Singleton, V. (2005). Object lessons. *Organization, 12*(3), 331–355. doi:10.1177/1350508405051270.

Law, J., and Urry, J. (2011). Enacting the social. *Economy and Society, 33*(3), 390–410. doi:10.1080/0308514042000225716.

Leach, M., Scoones, I., and Wynne, B. (Eds.). (2005). *Science and citizens: Globalization and the challenge of engagement.* London: Zed.

Leane, E. (2002). Chromodynamics: Science and colonialism in Kim Stanley Robinson's Mars trilogy. *Ariel, 33*(1), 83–104.

Leder, D. (1990). *The absent body.* Chicago: University of Chicago Press.

LeDoux, J. E. (2002). *Synaptic self: How our brains become who we are.* London: Macmillan.

Lefebvre, H. (1991). *Critique of everyday life*, vol. 1: *Introduction.* London: Verso.

Lengwiler, M. (2008). Participatory approaches in science and technology: Historical origins and current practices in critical perspective. *Science, Technology, and Human Values, 33*, 186–200.

Lenin, V. I. (1899). The development of capitalism in Russia. In V. I. Lenin (Ed.), *Collected works*, vol. 3 (pp. 21–608). Moscow: Foreign Languages.

Lenin, V. I. (1902). What is to be done? Burning questions of our movement. In V. I. Lenin (Ed.), *Collected works*, vol. 5 (pp. 347–530). Moscow: Foreign Languages.

Lenin, V. I. (1917). The state and revolution: The Marxist theory of the state and the tasks of the proletariat in the revolution. In V. I. Lenin (Ed.), *Collected works*, vol. 25 (pp. 381–492). Moscow: Foreign Languages.

Lenin, V. I. (1970). *Materialism and empirio-criticism: Critical comments on a reactionary philosophy.* Moscow: Progress. (Original work published 1970.)

Lewontin, R. C. (2000). *The triple helix: Gene, organism, and environment.* Cambridge, MA: Harvard University Press.

Lezaun, J. (2006). Creating a new object of government: Making genetically modified organisms traceable. *Social Studies of Science, 36*(4), 499–531. doi:10.1177/0306312706059461.

Lezaun, J. (2014). A reader's guide to the "ontological turn"—part 2. *Somatosphere: Science, Medicine, and Anthropology.* http://somatosphere.net/?p=6802.

Lezaun, J., and Montgomery, C. M. (2014). The pharmaceutical commons: Sharing and exclusion in global health drug development. *Science, Technology, and Human Values, 40*(1), 3–29. doi:10.1177/0162243914542349.

Lezaun, J., and Soneryd, L. (2007). Consulting citizens: Technologies of elicitation and the mobility of publics. *Public Understanding of Science, 16*(3), 279–297. doi:10.1177/0963662507079371.

Li, C. (Ed.). (2010). *China's emerging middle class: Beyond economic transformation.* Washington, DC: Brookings Institution Press.

Liben, L. S. (1999). Developing an understanding of external spatial representations. In I. E. Sigel (Ed.), *Development of mental representation: Theories and applications* (pp. 297–321). Mahwah, NJ: Lawrence Erlbaum.

Liberatore, A., and Funtowicz, S. (2003). Introduction: "Democratising" expertise, "expertising" democracy: What does this mean, and why bother? *Science and Public Policy, 30*(3), 146–150.

Liberona, A., and Myers, N. (2016). Becoming sensor in an oak savannah—cultivating new modes of attention in a 10,000-year-old happening. https://becomingsensor.com/

Liboiron, M. (2016). Care and solidarity are conditions for interventionist research. *Engaging Science, Technology, and Society, 2,* 67–72.

Liboiron, M., and Pine, K. H. (2015). The politics of measurement and action. In *Proceedings of the 33rd annual ACM Conference on Human Factors in Computing Systems* (pp. 3147–3156). doi:10.1145/2702123.2702298.

Lightfoot, G., and Harvie, D. (2016). Finance: Finding a philosophical fit? In R. Mir, H. Willmott, and M. Greenwood (Eds.), *The Routledge companion to philosophy in organization studies* (pp. 388–394). London: Routledge.

Lilley, S., Lightfoot, G., and Amaral, P. (2004). *Representing organization: Knowledge, management, and the information age.* Oxford: Oxford University Press.

Lindtner, S. (2015). Hacking with Chinese characteristics: The promises of the maker movement against China's manufacturing culture. *Science, Technology, and Human Values, 40*(5), 854–879.

Linebaugh, P. (2008). *The Magna Carta manifesto: Liberties and commons for all.* Berkeley: University of California Press.

Linebaugh, P., and Rediker, M. (2000). *The many-headed hydra: Sailors, slaves, commoners, and the hidden history of the revolutionary Atlantic.* London: Verso.

Littlefield, M. (2009). Constructing the organ of deceit. *Science, Technology, and Human Values, 34*(3), 365–392. doi:10.1177/0162243908328756.

Longino, H. E. (2002). *The fate of knowledge.* Princeton, NJ: Princeton University Press.

López, A. R., and Weinstein, B. (2012). *The making of the middle class: Toward a transnational history.* Durham. NC: Duke University Press.

López, D. (2010). The securitization of care spaces: Lessons from telecare. In M. Schillmeier and M. Domenech (Eds.), *New technologies and emerging spaces of care* (pp. 39–55). Farnham, UK: Ashgate.

Lovink, G., and Rossiter, N. (Eds.). (2011). *My creativity reader: A critique of creative industries.* Amsterdam: Institute of Network Cultures.

Lowe, L. (1996). *Immigrant acts: On Asian American cultural politics.* Durham, NC: Duke University Press.

Lucassen, J., and Lucassen, L. (1997). *Migration, migration history, history: Old paradigms and new perspectives.* Bern, Switzerland: International Institute of Social History.

Luckman, S. (2016). *Craft and the creative economy.* London: Palgrave.

Luhmann, N. (1995). *Social systems.* Stanford, CA: Stanford University Press.

Lukács, G. (1971). *History and class consciousness: Studies in Marxist dialectics.* London: Merlin.

Lury, C., Parisi, L., and Terranova, T. (2012). Introduction: The becoming topological of culture. *Theory, Culture and Society, 29*(4–5), 3–35. doi:10.1177/0263276412454552.

Lynch, M., and Cole, S. (2005). Science and technology studies on trial: Dilemmas of expertise. *Social Studies of Science, 35,* 269–312.

MacIntyre, A. (1981). *After virtue: A study in moral theory.* Notre Dame, IN: University of Notre Dame Press.

Mackenzie, A. (2013). From validating to verifying: Public appeals in synthetic biology. *Science as Culture, 22*(4), 476–496. doi:10.1080/14636778.2013.764067.

MacKenzie, D. (2009). Beneath all the toxic acronyms lies a basic cultural issue. *Financial Times,* p. 12.

MacKenzie, D. (2011). Evaluation cultures? On invoking "culture" in the analysis of behaviour in financial markets. http://www.sps.ed.ac.uk/__data/assets/pdf_file/0007/64564/EvalCults11.pdf.

MacKenzie, D. (2012). Knowledge production in financial markets: Credit default swaps, the ABX and the subprime crisis. *Economy and Society, 41*(3), 335–359. doi:10.1080/03085147.2012.661635.

MacKenzie, D., and Millo, Y. (2003). Constructing a market, performing theory: The historical sociology of a financial derivatives exchange. *American Journal of Sociology, 109*(1), 107–145.

Maddison, A. (2003). *The world economy: Historical statistics.* Paris: Development Centre of the Organisation for Economic Co-operation and Development.

Maddison, A. (2006). *The world economy,* vol. 1: *A millennial perspective,* vol. 2: *Historical statistics.* Paris: Development Centre of the Organisation for Economic Co-operation and Development.

Maddison, B. (2010). Radical commons discourse and the challenges of colonialism. *Radical History Review, 2010*(108), 29–48. doi:10.1215/01636545-2010-002.

Malabou, C. (2008). *What should we do with our brain?* New York: Fordham University Press.

Malafouris, L. (2004). The cognitive basis of material engagement: Where brain, body and culture conflate. In E. DeMarrais, C. Gosden, and C. Renfrew (Eds.), *Rethinking materiality: The engagement of mind with the material world* (pp. 53–62). Cambridge: McDonald Institute for Archaeological Research.

Malafouris, L. (2008). At the potter's wheel: An argument for material agency. In C. Knappett and L. Malafouris (Eds.), *Material agency: Towards a non-anthropocentric approach* (pp. 19–36). New York: Springer.

Malafouris, L. (2010). Metaplasticity and the human becoming: Principles of neuroarchaeology. *Journal of Anthropological Sciences, 88*(4), 49–72.

Maldonado-Torres, N. (2007). On the coloniality of being. *Cultural Studies, 21*(2), 240–270. doi:10.1080/09502380601162548.

Malewitz, R. (2014). *The practice of misuse: Rugged consumerism in contemporary American culture.* Stanford, CA: Stanford University Press.

Mander, J., and Tauli-Corpuz, V. (Eds.). (2006). *Paradigm wars: Indigenous peoples' resistance to globalization.* San Francisco: Sierra Club.

Maniglier, P. (2012). What is a problematic? *Radical Philosophy, 173,* 21–22.

Mann, A. (2011). No corn, no country: Mexican farmers opposing the North American Free Trade Agreement. *Journal of Iberian and Latin American Research, 17*(2), 197–211. doi:10.1080/13260219.2011.628286.

Marazzi, C. (2007). Rules for the incommensurable. *SubStance, 36*(1), 11–36.

Marazzi, C. (2010). *The violence of financial capitalism.* Los Angeles: Semiotext(e).

Marcuse, H. (1991). *One-dimensional man: Studies in the ideology of advanced industrial society.* Boston: Beacon.

Margulis, L. (1998). *Symbiotic planet: A new look at evolution.* New York: Basic Books.

Margulis, L., and Sagan, D. (2002). *Acquiring genomes: A theory of the origins of species.* New York: Basic Books.

Marres, N. (2009). Testing powers of engagement: Green living experiments, the ontological turn and the undoability of involvement. *European Journal of Social Theory, 12*(1), 117–133.

Marsh, P. (2012). *The new industrial revolution: Consumers, globalization and the end of mass production.* New Haven, CT: Yale University Press.

Marti, P., Frens, J., Hengeveld, B., and Levy, P. (2016). On making [Special issue]. *Interaction Design and Architecture(s) Journal (IxD&A), 30,* 1–112.

Martignoni, M. (2015). Postcolonial organising: An oral history of the Eritrean community in Milan. PhD diss., University of Leicester.

Martignoni, M., and Papadopoulos, D. (2014). Genealogies of autonomous mobility. In E. F. Isin and P. Nyers (Eds.), *Routledge handbook of global citizenship studies* (pp. 38–48). London: Routledge.

Martin, E. (1990). Towards an anthropology of immunology: The body as nation state. *Medical Anthropology Quarterly, 4*(4), 410–426.

Martin, E. (1992). The end of the body? *American Ethnologist, 19*(1), 121–140.

Martin, E. (1994). *Flexible bodies: Tracking immunity in American culture from the days of polio to the age of AIDS.* Boston: Beacon.

Martin, E. (2002). "Flexible Körper." Wissenschaft und Industrie im Zeitalter des flexiblen Kapitalismus. In B. Duden and D. Noeres (Eds.), *Auf den Spuren des Körpers in einer technogenen Welt* (pp. 29–54). Opladen, Germany: Leske + Budrich.

Martin, E. (2010). Self-making and the brain. *Subjectivity, 3*(4), 366–381.

Martin, R. (2002). *Financialization of daily life.* Philadelphia: Temple University Press.

Martin, R. (2011). *Under new management: Universities, administrative labor, and the professional turn.* Philadelphia: Temple University Press.

Marx, K. (1973). *Grundrisse: Foundations of the critique of political economy* (M. Nicolaus, Trans.). London: Penguin. (Original work published 1857.)

Marx, K. (1975). Economic and philosophical manuscripts of 1844. In K. Marx and F. Engels (Eds.), *Collected works,* vol. 3 (pp. 229–348). London: Lawrence and Wishart. (Original work published 1932.)

Marx, K. (1976). Critique of the Gotha programme. In K. Marx and F. Engels (Eds.), *Collected works,* vol. 24 (pp. 75–99). London: Lawrence and Wishart. (Original work published 1875.)

Marx, K., and Engels, F. (1976). The German ideology. In K. Marx and F. Engels (Eds.), *Collected works,* vol. 5. London: Lawrence and Wishart. (Original work published 1846.)

Masterpasqua, F. (2009). Psychology and epigenetics. *Review of General Psychology, 13*(3), 194–201.

Mattoni, A., and Doerr, N. (2007). Images within the precarity movement in Italy. *Feminist Review, 87*(1), 130–135. doi:10.2307/30140807.

Maxigas. (2012). Hacklabs and hackerspaces: Tracing two genealogies. *Journal of Peer Production, 2.*

Maxigas and Troxler, P. (2014). Shared machine shops [Special issue]. *Journal of Peer Production, 5.*

Mayer, S. (2003). Science out of step with the public: The need for public accountability of science in the UK. *Science and Public Policy, 30*(3), 177–182.

McCarthy, J. (2005). Commons as counterhegemonic projects. *Capitalism Nature Socialism, 16*(1), 9–24. doi:10.1080/1045575052000335348.

McClintock, A. (1995). *Imperial leather: Race, gender, and sexuality in the colonial contest.* New York: Routledge.

McCormick, S. (2007). Democratizing science movements: A new framework for mobilization and contestation. *Social Studies of Science, 37*(4), 609–624.

McCulloch, W., and Pitts, W. H. (1943). A logical calculus of the idea immanent in nervous activity. *Bulletin of Mathematical Biophysics, 5*(4), 115–133.

McKay, B., Nehring, R., and Walsh-Dilley, M. (2014). The "state" of food sovereignty in Latin America: Political projects and alternative pathways in Venezuela, Ecuador and Bolivia. *Journal of Peasant Studies, 41*(6), 1175–1200. doi:10.1080/03066150.2014.964217.

McLean, S. (2009). Stories and cosmogonies: Imagining creativity beyond "nature" and "culture." *Cultural Anthropology, 24*(2), 213–245.

McLean, S. (2011). Black goo: Forceful encounters with matter in Europe's muddy margins. *Cultural Anthropology, 26*(4), 589–619. doi:10.1111 /j.1548-1360.2009.01130.x/abstract.

McNeil, M. (2013). Between a rock and a hard place: The deficit model, the diffusion model and publics in STS. *Science as Culture, 22*(4), 589–608. doi:10.1080 /14636778.2013.764068.

McNeil, M., and Haran, J. (2013). Publics of bioscience. *Science as Culture, 22*(4), 433–608. doi:10.1080/09505431.2013.812383.

Mehta, L. (2010). *The limits to scarcity: Contesting the politics of allocation.* London: Earthscan.

Meillassoux, Q. (2009). *After finitude: An essay on the necessity of contingency.* London: Bloomsbury.

Meloni, M., Williams, S., and Martin, P. (Eds.). (2016). *Biosocial matters: Rethinking sociology-biology relations in the twenty-first century.* Hoboken, NJ: Wiley-Blackwell.

Melucci, A. (1996). *Challenging codes: Collective action in the information age.* Cambridge: Cambridge University Press.

Merchant, C. (1990). *The death of nature: Women, ecology, and the scientific revolution.* New York: HarperCollins.

Merleau-Ponty, M. (1966). *Phänomenologie der Wahrnehmung.* Berlin: De Gruyter.

Merriman, J. M. (2014). *Massacre: The life and death of the Paris Commune.* New York: Basic Books.

Metzger, G., and Breitwieser, S. (2005). *Gustav Metzger: History history.* Ostfildern-Ruit, Germany: Hatje Cantz/Generali Foundation.

Meyerhoff, E., Johnson, E., and Braun, B. (2011). Time and the university. *ACME: An International E-Journal for Critical Geographies, 10*(3), 483–507.

Mezzadra, S. (2001). *Diritto di fuga: Migrazioni, cittadinanza, globalizzazione.* Verona: Ombre Corte.

Mezzadra, S. (2011a). The gaze of autonomy: Capitalism, migration and social struggles. In V. Squire (Ed.), *The contested politics of mobility: Borderzones and irregularity* (pp. 121–143). London: Routledge.

Mezzadra, S. (2011b). The topicality of prehistory: A new reading of Marx's analysis of "so-called primitive accumulation." *Rethinking Marxism, 23*(3), 302–321. doi:10.1080/08935696.2011.582995.

Michael, M. (2006). *Technoscience and everyday life: The complex simplicities of the mundane.* Maidenhead, UK: Open University Press.

Middleton, D., and Brown, S. (2005). *The social psychology of experience: Studies in remembering and forgetting.* London: Sage.

Midnight Notes Collective. (1990). Introduction to the New Enclosures. *Midnight Notes, 10*(1–9).

Mignolo, W. (2002). The geopolitics of knowledge and the colonial difference. *South Atlantic Quarterly, 101*(1), 57–96.

Mignolo, W. (2007). Delinking. *Cultural Studies, 21*(2), 449–514. doi:10.1080/09502380601162647.

Mignolo, W. (2011). Geopolitics of sensing and knowing: On (de)coloniality, border thinking and epistemic disobedience. *Postcolonial Studies: Culture, Politics, Economy, 14*(3), 273–283. doi:10.1080/13688790.2011.613105.

Mignolo, W. (2013). On pluriversality. *Walter Mignolo.* http://waltermignolo.com/on-pluriversality/.

Mignolo, W., and Escobar, A. (Eds.). (2010). *Globalization and the decolonial option.* London: Routledge.

Millo, Y., and MacKenzie, D. (2009). The usefulness of inaccurate models: Towards an understanding of the emergence of financial risk management. *Accounting, Organizations and Society, 34*(5), 638–653.

Mirowski, P. (2011). *Science-mart: Privatizing American science.* Cambridge, MA: Harvard University Press.

Misa, T. J. (2011). *Leonardo to the Internet: Technology and culture from the Renaissance to the present* (2nd ed.). Baltimore: Johns Hopkins University Press.

Mishel, L. R., Bernstein, J., and Allegretto, S. A. (2007). *The state of working America 2006/2007.* Ithaca, NY: Cornell University Press.

Mitchell, A. R., Jeppesen, P., Nicol, L., Morrison, H., and Kipling, D. (1996). Epigenetic control of mammalian centromere protein binding: Does DNA methylation have a role? *Journal of Cell Science, 109*(9), 2199–2206.

Moeran, B., and Pedersen, J. S. (Eds.). (2011). *Negotiating values in the creative industries: Fairs, festivals and competitive events.* Cambridge: Cambridge University Press.

Mol, A. (1999). Ontological politics: A word and some questions. *Sociological Review, 47*(S1), 74–89. doi:10.1111/1467-954X.46.s.5.

Mol, A. (2002). *The body multiple: Ontology in medical practice.* Durham, NC: Duke University Press.

Mol, A., and Law, J. (2004). Embodied action, enacted bodies: The example of hypoglycaemia. *Body and Society, 10*(2–3), 43–62. doi:10.1177/1357034X04042932.

Mollison, B. (1988). *Permaculture: A designers' manual* (2nd ed.). Sisters Creek, Tasmania: Tagari.

Mollison, B., and Holmgren, D. (1990). *Permaculture one: A perennial agriculture for human settlements.* Sisters Creek, Tasmania: Tagari.

Mollison, B., and Slay, R. M. (1991). *Introduction to permaculture.* Sisters Creek, Tasmania: Tagari.

Moore, A. W. (2015). *Occupation culture: Art and squatting in the city from below.* London: Minor Compositions.

Mora, C., Tittensor, D. P., Adl, S., Simpson, A. G., and Worm, B. (2011). How many species are there on Earth and in the ocean? *PLOS Biology, 9*(8), e1001127. doi:10.1371/journal.pbio.1001127.

Mora-Gámez, F. (2016). Reparation beyond statehood: Assembling rights restitution in post-conflict Colombia. PhD diss., University of Leicester.

Moreira, T. (2006). Heterogeneity and coordination of blood pressure in neurosurgery. *Social Studies of Science, 36*(1), 69–97.

Morini, C. (2007). The feminization of labour in cognitive capitalism. *Feminist Review, 87*(1), 40–59. doi:10.2307/30140799.

Morini, C., and Fumagalli, A. (2010). Life put to work: Towards a life theory of value. *Ephemera: Theory and Politics in Organization, 10*(3/4), 234–252.

Morita, A. (2013). Traveling engineers, machines, and comparisons: Intersecting imaginations and journeys in the Thai local engineering industry. *East Asian Science, Technology and Society, 7*(2), 221–241. doi:10.1215/18752160-2145403.

Morris, S. J., and Gotel, O. C. Z. (2006). Flow diagrams: Rise and fall of the first software engineering notation. In D. Barker-Plummer, R. Cox, and N. Swoboda (Eds.), *Diagrammatic representation and inference: Proceedings of the 4th international conference, Diagrams 2006, Stanford, CA, June 28–30, 2006* (pp. 130–144). New York: Springer.

Morse, J. M. (2006). The Politics of evidence. *Qualitative Health Research, 16*(3), 395–404. doi:10.1177/1049732305285482.

Moser, I. (2008). Making Alzheimer's disease matter: Enacting, interfering and doing politics of nature. *Geoforum, 39*(1), 98–110.

Moses, M. V. (2001). Magical realism at world's end. *Literary Imagination, 3*(1), 105–133.

Mouffe, C. (2000). *The democratic paradox.* London: Verso.

Moulier Boutang, Y. (1998). *De l'esclavage au salariat: Economie historique du salariat bridé.* Paris: Presses Universitaires de France.

Moulier Boutang, Y. (2011). *Cognitive capitalism.* Cambridge: Polity.

Moulier Boutang, Y., and Grelet, S. (2001). The art of flight: An interview with Yann Moulier-Boutang. *Rethinking Marxism, 13*(3–4), 227–235.

Müller, G. B. (2007). Evo-devo: Extending the evolutionary synthesis. *Nature Reviews Genetics, 8*(12), 943–949.

Müller, H. (2000). Kinder, denkt an die Zwangsläufigkeit. Freiheit, Korruption, Konterrevolution: Ein Gespräch zwischen Sascha Anderson, Heiner Müller und A. R. Penck vor zehn Jahren. *Frankfurter Allgemeine Zeitung, BS* 3, January 20.

Müller, R., and Kenney, M. (2014). Agential conversations: Interviewing postdoctoral life scientists and the politics of mundane research practices. *Science as Culture, 23*(4), 537–559.

Muniesa, F. (2007). Market technologies and the pragmatics of prices. *Economy and Society, 36*(3), 377–395. doi:10.1080/03085140701428340.

Munro, R. (2004). Just waiting: Endless deferral and the social injustice of "suspending" participants between bidding and evaluation. In R. Lippens (Ed.), *Imaginary boundaries of justice: Social and legal justice across disciplines* (pp. 51–67). Oxford: Hart.

Munro, R., and Mouritsen, J. (Eds.). (1996). *Accountability: Power, ethos, and the technologies of managing.* London: International Thomson Business Press.

Munster, A. (2013). *An aesthesia of networks: Conjunctive experience in art and technology.* Cambridge, MA: MIT Press.

Murgia, A., and Selmi, G. (2012). "Inspire and conspire": Italian precarious workers between self-organization and self-advocacy. *Interface: A Journal for and about Social Movements, 4*(2), 181–196.

Murphy, M. (2012). *Seizing the means of reproduction: Entanglements of feminism, health, and technoscience.* Durham, NC: Duke University Press.

Murphy, M. (2013). Economization of life: Calculative infrastructures of population and economy. In P. Rawes (Ed.), *Relational architectural ecologies: Architecture, nature and subjectivity* (pp. 139–155). London: Routledge.

Musiani, F. (2013). Decentralizing DNS: Peers, infrastructure, and Internet governance. *Georgetown Journal of International Affairs,* 111–118.

Musiani, F. (2015). Practice, plurality, performativity, and plumbing: Internet governance research meets science and technology studies. *Science, Technology, and Human Values, 40*(2), 272–286. doi:10.1177/0162243914553803.

Myers, N. (2008). Molecular embodiments and the body-work of modeling in protein crystallography. *Social Studies of Science, 38*(2), 163–199.

Myers, N. (2012). Dance your PhD: Embodied animations, body experiments, and the affective entanglements of life science research. *Body and Society, 18*(1), 151–189. doi:10.1177/1357034x11430965.

Myers, N. (2014). Sensing botanical sensoria: A kriya for cultivating your inner plant. *Centre for Imaginative Ethnography, Imaginings Series.* http://imaginativeethnography.org/imaginings/affect/sensing-botanical -sensoria/.

Myers, N. (2015). Conversations on plant sensing: Notes from the field. *NatureCulture, 3,* 35–66.

Myers, N., and Dumit, J. (2011). Haptic creativity and the mid-embodiments of experimental life. In F. E. Mascia-Lees (Ed.), *A companion to the anthropology of the body and embodiment* (pp. 239–261). Oxford: Wiley-Blackwell.

Nancy, J.-L. (1991). *The inoperative community.* Minneapolis: University of Minnesota Press.

Nancy, J.-L. (2000). *Being singular plural.* Stanford, CA: Stanford University Press.

Neeson, J. M. (1993). *Commoners: Common right, enclosure and social change in common-field England, 1700–1820.* Cambridge: Cambridge University Press.

Negri, A. (1988). *Revolution retrieved: Writings on Marx, Keynes, capitalist crisis, and new social subjects (1967–83).* London: Red Notes.

Negri, A. (2005a). *Books for burning: Between civil war and democracy in 1970s Italy.* London: Verso.

Negri, A. (2005b). *The politics of subversion: A manifesto for the twenty-first century.* Cambridge: Polity.

Negri, A. (2010). Postface: A reflection on income in the "great crisis" of 2007 and beyond. In A. Fumagalli and S. Mezzadra (Eds.), *Crisis in the global economy: Financial markets, social struggles, and new political scenarios* (pp. 263–271). New York: Semiotext(e).

Neidich, W. (Ed.) (2014). *The psychopathologies of cognitive capitalism: Part two.* Berlin: Archive Books.

Newfield, C. (2008). *Unmaking the public university: The forty-year assault on the middle class.* Cambridge, MA: Harvard University Press.

Newfield, C. (2011). Devolving public universities: Lessons from America. *Radical Philosophy, 169,* 36–42.

Newfont, K. (2012). *Blue Ridge commons: Environmental activism and forest history in western North Carolina.* Athens: University of Georgia Press.

Neyland, D., and Woolgar, S. (2002). Accountability in action? The case of a database purchasing decision. *British Journal of Sociology, 53*(2), 259–274.

Nichols, J. (1800). *The history and antiquities of the county of Leicester,* vol. 3, part 1. Leicester: John Nichols.

Nichols, T. (Ed.) (1980). *Capital and labour: Studies in the capitalist labour process.* London: Fontana.

Niewohner, J. (2011). Epigenetics: Embedded bodies and the molecularisation of biography and milieu. *BioSocieties, 6*(3), 279–298.

Noorani, T., Blencowe, C., and Brigstocke, J. (2013). *Problems of participation: Reflections on authority, democracy, and the struggle for common life.* Lewes, UK: ARN Press.

Nordhaus, W. D. (2004). *Retrospective on the 1970s productivity slowdown* (Working Paper 109501). Cambridge, MA: National Bureau of Economic Research.

North, L. L., and Grinspun, R. (2016). Neo-extractivism and the new Latin American developmentalism: The missing piece of rural transformation. *Third World Quarterly, 37*(8), 1483–1504. doi:10.1080/01436597.2016.1159508.

Nowotny, H., and Rose, H. (1979). *Counter-movements in the sciences: The sociology of the alternatives to big science.* Dordrecht, Netherlands: Reidel.

Nowotny, H., Scott, P., and Gibbons, M. (2001). *Re-thinking science: Knowledge and the public in an age of uncertainty.* Cambridge: Polity.

Nunes, R. (2012). The lessons of 2011: Three theses on organisation. *Mute.* http://www.metamute.org/editorial/articles/lessons-2011-three-theses -organisation#.

Nunes, R. (2014). *Organisation of the organisationless: Collective action after networks.* London: Mute Books.

Nyers, P. (2006). The accidental citizen: Acts of sovereignty and (un)making citizenship. *Economy and Society, 35*, 22–41.

Nyers, P. (2010). Forms of irregular citizenship. In V. Squire (Ed.), *The contested politics of mobility: Borderzones and irregularity* (pp. 184–198). New York: Routledge.

Nyers, P. (2015). Migrant citizenships and autonomous mobilities. *Migration, Mobility, and Displacement, 1*(1), 23–39.

Nyers, P. (Ed.) (2009). *Securitizations of citizenship.* London: Routledge.

Nyers, P., and Rygiel, K. (2012). *Citizenship, migrant activism and the politics of movement.* London: Routledge.

O'Connell Davidson, J. (2010). New slavery, old binaries: Human trafficking and the borders of "freedom." *Global Networks, 10*(2), 244–261.

O'Connell Davidson, J. (2011). Moving children? Child trafficking, child migration, and child rights. *Critical Social Policy, 31*(3), 454–477.

O'Connell Davidson, J. (2015). *Modern slavery: The margins of freedom.* New York: Palgrave Macmillan.

O'Riordan, K. (2013). Biodigital publics: Personal genomes as digital media artefacts. *Science as Culture, 22*(4), 516–539. doi:10.1080/14636778.2013.764069.

Olivera, O., and Lewis, T. (2004). *Cochabamba! Water war in Bolivia.* Cambridge, MA: South End.

Olwig, K. R. (2013). Globalism and the enclosure of the landscape commons. In I. D. Rotherham (Ed.), *Cultural severance and the environment: The ending of traditional and customary practice on commons and landscapes managed in common* (pp. 31–46). Dordrecht, Netherlands: Springer.

Orhangazi, O. (2008). *Financialization and the US economy.* Cheltenham, UK: Edward Elgar.

Orr, J. (2006). *Panic diaries: A genealogy of panic disorder.* Durham, NC: Duke University Press.

Orr, J. (2012). Punk justice. *Scholar and Feminist Online, 11*(1).

Orr, J. (2015). Enchanting catastrophe: Magical subrealism and BP's Macondo. *Catalyst: Feminism, Theory, Technoscience, 1*(1).

Orser Jr., C. E., and Funari, P. P. A. (2001). Archaeology and slave resistance and rebellion. *World Archaeology, 33*(1), 61–72. doi:10.1080/00438240120047636.

Ostrom, E. (1990). *Governing the commons: The evolution of institutions for collective action.* Cambridge: Cambridge University Press.

Overton, W. F. (1998). The arrow of time and cycles of time: Concepts of change, cognition and embodiment. *Psychological Inquiry, 5*, 215–237.

Oyama, S. (2000). *The ontogeny of information: Developmental systems and evolution.* Durham, NC: Duke University Press.

Ozarow, D., and Croucher, R. (2014). Workers' self-management, recovered companies and the sociology of work. *Sociology, 48*(5), 989–1006. doi:10.1177/0038038514539064.

Palmer, R. (1979). *A ballad history of England from 1588 to the present day.* London: Batsford.

Palmer, R. (1989). Ay up, me duck; Or, some Leicestershire ballads and songs. *Leicestershire Historian, 3*(7), 14–29.

Panagiotidis, E., and Tsianos, V. (2007). Denaturalising "camps": Überwachen und Entschleunigen in der Schengener Ägäis-Zone. In Transit Migration Forschungsgruppe (Ed.), *Turbulente Ränder: Neue Perspektiven auf Migration an den Grenzen Europas* (pp. 57–85). Bielefeld, Germany: Transcript.

Pandian, A. (2009). *Crooked stalks: Cultivating virtue in South India.* Durham, NC: Duke University Press.

Papadopoulos, D. (2002). Dialectics of subjectivity: North-Atlantic certainties, neo-liberal rationality, and liberation promises. *International Journal of Critical Psychology, 6,* 99–122.

Papadopoulos, D. (2003). The ordinary superstition of subjectivity: Liberalism and technostructural violence. *Theory and Psychology, 13*(1), 73–93.

Papadopoulos, D. (2006). World 2: On the significance and impossibility of articulation. *Culture, Theory and Critique, 47*(2), 165–179.

Papadopoulos, D. (2008). In the ruins of representation: Identity, individuality, subjectification. *British Journal of Social Psychology, 47*(1), 139–165. doi:10.1348/014466607X187037.

Papadopoulos, D. (2010). *L. S. Wygotski: Werk und Rezeption* (2nd ed.). Berlin: Lehmanns Media.

Papadopoulos, D. (2014). Generation M. Matter, makers, microbiomes: Compost for Gaia. *Teknokultura: Journal of Digital Culture and Social Movements, 11*(3), 637–645.

Papadopoulos, D. (2017). The two endings of the precarious movement. In A. Bove, E. Armano, and A. Murgia (Eds.), *Mapping precariousness: Subjectivities and resistance* (pp. 137–148). London: Routledge.

Papadopoulos, D., and Stephenson, N. (2007). Traveling experience: Rethinking the limits of empirical evidence in social science. In V. van Deventer, M. Terre Blanche, E. Fourie, and P. Segalo (Eds.), *Citizen city: Between constructing agent and constructed agency* (pp. 128–137). Toronto: Captus.

Papadopoulos, D., Stephenson, N., and Tsianos, V. (2008). *Escape routes: Control and subversion in the 21st century.* London: Pluto.

Papadopoulos, D., and Tsianos, V. (2007a). The autonomy of migration: The animals of undocumented mobility. In A. Hickey-Moody and P. Malins (Eds.), *Deleuzian encounters: Studies in contemporary social issues* (pp. 223–235). Basingstoke, UK: Palgrave Macmillan.

Papadopoulos, D., and Tsianos, V. (2007b). How to do sovereignty without people? The subjectless condition of postliberal power. *Boundary 2: International Journal of Literature and Culture, 34*(1), 135–172.

Papadopoulos, D., Tsianos, V., and Tsomou, M. (2015). Athens: Metropolitan blockade—real democracy. In P. Weibel (Ed.), *Global activism: Art and conflict in the 21st century* (pp. 225–232). Cambridge, MA: MIT Press.

Parisi, L. (2012). Digital design and topological control. *Theory, Culture and Society, 29*(4–5), 165–192. doi:10.1177/0263276412443568.

Parisi, L. (2013). *Contagious architecture: Computation, aesthetics, and space.* Cambridge, MA: MIT Press.

Parisi, L., and Terranova, T. (2000). Heat-death: Emergence and control in genetic engineering and artificial life. *CTheory.* http://www.ctheory.net/articles.aspx?id=127.

Parker, M. (Ed.) (2002). *Utopia and organization.* Oxford: Blackwell.

Parker, M., Cheney, G., Fournier, V., and Land, C. (Eds.). (2014). *The Routledge companion to alternative organization.* London: Routledge.

Parker, M., Fournier, V., and Reedy, P. (Eds.). (2007). *The dictionary of alternatives: Utopianism and organization.* London: Zed Books.

Passoth, J.-H., Peuker, B. M., and Schillmeier, M. (2012). *Agency without actors? New approaches to collective action.* London: Routledge.

Passoth, J.-H., and Rowland, N. J. (2010). Actor-network state: Integrating actor-network theory and state theory. *International Sociology, 25*(6), 818–841. doi:10.1177/0268580909351325.

Patton, C. (1985). *Sex and germs: The politics of AIDS.* Boston: South End Press.

Patton, C. (1990). *Inventing AIDS.* New York: Routledge.

Patton, C., and Kelly, J. (1987). *Making it: A woman's guide to sex in the age of AIDS.* Ithaca, NY: Firebrand Books.

Patton, P. (1996). Sovereignty, law and difference in Australia: After the Mabo case. *Alternatives: Global, Local, Political, 21*(2), 149–170.

Patton, P. (2000). *Deleuze and the political.* London: Routledge.

Paulani, L. M. (2009). The crisis of the finance-led regime of accumulation and the situation of Brazil. *Estudos Avançados, 23*(66), 25–39.

Peck, J., and Tickell, A. (2002). Neoliberalizing space. *Antipode, 34*(3), 380–404.

Pellandini-Simányi, L., Hammer, F., and Vargha, Z. (2015). The financialization of everyday life or the domestication of finance? How mortgages engage with borrowers' temporal horizons, relationships and rationality in Hungary. *Cultural Studies, 29*(5–6), 733–759. doi:10.1080/09502386.2015.1017142.

Pérez-Bustos, T. (2016). Embodying a caring science: An ethnographic analysis of the communicative practices of a Colombian trans-woman scientist in the media. *Universitas Humanística, 82,* 429–459. doi:10.11144/Javeriana.uh82.ecse.

Perrier, M., and Withers, D. (2016). An archival feminist pedagogy: Unlearning and objects as affective knowledge companions. *Continuum: Journal of Media and Cultural Studies, 30*(3), 355–366. doi:10.1080/10304312.2016.1166561.

Pickerill, J., and Chatterton, P. (2006). Notes towards autonomous geographies: Creation, resistance and self-management as survival tactics. *Progress in Human Geography, 30*(6), 730–746.

Pickering, A. (1994). After representation: Science studies in the performative idiom. *PSA: Proceedings of the biennal meeting of the Philosophy of Science Association, 2,* 413–419.

Pickering, A. (1995). *The mangle of practice: Time, agency, and science.* Chicago: University of Chicago Press.

Pickering, A. (2008). New ontologies. In A. Pickering and K. Guzik (Eds.), *The mangle in practice: Science, society, and becoming* (pp. 1–14). Durham, NC: Duke University Press.

Pickering, A. (2010). *The cybernetic brain: Sketches of another future.* Chicago: University of Chicago Press.

Pickering, A. (2011). H-: Brains, selves and spirituality in the history of cybernetics. *Metanexus.* http://metanexus.net/print/essay/h-brains-selves-and-spirituality-history-cybernetics.

Pickering, A. (Ed.) (1992). *Science as practice and culture.* Chicago: University of Chicago Press.

Pickering, A., and Guzik, K. (2008). *The mangle in practice: Science, society, and becoming.* Durham, NC: Duke University Press.

Pickersgill, M. (2013). The social life of the brain: Neuroscience in society. *Current Sociology, 61*(3), 322–340. doi:10.1177/0011392113476464.

Pickersgill, M., Martin, P., and Cunningham-Burley, S. (2015). The changing brain: Neuroscience and the enduring import of everyday experience. *Public Understanding of Science, 24*(7), 878–892.

Pinker, S. (1997). *How the mind works.* New York: Norton.

Pitts-Taylor, V. (2010). The plastic brain: Neoliberalism and the neuronal self. *Health: Interdisciplinary Studies in Health, Illness and Medicine, 14*(6), 635–652. doi:10.1177/1363459309360796.

Pitts-Taylor, V. (2016). *The brain's body: Neuroscience and corporeal politics.* Durham, NC: Duke University Press.

Pizio, D. (2017). Organizing privacy: Anonymous infrastructures and the emergence of the hidden Internet. PhD diss., University of Leicester. Manuscript in preparation.

Pollert, A., and Charlwood, A. (2009). The vulnerable worker in Britain and problems at work. *Work, Employment and Society, 23*(2), 343–362. doi:10.1177/0950017009106771.

Porter, G., and Tiusanen, J. (2006). Performing resistance to the new rural order: An unpublished ballad opera and the green song. *Eighteenth Century, 47*(2/3), 203–232.

Poulantzas, N. (1978). *State, power, socialism.* London: New Left Books.

Pozzolini, A. (1970). *Antonio Gramsci: An introduction to his thought.* London: Pluto.

Pratt, M. L. (1992). *Imperial eyes: Travel writing and transculturation.* London: Routledge.

Precarious Workers Brigade and Carrot Workers Collective. (2014). Free labour syndrome: Volunteer work and unpaid overtime in the creative and cultural sector. In M. Kozłowski, A. Kurant, J. Sowa, K. Szadkowski, and J. Szreder (Eds.), *Joy forever: The political economy of social creativity* (pp. 211–226). London: Mayfly Books.

Price, R. (Ed.). (1979). *Maroon societies: Rebel slave communities in the Americas.* Baltimore: Johns Hopkins University Press.

Pryke, M. (2010). Money's eyes: The visual preparation of financial markets. *Economy and Society, 39*(4), 427–459. doi:10.1080/03085147.2010.510679.

Puig de la Bellacasa, M. (2009). Touching technologies, touching visions: The reclaiming of sensorial experience and the politics of speculative thinking. *Subjectivity, 28*(1), 297–315. doi:10.1057/sub.2009.17.

Puig de la Bellacasa, M. (2010). Ethical doings in naturecultures. *Ethics, Place and Environment: A Journal of Philosophy and Geography, 13*(2), 151–169.

Puig de la Bellacasa, M. (2011). Matters of care in technoscience: Assembling neglected things. *Social Studies of Science, 41*(1), 85–106.

Puig de la Bellacasa, M. (2012). "Nothing comes without its world": Thinking with care. *Sociological Review, 60*(2), 197–216.

Puig de la Bellacasa, M. (2013). *Politiques féministes et construction des savoirs: "Penser nous devons!"* Paris: Harmattan.

Puig de la Bellacasa, M. (2014a). Ecological thinking, material spirituality, and the poetics of infrastructure. In G. Bowker, S. Timmermans, A. E. Clarke, and E. Balka (Eds.), *Boundaries objects and beyond: Working with Leigh Star* (pp. 47–68). Cambridge, MA: MIT Press.

Puig de la Bellacasa, M. (2014b). Encountering bioinfrastructure: Ecological struggles and the sciences of soil. *Social Epistemology, 28*(1), 26–40. doi:10.1080/02 691728.2013.862879.

Puig de la Bellacasa, M. (2014c). *Les savoirs situés de Sandra Harding et Donna Haraway: Science et épistémologies féministes.* Paris: Harmattan.

Puig de la Bellacasa, M. (2015). Making time for soil: Technoscientific futurity and the pace of care. *Social Studies of Science, 45*(5), 691–716. doi:10.1177 /0306312715599851.

Puig de la Bellacasa, M. (2017). *Matters of care: Speculative ethics in more than human worlds.* Minneapolis: University of Minnesota Press.

Pulido Martinez, H. C. (2013). Outlining critical psychology of work in Latin America. *Annual Review of Critical Psychology, 10,* 672–689.

Quijano, A. (2000). Coloniality of power, Eurocentrism, and Latin America. *Nepantla: Views from South, 1*(3), 533–580.

Quijano, A. (2007). Coloniality and modernity/rationality. *Cultural Studies, 21*(2), 168–178. doi:10.1080/09502380601164353.

Quijano, A., and Wallerstein, I. M. (1992). Americanity as a concept, or the Americas in the modern world-system. *International Social Science Journal, 44,* 549–557.

Rabinow, P. (1996). *Essays on the anthropology of reason.* Princeton, NJ: Princeton University Press.

Raffles, H. (2002). *In Amazonia: A natural history.* Princeton, NJ: Princeton University Press.

Rancière, J. (1998). *Disagreement: Politics and philosophy.* Minneapolis: University of Minnesota Press.

Rapp, R. (2000). *Testing women, testing the fetus: The social impact of amniocentesis in America.* New York: Routledge.

Rappert, B. (2014). Present absences: Hauntings and whirlwinds in "-graphy." *Social Epistemology, 28*(1), 41–55. doi:10.1080/02691728.2013.862876.

Rappert, B., and Bauchspies, W. K. (2014). Introducing absence. *Social Epistemology, 28*(1), 1–3. doi:10.1080/02691728.2013.862875.

Raqs Media Collective. (2003). X notes on practice: Stubborn structures and insistent seepage in a networked world. In V. Marina and G. Melanie (Eds.), *Immaterial labour: Work, research and art* (pp. 213–231). London: Black Dog.

Ratner, C. (2013). *Cooperation, community, and co-ops in a global era.* New York: Springer.

Ratner, C. (2015). *The politics of cooperation and co-ops: Forms of cooperation and co-ops, and the politics that shape them.* Hauppauge, NY: Nova Science.

Rato, M., and Ree, R. (2012). Materializing information: 3D printing and social change. *First Monday, 17*(7).

Raunig, G. (2006). Instituent practices: Fleeing, instituting, transforming. *Transversal—European Institute for Progressive Cultural Policies Journal, 01.2006.* http://eipcp.net/transversal/0106/raunig/en.

Raunig, G. (2007). Instituent practices, No. 2. Institutional critique, constituent power, and the persistence of instituting. *Transversal—European Institute for Progressive Cultural Policies Journal, 01.2007.* http://eipcp.net/transversal/0507/raunig/en.

Ravetz, J. R. (1971). *Scientific knowledge and its social problems.* Oxford: Clarendon.

Ravetz, J. R. (1990). *The merger of knowledge with power: Essays in critical science.* London: Mansell.

Ravetz, J. R. (2006). *The no-nonsense guide to science.* Oxford: New Internationalist.

Rawls, J. (1971). *A theory of justice.* Cambridge, MA: Belknap Press of Harvard University Press.

Read, J. (2011). The production of subjectivity: From transindividuality to the commons. *New Formations, 70*(1), 113–131. doi:10.3898/newf.70.07.2010.

Reardon, J. (2012). The democratic, anti-racist genome? Technoscience at the limits of liberalism. *Science as Culture, 21*(1), 25–47. doi:10.1080/09505431.2011.565322.

Rediker, M. (1987). *Between the devil and the deep blue sea: Merchant seamen, pirates and the Anglo-American maritime world, 1700–1750.* Cambridge: Cambridge University Press.

Reid, H. G., and Taylor, B. (2010). *Recovering the commons: Democracy, place, and global justice.* Urbana: University of Illinois Press.

Renaut, A. (1997). *The era of the individual: A contribution to a history of subjectivity.* Princeton, NJ: Princeton University Press.

Resnick, S. A., and Wolff, R. D. (2006). *New departures in Marxian theory.* London: Routledge.

Reynolds, L. (2013). The contested publics of the UK GM controversy: A tale of entanglement and purification. *Science as Culture, 22*(4), 452–475. doi:10.1080/14636778.2013.764070.

Rheinberger, H.-J. (1997). *Toward a history of epistemic things: Synthesizing proteins in the test tube.* Stanford, CA: Stanford University Press.

Rhodes, R. A. W. (1997). *Understanding governance: Policy networks, governance, reflexivity and accountability.* Buckingham, UK: Open University Press.

Ribes, D., and Polk, J. B. (2015). Organizing for ontological change: The kernel of an AIDS research infrastructure. *Social Studies of Science, 45*(2), 214–241. doi:10.1177/0306312714558136.

Ricoveri, G. (2013). *Nature for sale: The commons versus commodities.* London: Pluto.

Rifkin, J. (2011). *The third industrial revolution: How lateral power is transforming energy, the economy, and the world.* New York: Palgrave Macmillan.

Rigby, J., and Schlembach, R. (2013). Impossible protest: Noborders in Calais. *Citizenship Studies, 17*(2), 157–172. doi:10.1080/13621025.2013.780731.

Robbins, B. (2011). The worlding of the American novel. In L. Cassuto (Ed.), *The Cambridge history of the American novel* (pp. 1097–1107). Cambridge: Cambridge University Press.

Robert, J. S. (2004). *Embryology, epigenesis, and evolution: Taking development seriously.* Cambridge: Cambridge University Press.

Roberts, C., Kippax, S., Spongberg, M., and Crawford, J. (1996). "Going down": Oral sex, imaginary bodies and HIV. *Body and Society, 2,* 107–124.

Robinson, C. J. (2000). *Black Marxism: The making of the black radical tradition.* Chapel Hill: University of North Carolina Press.

Robinson, K. S. (1992). *Red Mars.* London: HarperCollins.

Robinson, K. S. (1993). *Green Mars.* London: HarperCollins.

Robinson, K. S. (1996). *Blue Mars.* London: HarperCollins.

Roden, D. (2014). *Posthuman life: Philosophy at the edge of the human.* London: Routledge.

Rodgers, C. P., Straughton, E., Winchester, A. J. L., and Pieraccini, M. (2011). *Contested common land: Environmental governance past and present.* Abingdon, UK: Earthscan.

Rodríguez, N. (1996). The battle for the border: Notes on autonomous migration, transnational communities, and the state. *Social Justice, 23*(3), 21–37.

Rodríguez Giralt, I. (2015). Birds as lines: The production of alternative regimes of environmental management in the aftermath of a toxic disaster. *Geoforum, 66,* 156–166. doi:10.1016/j.geoforum.2015.05.002.

Rona-Tas, A., and Hiss, S. (2011). Forecasting as valuation: The role of ratings and predictions in the subprime mortgage crisis in the United States. In J. Beckert and P. Aspers (Eds.), *The worth of goods: Valuation and pricing in the economy* (pp. 223–246). Oxford: Oxford University Press.

Root, R. B. (1967). The niche exploitation pattern of the blue-gray gnatcatcher. *Ecological Monographs, 37,* 317–350.

Rorty, R. (1967). *The linguistic turn: Essays in philosophical method.* Chicago: University of Chicago Press.

Rosas, G. (2012). *Barrio libre: Criminalizing states and delinquent refusals of the new frontier.* Durham, NC: Duke University Press.

Rose, D. B. (2004). *Reports from a wild country: Ethics for decolonisation.* Sydney: University of New South Wales Press.

Rose, D. B., and van Dooren, T. (2011). Unloved others: Death of the disregarded in the time of extinctions. *Australian Humanities Review, 50.*

Rose, G. (1997). *Love's work.* London: Vintage.

Rose, H. (1983). Hand, brain, and heart: A feminist epistemology for the natural sciences. *Signs: Journal of Women in Culture and Society, 9*(1), 73–90.

Rose, H. (1994). *Love, Power and Knowledge.* Cambridge: Polity.

Rose, H., and Rose, S. (1969). *Science and society.* London: Penguin.

Rose, H., and Rose, S. (1976a). *The political economy of science: Ideology of/in the natural sciences.* London: Macmillan.

Rose, H., and Rose, S. (1976b). *The radicalisation of science: Ideology of/in the natural sciences.* London: Macmillan.

Rose, N. (2001). The politics of life itself. *Theory, Culture and Society, 18*(6), 1–30.

Rose, N. (2013). The human sciences in a biological age. *Theory, Culture and Society, 30*(1), 3–34. doi:10.1177/0263276412456569.

Rose, S. (1998). *Lifelines: Biology beyond determinism.* Oxford: Oxford University Press.

Rosenau, J. N., and Czempiel, E. O. (1992). *Governance without government: Order and change in world politics.* Cambridge: Cambridge University Press.

Rosengarten, M. (2004). Consumer activism in the pharmacology of HIV. *Body and Society, 10*(1), 91–107.

Rosengarten, M. (2009). *HIV interventions: Biomedicine and the traffic between information and flesh.* Seattle: University of Washington Press.

Rosengarten, M., and Michael, M. (2009). Rethinking the bioethical enactment of medically drugged bodies: Paradoxes of using anti-HIV drug therapy as a technology for prevention. *Science as Culture, 18*(2), 183–199. doi:10.1080/09505430902885565.

Ross, A. (2009). *Nice work if you can get it: Life and labor in precarious times.* New York: New York University Press.

Ross, K. (2015). *Communal luxury: The political imaginary of the Paris Commune.* London: Verso.

Roth, K. H. (2008). Global crisis—global proletarianisation—counter-perspectives. *Wildcat.* http://www.wildcat-www.de/en/actual/e068roth_crisis.html.

Roth, W.-M., and Barton, A. C. (2004). *Rethinking scientific literacy.* London: Routledge.

Rowe, G., and Frewer, L. J. (2004). Evaluating public-participation exercises: A research agenda. *Science, Technology, and Human Values, 29,* 512–557.

Rowland, N. J., and Passoth, J.-H. (2014). Infrastructure and the state in science and technology studies. *Social Studies of Science, 45*(1), 137–145. doi:10.1177/0306312714537566.

Rowlinson, M., and Hassard, J. (2001). Marxist political economy, revolutionary politics, and labour process theory. *International Studies of Management and Organization, 30*(4), 85–112.

Rucht, D. (1991). *Research on social movements: The state of the art in Western Europe and the USA*. Frankfurt am Main: Campus.

Rumford, C. (2008). Introduction: Citizens and borderwork in Europe. *Space and Polity, 12*(1), 1–12. doi:10.1080/13562570801969333.

Russell, T. (2010). *A renegade history of the United States*. New York: Free Press.

Ryan, K. E., and Hood, L. K. (2004). Guarding the castle and opening the gates. *Qualitative Inquiry, 10*(1), 79–95. doi:10.1177/1077800403259483.

Rygiel, K. (2011). Bordering solidarities: Migrant activism and the politics of movement and camps at Calais. *Citizenship Studies, 15*(1), 1–19. doi:10.1080/136210 25.2011.534911.

Sahlins, M. D. (2002). *Waiting for Foucault, still*. Chicago: Prickly Paradigm.

Sakai, N. (1997). *Translation and subjectivity: On "Japan" and cultural nationalism*. Minneapolis: University of Minnesota Press.

Salvini, F. (2008). The moons of Jupiter: Networked institutions in the productive transformations of Europe. *Transversal—European Institute for Progressive Cultural Policies Journal*. http://eipcp.net/transversal/0508/salvini/en.

Salvini, F. (2013). Struggles for the right to the city: Assembling politics on the streets of Barcelona. PhD diss., Queen Mary University of London.

Sampson, E. E. (1996). Establishing embodiment in psychology. *Theory and Psychology, 6*(4), 601–624.

Sánchez Criado, T., Rodríguez Giralt, I., and Mencaroni, A. (2016). Care in the (critical) making: Open prototyping, or the radicalisation of independent-living politics. ALTER—*European Journal of Disability Research, 10*(1), 24–39.

Sandbothe, M. (1998). *Die Verzeitlichung der Zeit: Grundtendenzen der modernen Zeitdebatte in Philosophie und Wissenschaft*. Darmstadt, Germany: Wissenschaftliche Buchgesellschaft.

Sandoval, C. (1991). U.S. third world feminism: The theory and method of oppositional consciousness in the postmodern world. *Genders, 10*, 1–24.

Sassen, S. (2004). The repositioning of citizenship: Emergent subjects and spaces for politics. In P. A. Passavant and J. Dean (Eds.), *Empire's new clothes: Reading Hardt and Negri* (pp. 177–201). New York: Routledge.

Savage, J. (2015). *1966: The year the decade exploded*. London: Faber and Faber.

Savage, M. (2002). Individuality and class: The rise and fall of the gentlemanly social contract in Britain. In O. Zunz, L. J. Schoppa, and N. Hiwatari (Eds.), *Social contracts under stress: The middle classes of America, Europe, and Japan at the turn of the century* (pp. 47–65). New York: Russell Sage.

Savransky, M. (2017). A decolonial imagination: Sociology, anthropology and the politics of reality. *Sociology, 51*(1), 11–26.

Savvides, L. (2018). 3D printing: Politics, material hacking and grassroots innovation. PhD diss., University of Leicester.

Sayers, D. O. (2014a). *A desolate place for a defiant people: The archaeology of maroons, indigenous Americans, and enslaved laborers in the Great Dismal Swamp*. Gainesville: University Press of Florida.

Sayers, D. O. (2014b). Maroon and leftist praxis in historical archaeology. In P. P. A. Funari and C. E. Orser Jr. (Eds.), *Current perspectives on the archaeology of african slavery in Latin America* (pp. 5–21). New York: Springer.

Scanlon, T. (1998). *What we owe to each other.* Cambridge, MA: Harvard University Press.

Scarry, E. (1985). *The body in pain: The making and unmaking of the world.* New York: Oxford University Press.

Schillmeier, M. (2010). *Rethinking disability: Bodies, senses, and things.* London: Routledge.

Schillmeier, M. (2015). Caring for social complexity in nanomedicine. *Nanomedicine, 10*(20), 3181–3193.

Schillmeier, M. (2016). *Eventful bodies: The cosmopolitics of illness.* Abingdon, UK: Routledge.

Schillmeier, M., and Domenech, M. (2010). *New technologies and emerging spaces of care.* Aldershot, UK: Ashgate.

Scholl, B. J., and Leslie, A. M. (1999). Modularity, development, and "Theory of Mind." *Mind and Language, 14*(1), 131–153.

Schrader, A. (2010). Responding to *Pfiesteria piscicida* (the fish killer): Phantomatic ontologies, indeterminacy, and responsibility in toxic microbiology. *Social Studies of Science, 40*(2), 275–306. doi:10.1177/0306312709344902.

Schraube, E. (2005). "Torturing things until they confess": Günther Anders's critique of technology. *Science as Culture, 14*(1), 77–85.

Schraube, E. (2009). Technology as materialized action and its ambivalences. *Theory and Psychology, 19*(2), 296–312. doi:10.1177/0959354309103543.

Schraube, E. (2013). First-person perspective and sociomaterial decentering: Studying technology from the standpoint of the subject. *Subjectivity, 6*(1), 12–32. doi:10.1057/sub.2012.28.

Schraube, E., and Højholt, C. (2016). *Psychology and the conduct of everyday life.* New York: Routledge.

Schull, N., and Zaloom, C. (2011). The shortsighted brain: Neuroeconomics and the governance of choice in time. *Social Studies of Science, 41*(4), 515–538. doi:10.1177/0306312710397689.

Schuster, L. (2003). *The use and abuse of political asylum in Britain and Germany.* London: Frank Cass.

Scott, J. C. (2009). *The art of not being governed: An anarchist history of upland Southeast Asia.* New Haven, CT: Yale University Press.

Scott, J. C. (2012). *Two cheers for anarchism: Six easy pieces on autonomy, dignity, and meaningful work and play.* Princeton, NJ: Princeton University Press.

Scott, J. W. (1991). The evidence of experience. *Critical Inquiry, 17*(4), 773–797.

Scott, J. W. (2001). Die Zukunft von *gender*: Fantasien zur Jahrtausendwende. In C. Honegger and A. Caroline (Eds.), *Gender—die Tücken einer Kategorie: Joan W. Scott, Geschichte und Politik; Beiträge zum Symposion anlässlich der Verleihung des Hans-Sigrist-Preises 1999 der Universität Bern an Joan W. Scott* (pp. 39–63). Zurich: Chronos.

Sedgwick, E. K. (1990). *Epistemology of the closet*. Berkeley: University of California Press.

Selmon, S. (1988). Magic realism as post-colonial discourse. *Canadian Literature, 116*, 9–24.

Sennett, R. (2008). *The craftsman*. London: Penguin.

Shapin, S. (2008). *The scientific life: A moral history of a late modern vocation*. Chicago: University of Chicago Press.

Shapin, S., and Schaffer, S. (1985). *Leviathan and the air-pump: Hobbes, Boyle, and the experimental life*. Princeton, NJ: Princeton University Press.

Sharma, N. (2009). Escape artists: Migrants and the politics of naming. *Subjectivity, 29*(1), 467–476.

Shaw, D. B. (2008). *Technoculture: The key concepts*. Oxford: Berg.

Shenk, D. (2010). *The genius in all of us: New insights into genetics, talent, and IQ*. New York: Anchor.

Shilts, R. (1987). *And the band played on: Politics, people, and the AIDS epidemic*. New York: St. Martin's Griffin.

Shiva, V. (2005). *Earth democracy: Justice, sustainability and peace*. Berkeley, CA: North Atlantic Books.

Shoemaker, R. B. (2007). *The London mob: Violence and disorder in eighteenth-century England*. London: Continuum.

Shukaitis, S. (2013). Recomposing precarity: Notes on the laboured politics of class composition. *Ephemera: Theory and Politics in Organization, 13*(3), 641–658.

Shukaitis, S. (2014). Learning not to labor. *Rethinking Marxism, 26*(2), 193–205. doi:10.1080/08935696.2014.888835.

Shukaitis, S., Graeber, D., and Biddle, E. (2007). *Constituent imagination: Militant investigations, collective theorization*. Oakland, CA: AK Press.

Simberloff, D., and Dayan, T. (1991). The guild concept and the structure of ecological communities. *Annual Review of Ecology and Systematics, 22*, 115–143.

Simon, N. (2010). *The participatory museum*: Museum 2.0. https://museumtwo .blogspot.gr/2014/07/new-approach-historic-mission-remaking.html.

Simpson, L. B. (2014). Land as pedagogy: Nishnaabeg intelligence and rebellious transformation. *Decolonization: Indigeneity, Education and Society, 3*(3), 1–25.

Sims, B. (2007). "The day after the hurricane": Infrastructure, order, and the New Orleans police department's response to Hurricane Katrina. *Social Studies of Science, 37*(1), 111–118. doi:10.1177/0306312706069432.

Sismondo, S. (2009). Ghosts in the machine: Publication planning in the medical sciences. *Social Studies of Science, 39*(2), 171–198. doi:10.1177 /0306312708101047.

Sismondo, S. (2015). Ontological turns, turnoffs and roundabouts. *Social Studies of Science, 45*(3), 441–448.

Sitrin, M. (2006). *Horizontalism: Voices of popular power in Argentina*. Oakland, CA: AK Press.

Sitrin, M. (2012). *Everyday revolutions: Horizontalism and autonomy in Argentina*. London: Zed.

Sivek, S. C. (2011). "We need a showing of all hands": Technological utopianism in MAKE magazine. *Journal of Communication Inquiry, 35*(3), 187–209. doi:10.1177/0196859911410317.

Slack, J. D. (1996). The theory and method of articulation in cultural studies. In D. Morley and K.-H. Chen (Eds.), *Stuart Hall: Critical dialogues in cultural studies* (pp. 112–127). London: Routledge.

Smith, C. W. (1999). *Success and survival on Wall Street: Understanding the mind of the market.* Lanham, MD: Rowman and Littlefield.

Smith, C. W. (2011). Coping with contingencies in equity option markets: The "rationality" of pricing. In J. Beckert and P. Aspers (Eds.), *The worth of goods: Valuation and pricing in the economy* (pp. 272–294). Oxford: Oxford University Press.

Smith, D. E. (1987). *The everyday world as problematic: A feminist sociology.* Boston: Northeastern University Press.

Smith, D. W. (2006). Axiomatics and problematics as two modes of formalization: Deleuze's epistemology of mathematics. In S. Duffy (Ed.), *Virtual mathematics: The logic of difference* (pp. 145–168). Manchester, UK: Clinamen.

Smith, D. W. (2012). *Essays on Deleuze.* Edinburgh: Edinburgh University Press.

Smith, M. R., and Marx, L. (Eds.). (1994). *Does technology drive history? The dilemma of technological determinism.* Cambridge, MA: MIT Press.

Smith, P. H. (2004). *The body of the artisan: Art and experience in the scientific revolution.* Chicago: University of Chicago Press.

Smith, W., Higgins, M., Kokkinidis, G., and Parker, M. (2015). Becoming invisible: The ethics and politics of imperceptibility. *Culture and Organization,* 1–20. doi:10.1080/14759551.2015.1110584.

Söderberg, J., and Delfanti, A. (2015). Hacking hacked! The life cycles of digital innovation [Special section]. *Science, Technology, and Human Values, 40*(5), 793–798. doi:10.1177/0162243915595091.

Sonigo, P. (2005). The robot and the forest. *Current Sociology, 53*(2), 311–322.

Spinoza, B. (1994). *A Spinoza reader: The ethics and other works* (E. M. Curley, Trans.). Princeton, NJ: Princeton University Press.

Spivak, G. C. (1999). *A critique of postcolonial reason: Toward a history of the vanishing present.* Cambridge, MA: Harvard University Press.

Staeheli, U. (2012). Listing the global: Dis/connectivity beyond representation? *Distinktion: Scandinavian Journal of Social Theory, 13*(3), 233–246. doi:10.1080/1600910x.2012.724646.

Staeheli, U. (2013). Entnetzt euch! Praktiken und Ästhetiken der Anschlusslosigkeit. *Mittelweg 36, 4,* 3–28.

Stallman, R. (2013). Why "open source" misses the point of free software. *GNU Operating System, Free Software Foundation.* http://www.gnu.org/philosophy/open-source-misses-the-point.html.

Stam, H. J. (1996). The body's psychology and the psychology's body [Special issue]. *Theory and Psychology, 6*(4), 555–764.

Stamets, P. (2005). *Mycelium running: How mushrooms can help save the world.* Berkeley, CA: Ten Speed.

Stamos, D. N. (2003). *The species problem: Biological species, ontology, and the metaphysics of biology.* Lanham, MD: Lexington Books.

Standing, G. (2011). *The precariat: The new dangerous class.* London: Bloomsbury Academic.

Star, S. L. (1983). Simplification in scientific work: An example from neuroscience research. *Social Studies of Science, 13*(2), 205–228.

Star, S. L. (1989). *Regions of the mind: Brain research and the quest for scientific certainty.* Stanford, CA: Stanford University Press.

Star, S. L. (1991). The sociology of the invisible: The primacy of work in the writings of Anselm Strauss. In D. R. Maines (Ed.), *Social organization and social process: Essays in honor of Anselm Strauss* (pp. 265–283). New York: De Gruyter.

Star, S. L. (1993). Cooperation without consensus in scientific problem solving: Dynamics of closure in open systems. In S. Easterbrook (Ed.), cscw*: Cooperation or conflict?* (pp. 93–106). London: Springer.

Star, S. L. (2010). This is not a boundary object: Reflections on the origin of a concept. *Science, Technology, and Human Values, 35*(5), 601–617. doi:10.1177/0162243910377624.

Star, S. L. (Ed.) (1995). *Ecologies of knowledge: Work and politics in science and technology.* Albany: State University of New York Press.

Starhawk. (1982). *Dreaming the dark: Magic, sex, and politics.* Boston: Beacon.

Starhawk. (2002). *Webs of power: Notes from the global uprising.* Gabriola Island, Canada: New Society.

Stark, D. (2009). *The sense of dissonance: Accounts of worth in economic life.* Princeton, NJ: Princeton University Press.

Starosielski, N. (2015). *The undersea network.* Durham, NC: Duke University Press.

Steels, L., and Brooks, R. A. (1995). *The artificial life route to artificial intelligence: Building embodied, situated agents.* Hillsdale, NJ: Lawrence Erlbaum.

Stein, J. A. (2017). The political imaginaries of 3D printing: Prompting mainstream awareness of design and making. *Design and Culture, 9*(1), 3–27. doi:10.1080/1 7547075.2017.1279941.

Steinfeld, R. J. (2001). *Coercion, contract, and free labor in the nineteenth century.* Cambridge: Cambridge University Press.

Stengers, I. (2000). *The invention of modern science.* Minneapolis: University of Minnesota Press.

Stengers, I. (2005). The cosmopolitical proposal. In B. Latour and P. Weibel (Eds.), *Making things public: Atmospheres of democracy* (pp. 994–1003). Cambridge, MA: mit Press.

Stengers, I. (2010). *Cosmopolitics I: I. The science wars, II. The invention of mechanics, III. Thermodynamics.* Minneapolis: University of Minnesota Press.

Stephenson, N. (2003). Interrupting neo-liberal subjectivities. *Continuum: Journal of Media and Cultural Studies, 17*(2), 135–146.

Stephenson, N. (2010). Emerging infectious disease/emerging forms of biological sovereignty. *Science, Technology, and Human Values, 36*(5), 616–637. doi:10.1177/0162243910388023.

Stephenson, N., and Papadopoulos, D. (2006). *Analysing everyday experience: Social research and political change.* London: Palgrave Macmillan.

Sterling, B. (1977). *Involution ocean.* London: Barnard's Inn.

Sterling, B. (2005). *Shaping things.* Cambridge, MA: MIT Press.

Stewart, M. (2013). Smooth effects: The erasure of labour and production of police as experts through augmented objects. *M/C Journal: A Journal of Media and Culture, 16*(6). http://journal.media-culture.org.au/index.php/mcjournal/article/view/746.

Stewart-Harawira, M. (2013). Challenging knowledge capitalism: Indigenous research in the 21st century. *Socialist Studies, 9*(1), 39–51.

Stockdill, B. C. (2003). *Activism against AIDS: At the intersections of sexuality, race, gender, and class.* Boulder, CO: Lynne Rienner.

Strang, V. (1997). *Uncommon ground: Cultural landscapes and environmental values.* Oxford: Berg.

Strathern, M. (1991). *Partial connections.* Lanham, MD: Rowman and Littlefield.

Strathern, M. (2005). *Kinship, law and the unexpected: Relatives are always a surprise.* Cambridge: Cambridge University Press.

Straw, W., and Warner, N. (2014). *The march of the modern makers: An industrial strategy for the creative industries.* London: Institute for Public Policy Research.

Stubbe, J. (2016). Material practice as a form of critique. *Interaction Design and Architecture(s) Journal (IxD&A), 30,* 30–46.

Suchman, L., and Bishop, L. (2000). Problematizing "innovation" as a critical project. *Technology Analysis and Strategic Management, 12*(3), 327–333. doi:10.1080/713698477.

Suchman, L. A. (2007). *Human-machine reconfigurations: Plans and situated actions* (2nd ed.). Cambridge: Cambridge University Press.

Sullivan, T. A., Warren, E., and Westbrook, J. L. (2000). *The fragile middle class: Americans in debt.* New Haven, CT: Yale University Press.

Sultan, S. E. (2007). Development in context: The timely emergence of eco-devo. *Trends in Ecology and Evolution, 22*(11), 575–582.

Sundberg, J. (2014). Decolonizing posthumanist geographies. *Cultural Geographies, 21*(1), 33–47. doi:10.1177/1474474013486067.

Sunder Rajan, K. (2006). *Biocapital: The constitution of postgenomic life.* Durham, NC: Duke University Press.

Suvin, D. (1979). *Metamorphoses of science fiction: On the poetics and history of a literary genre.* New Haven, CT: Yale University Press.

Svampa, M. (2015). The "commodities consensus" and valuation languages in Latin America. *Alternautas, 2*(1), 45–59.

Svirsky, M. (2010). The production of terra nullius and the Zionist-Palestinian conflict. In S. Bignall and P. Patton (Eds.), *Deleuze and the postcolonial* (pp. 220–250). Edinburgh: Edinburgh University Press.

Tabarés-Gutiérrez, R. (2016). Approaching maker's phenomenon. *Interaction Design and Architecture(s) Journal (IxD&A), 30,* 19–29.

Tadajewski, M., Maclaran, P., Parsons, E., and Parker, M. (Eds.). (2011). *Key concepts in critical management studies.* London: Sage.

Taks, J. (2008). "El agua es de todos/water for all/": Water resources and development in Uruguay. *Development, 51*(1), 17–22.

Tarì, M., and Vanni, I. (2005). On the life and deeds of San Precario, patron saint of precarious workers and lives. *Fibreculture Journal, 5.*

Taylor, C. (1986). Sprache und Gesellschaft. In A. Honneth and H. Joas (Eds.), *Kommunikatives Handeln* (pp. 35–52). Frankfurt am Main: Suhrkamp.

Taylor, C. (1989). *Sources of the self: The making of modern identity.* Cambridge, MA: Harvard University Press.

Taylor, C. (1991). *The malaise of modernity.* Concord, Canada: Anansi.

Taylor, C. (2007). *A secular age.* Cambridge, MA: Belknap Press of Harvard University Press.

Teo, T. (2015). Critical psychology: A geography of intellectual engagement and resistance. *American Psychologist, 70*(3), 243–254. doi:10.1037/a0038727.

Terranova, T. (2000). Free labor: Producing culture for the digital economy. *Social Text, 18*(2), 33–58.

Terranova, T. (2004). *Network culture: Politics for the information age.* London: Pluto.

Terranova, T. (2014). Red stack attack! Algorithms, capital and the automation of the common. *EuroNomade.* http://www.euronomade.info/?p=2268.

Thelen, E., and Smith, L. B. (1994). *A dynamic systems approach to the development of cognition and action.* Cambridge, MA: MIT Press.

Thomas, G. M., and Latimer, J. (2015). In/exclusion in the clinic: Down's syndrome, dysmorphology and the ethics of everyday medical work. *Sociology, 49*(5), 937–954. doi:10.1177/0038038515588470.

Thomas, N. (2010). *Islanders: The Pacific in the age of empire.* New Haven, CT: Yale University Press.

Thompson, A. O. (2006). *Flight to freedom: African runaways and maroons in the Americas.* Kingston, Jamaica: University of the West Indies Press.

Thompson, C. (2005). *Making parents: The ontological choreography of reproductive technologies.* Cambridge, MA: MIT Press.

Thompson, E., and Varela, F. J. (2001). Radical embodiment: Neural dynamics and consciousness. *Trends in Cognitive Sciences, 5*(10), 418–425.

Thompson, E. P. (1971). The moral economy of the English crowd in the eighteenth century. *Past and Present, 50*(1), 76–136. doi:10.1093/past/50.1.76.

Thompson, E. P. (1991). *Customs in common.* London: Merlin.

Thompson, M. (1994). *Long road to freedom: The advocate history of the gay and lesbian movement.* New York: St. Martin's.

Thompson, P. (2005). Foundation and empire: A critique of Hardt and Negri. *Capital and Class, 29*(2), 73–98. doi:10.1177/030981680508600105.

Thompson, P., and Smith, C. (2001). Follow the redbrick road. Reflections on pathways in and out of the labor process debate. *International Studies of Management and Organization, 30*(4), 40–67.

Thornton, D. J. (2011). *Brain culture: Neuroscience and popular media.* New Brunswick, NJ: Rutgers University Press.

Thorpe, C., and Gregory, J. (2010). Producing the post-Fordist public: The political economy of public engagement with science. *Science as Culture, 19*(3), 273–301. doi:10.1080/09505430903194504.

Thurnell-Read, T. (2014). Craft, tangibility and affect at work in the microbrewery. *Emotion, Space and Society, 13*, 46–54. doi:10.1016/j.emospa.2014.03.001.

Ticktin, M. (2016). Calais: Containment politics in the "jungle." *Funambulist: Politics of Space and Bodies, 5*, 28–33.

Tilly, C. (2004). *Social movements, 1768–2004.* Boulder, CO: Paradigm.

Timmer, M. P., and de Vries, G. J. (2007). *A cross-country database for sectoral employment and productivity in Asia and Latin America, 1950–2005: Groningen Growth and Development Centre Research Memorandum GD-98.* Groningen, Netherlands: University of Groningen.

Todorov, T. (1984). *The conquest of America: The question of the other.* New York: Harper and Row.

Tolman, E. C. (1954). A psychological model. In T. Parsons and E. A. Shils (Eds.), *Toward a general theory of action* (pp. 277–361). Cambridge, MA: Harvard University Press.

Tomba, M. (2009). Another type of Gewalt: Beyond law. Re-reading Benjamin. *Historical Materialism, 17*(1), 126–144.

Touraine, A. (1981). *The voice and the eye: An analysis of social movements.* Cambridge: Cambridge University Press.

Transit Migration Forschungsgruppe (Ed.). (2006). *Turbulente Ränder: Neue Perspektiven auf Migration an den Grenzen Europas.* Bielefeld, Germany: Transcript.

Treichler, P. A. (1999). *How to have theory in an epidemic: Cultural chronicles of AIDS.* Durham, NC: Duke University Press.

Tremayne, W. J. (2013). *The good life lab: Radical experiments in hands-on living.* North Adams, MA: Storey.

Trimikliniotis, N., Parsanoglou, D., and Tsianos, V. (2015a). Mobile commons and/in precarious spaces: Mapping migrant struggles and social resistance. *Critical Sociology, 42*, 7–8, 1035–1049.

Trimikliniotis, N., Parsanoglou, D., and Tsianos, V. (2015b). *Mobile commons, migrant digitalities and the right to the city.* Basingstoke, UK: Palgrave Macmillan.

Tronti, M. (2005). The strategy of refusal. Libcom.org. https://libcom.org/library/strategy-refusal-mario-tronti.

Troutt, D. D. (2006). *After the storm: Black intellectuals explore the meaning of Hurricane Katrina.* New York: New Press.

Truscello, M., and Gordon, U. (2013). Whose streets? Anarchism, technology and the petromodern state. *Anarchist Studies, 21*(1), 9–27.

Tsianos, V. (2015a). Die Grenzschutzpolitik der EU muss sich ändern. *Mediendienst Integration.* http://mediendienst-integration.de/artikel/vassilis-tsianos-frontex-grenzschutz-im-mittelmeer-moratorium.html.

Tsianos, V. (2015b). The militaristic tackling of immigration has been decided. *AnalyzeGreece!* http://www.analyzegreece.gr/topics/immigrants-rights-and -racism/item/201-v-tsianos-the-militaristic-tackling-of-immigration-has -been-decided.

Tsianos, V. (2017). Die (Un-)Durchlässigkeit der europäischen Außengrenzen für Flüchtlinge: Der Fall Eurodac. *Soziale Probleme, 26*(2), 189–204.

Tsianos, V., and Papadopoulos, D. (2012). Crisis, migration and the death drive of capitalism. *Afterall: A Journal of Art, Context, and Enquiry*(31), 4–11.

Tsianos, V., Papadopoulos, D., and Stephenson, N. (2012). This is class war from above and they are winning it: What is to be done? *Rethinking Marxism, 24*(3), 448–457. doi:10.1080/08935696.2012.685285.

Tsianos, V., and Pieper, M. (2011). Postliberale Assemblagen. Rassismus in Zeiten der Gleichheit. In S. Friedrich (Ed.), *Rassismus in der Leistungsgesellschaft: Analysen und kritische Perspektiven zu den rassistischen Normalisierungsproz-essen der "Sarrazindebatte"* (pp. 114–133). Münster, Germany: Edition Assemblage.

Tsing, A. (2000). The global situation. *Cultural Anthropology, 15*(3), 327–360.

Tsing, A. (2005). *Friction: An ethnography of global connection.* Princeton, NJ: Princeton University Press.

Tsing, A. L. (2015). *The mushroom at the end of the world: On the possibility of life in capitalist ruins.* Princeton, NJ: Princeton University Press.

Tuck, E., and Yang, K. W. (2012). Decolonization is not a metaphor. *Decoloniza-tion: Indigeneity, Education and Society, 1*, 1–40.

Turnbull, D. (2000). *Masons, tricksters, and cartographers: Comparative studies in the sociology of scientific and indigenous knowledge.* Amsterdam: Harwood Academic.

Turnbull, D. (2002). Travelling knowledge: Narratives, assemblage and encounters. In M.-N. Bourguet, C. Licoppe, and H. O. Sibum (Eds.), *Instruments, travel and science: Itineraries of precision from the seventeenth to the twentieth century* (pp. 273–294). London: Routledge.

Turner, B. (1984). *The body and society: Explorations in social theory.* Oxford: Blackwell.

Turner, B. (2014). Food waste, intimacy and compost: The stirrings of a new ecol-ogy? *Scan: Journal of Media Arts Culture, 11*(1).

Turner, F. J. (1993). *History, frontier, and section: Three essays.* Albuquerque: University of New Mexico Press.

Turner, S. P. (2003). *Liberal democracy 3.0: Civil society in an age of experts.* London: Sage.

Tutton, R. (2007). Constructing participation in genetic databases: Citizenship, gov-ernance, and ambivalence. *Science, Technology, and Human Values, 32*, 172–195.

Tyler, I. (2010). Designed to fail: A biopolitics of British citizenship. *Citizenship Studies, 14*(1), 61–74.

Tyler, I. (2013). *Revolting subjects: Social abjection and resistance in neoliberal Britain.* London: Zed.

Universidad Nómada. (2008). Mental prototypes and monster institutions: Some notes by way of an introduction. *Transversal—European Institute for Progressive Cultural Policies Journal, 05.2008.* http://eipcp.net/transversal/0508/universidadnomada/en.

Urbonas, N., and Urbonas, G. (2008). *Devices for action.* Barcelona: Museu d'Art Contemporani de Barcelona (MACBA).

van de Sande, M. (2013). The prefigurative politics of Tahrir Square—an alternative perspective on the 2011 revolutions. *Res Publica, 19*(3), 223–239. doi:10.1007/s11158-013-9215-9.

van Dooren, T. (2010). Biopatents and the problem/promise of genetic leaks: Farming canola in Canada. *Capitalism Nature Socialism, 21*(2), 43–63.

van Dooren, T. (2011). Vultures and their people in India: Equity and entanglement in a time of extinctions. *Australian Humanities Review, 50.* http://www.australianhumanitiesreview.org/archive/Issue-May-2011/vandooren.html.

van Dooren, T. (2014). *Flight ways: Life and loss at the edge of extinction.* New York: Columbia University Press.

van Gunsteren, H. (1998). *A theory of citizenship: Organizing plurality in contemporary democracies.* Boulder, CO: Westview.

van Speybroeck, L., van de Vijver, G., and de Waele, D. (Eds.). (2002). *From epigenesis to epigenetics: The genome in context. Annals of the New York Academy of Sciences, Volume 981.* New York: New York Academy of Sciences.

Vaneigem, R. (1996). Postface: Observations sur le manifeste. In K. Marx and F. Engels (Eds.), *Manifeste du Parti Communiste.* Paris: Mille et une nuits.

Vaneigem, R., and Obrist, H. U. (2009). Hans Ulrich Obrist in conversation with Raoul Vaneigem. *e-flux, 6.* http://www.e-flux.com/journal/in-conversation-with-raoul-vaneigem/

Varela, F. J., Thompson, E., and Rosch, E. (1991). *The embodied mind: Cognitive science and human experience.* Cambridge, MA: MIT Press.

Vargha, Z. (2015). Insurance after markets. *Economic Sociology, 17*(1), 1–65.

Vasta, E. (2011). Immigrants and the paper market: Borrowing, renting and buying identities. *Ethnic and Racial Studies, 34*(2), 187–206.

Vazquez, R. (2011). Translation as erasure: Thoughts on modernity's epistemic violence. *Journal of Historical Sociology, 24*(1), 27–44. doi:10.1111/j.1467-6443.2011.01387.x.

Vercellone, C. (2007). From formal subsumption to general intellect: Elements for a Marxist reading of the thesis of cognitive capitalism. *Historical Materialism, 15*(1), 13–36.

Vercellone, C. (2010). The crisis of the law of value and the becoming-rent of profit. In A. Fumagalli and S. Mezzadra (Eds.), *Crisis in the global economy: Financial markets, social struggles, and new political scenarios* (pp. 85–118). New York: Semiotext(e).

Verran, H. (2001). *Science and an African logic.* Chicago: University of Chicago Press.

Verran, H. (2002). A postcolonial moment in science studies: Alternative firing regimes of environmental scientists and aboriginal landowners. *Social Studies of Science, 32*(5/6), 729–762.

Vieta, M. (2009). The social innovations of autogestion in Argentina's worker-recuperated enterprises: Cooperatively reorganizing productive life in hard times. *Labor Studies Journal, 35*(3), 295–321. doi:10.1177/0160449x09337903.

Virilio, P. (1995). *The art of the motor.* Minneapolis: University of Minnesota Press.

Virilio, P. (1998). Critical space. In J. Der Derian (Ed.), *The Virilio reader* (pp. 58–72). Malden, MA: Blackwell.

Virno, P. (2007). Anthropology and theory of institutions. *Transversal—European Institute for Progressive Cultural Policies Journal, 05.2007.*, http://eipcp.net /transversal/0407/virno/en.

Vis, F. (2013). A critical reflection on Big Data: Considering APIs, researchers and tools as data makers. *First Monday, 18*(10). doi:10.5210/fm.v18i10.4878.

Viveiros de Castro, E. (2004). Perspectival anthropology and the method of controlled equivocation. *Tipití: Journal of the Society for the Anthropology of Lowland South America, 2*(1).

von Oswald, A. (2002). Volkswagen, Wolfsburg und die italienischen "Gastarbe-iter." *Archiv für Sozialgeschichte, 42,* 55–79.

Vora, K. (2009). Indian transnational surrogacy and the commodification of vital energy. *Subjectivity, 28,* 266–278.

Vygotsky, L. S. (1987). *The collected works of L. S. Vygotsky,* vol. 1: *Problems of general psychology.* New York: Plenum.

Wade, R. (1988). *Village republics: Economic conditions for collective action in South India.* Cambridge: Cambridge University Press.

Wainwright, P. C., and Reilly, S. M. (Eds.). (1994). *Ecological morphology: Integrative organismal biology.* Chicago: University of Chicago Press.

Wall, D. (1999). *Earth First! and the anti-roads movement: Radical environmentalism and comparative social movements.* London: Routledge.

Wall, D. (2014). *The commons in history: Culture, conflict, and ecology.* Cambridge, MA: MIT Press.

Wallerstein, I. M. (1995). *After liberalism.* New York: New Press.

Wallerstein, I. M. (1998). *Utopistics: Or, historical choices of the twenty-first century.* New York: New Press.

Walsh, J. (2010). From border control to border care: The political and ethical potential of surveillance. *Surveillance and Society, 8*(2), 113–130.

Walsh, J. (2013). Remapping the border: Geospatial technologies and border activism. *Environment and Planning D: Society and Space, 31*(6), 969–987. doi:10.1068/d18112.

Walter-Herrmann, J., and Büching, C. (Eds.). (2013). *FabLab: Of machines, makers and inventors.* Bielefeld, Germany: Transcript.

Warhurst, C., Thompson, P., and Nickson, D. (2009). Labor process theory: Putting the materialism back into the meaning of service work. In M. Korczynski and C. L. Macdonald (Eds.), *Service work: Critical perspectives* (pp. 91–112). London: Routledge.

Warner, M. (1999). *The trouble with normal: Sex, politics, and the ethics of queer life*. New York: Free Press.

Waterman, P., Mattoni, A., Humphrys, E., Cox, L., and Esteves, A. M. (2012). For the global emancipation of labour: New movements and struggles around work, workers and precarity [Special Issue]. *Interface: A Journal for and about Social Movements, 4*(2), 1–368.

Waterton, C., Ellis, R., and Wynne, B. (2013). *Barcoding nature: Shifting cultures of taxonomy in an age of biodiversity loss*. London: Routledge.

Watney, S. (1997). *Policing desire: Pornography, AIDS, and the media* (3rd ed.). London: Cassell.

Watson, J. B. (1913). Psychology as the behaviorist views it. *Psychological Review, 20*, 158–177.

Watson, S. (1998). The neurobiology of sorcery: Deleuze and Guattari's brain. *Body and Society, 4*(4), 23–45. doi:10.1177/1357034x98004004002.

Weber, J. (2010). Making worlds: Epistemological, ontological and political dimensions of technoscience. *Poiesis and Praxis, 7*(1–2), 17–36. doi:10.1007/s10202-010-0076-4.

Webster, A. (2007). Crossing boundaries: Social science in the policy room. *Science, Technology, and Human Values, 32*, 458–478.

Weeks, J. (1995). *Invented moralities: Sexual values in the age of uncertainty*. Cambridge: Polity.

Weeks, K. (2007). Life within and against work: Affective labor, feminist critique, and post-Fordist politics. *Ephemera: Theory and Politics in Organization, 7*(1), 233–249.

Weeks, K. (2011). *The problem with work: Feminism, Marxism, antiwork politics, and postwork imaginaries*. Durham, NC: Duke University Press.

Weik, T. (1997). The archaeology of maroon societies in the Americas: Resistance, cultural continuity, and transformation in the African diaspora. *Historical Archaeology, 31*(2), 81–92.

Weinel, M. (2007). Primary source knowledge and technical decision-making: Mbeki and the AZT debate. *Studies in History and Philosophy of Science, 38*(4), 748–760.

Wellmer, A. (1977). Kommunikation und Emanzipation: Überlegungen zur "sprachanalytischen Wende" der kritischen Theorie. In U. Jaeggi and A. Honneth (Eds.), *Theorien des Historischen Materialismus* (pp. 465–499). Frankfurt am Main: Suhrkamp.

Welsh, I. (2000). New social movements. *Developments in Sociology, 16*, 43–60.

Welsh, I. (2007). Participation, innovation, and efficiency: Social movements and the new genetics in Germany and the UK. In I. Blühdorn and U. Jun (Eds.), *Economic efficiency-democratic empowerment: Contested modernization in Britain and Germany* (pp. 275–295). Lanham, MD: Lexington Books.

Welsh, I., Plows, A., and Evans, R. (2007). Human rights and genomics: Science, genomics and social movements at the 2004 London Social Forum. *New Genetics and Society, 26*(2), 123–135.

Welsh, I., and Wynne, B. (2013). Science, scientism and imaginaries of publics in the UK: Passive objects, incipient threats. *Science as Culture, 22*(4), 540–566. doi:10.1080/14636778.2013.764072.

Wenzel, J. (2006). Petro-magic-realism: Toward a political ecology of Nigerian literature. *Postcolonial Studies: Culture, Politics, Economy, 9*(4), 449–464. doi:10.1080/13688790600993263.

West-Eberhard, M. J. (2003). *Developmental plasticity and evolution.* Oxford: Oxford University Press.

Weston, K. (2013). Biosecuritization: The quest for synthetic blood and the taming of kinship. In C. H. Johnson, B. Jussen, D. W. Sabean, and S. Teuscher (Eds.), *Blood and kinship: Matter for metaphor from ancient Rome to the present* (pp. 244–265). New York: Berghahn.

Wexler, B. E. (2006). *Brain and culture: Neurobiology, ideology, and social change.* Cambridge, MA: MIT Press.

White, K. (1992). Elements of geopoetics. *Edinburgh Review, 88*, 163–178.

Whitehead, A. N. (1979). *Process and reality: An essay in cosmology* (D. R. Griffin and D. W. Sherburne, Eds.). New York: Free Press.

Wildcat, M., Irlbacher-Fox, S., Coulthard, G., and McDonald, M. (2014). Indigenous land-based education [Special issue]. *Decolonization: Indigeneity, Education and Society, 3*(3), i–xv, 1–186.

Williams, R. (1977). *Marxism and literature.* Oxford: Oxford University Press.

Williams, R. (1980). *Problems in materialism and culture: Selected essays.* London: Verso.

Williams, S. E., and Hero, J. M. (1998). Rainforest frogs of the Australian Wet Tropics: Guild classification and the ecological similarity of declining species. *Proceedings of the Royal Society B, 265*(1396), 597–602.

Wilson, A. (2008). Gustav Metzger's auto-destructive/auto-creative art. *Third Text, 22*(2), 177–194. doi:10.1080/09528820802012844.

Wilson, E. A. (1998). *Neural geographies: Feminism and the microstructure of cognition.* New York: Routledge.

Wilson, E. A. (2004). *Psychosomatic: Feminism and the neurological body.* Durham, NC: Duke University Press.

Wilson, R. (2015). Toward an ecopoetics of Oceania worlding the Asia-Pacific region as space-time ecumene. In Y. Shu and D. E. Pease (Eds.), *American studies as transnational practice: Turning toward the transpacific* (pp. 213–236). Hanover, NH: Dartmouth College Press.

Wilson, R., and Connery, C. (Eds.). (2005). *Worldings: World literature, field imaginaries, future practices. Doing cultural studies inside the U.S. war-machine.* Santa Cruz, CA: New Pacific Press.

Wimmer, R. (1980). *Universalisierung in der Ethik: Analyse, Kritik und Rekonstruktion ethischer Rationalitätsansprüche.* Frankfurt am Main: Suhrkamp.

Winner, L. (1986). *The whale and the reactor: A search for limits in an age of high technology.* Chicago: University of Chicago Press.

Winner, L. (1993). Upon opening the black box and finding it empty: Social constructivism and the philosophy of technology. *Science, Technology, and Human Values, 18*(3), 362.

Wired. (2011). How to make stuff. April 19. https://www.wired.com/magazine/19-04/.

Withers, D. (2013). Re-enacting process: Temporality, historicity and the Women's Liberation Music Archive. *International Journal of Heritage Studies, 20*(7–8), 688–701. doi:10.1080/13527258.2013.794745.

Withers, D. (2015). *Feminism, digital culture and the politics of transmission: Theory, practice and cultural heritage.* London: Rowman and Littlefield.

Withers, D. (2016). Theorising the women's liberation movement as cultural geritage. *Women's History Review, 25*(5), 847–862. doi:10.1080/09612025.2015.1132871.

Wittgenstein, L. (1958). *Philosophical investigations.* New York: Prentice Hall.

Wolf, P., and Troxler, P. (2016). Community-based business models: Insights from an emerging maker economy. *Interaction Design and Architecture(s) Journal (IxD&A), 30*, 75–94.

Wolfe, C. (2010). *What is posthumanism?* Minneapolis: University of Minnesota Press.

Wolfe, P. (2006). Settler colonialism and the elimination of the native. *Journal of Genocide Research, 8*(4), 387–409.

Wolff, R. D. (2008). Capitalist crisis, Marx's shadow. *Monthly Review.* http://monthly review.org/mrzine/wolff260908.html.

Woodhouse, E., Hess, D. J., Breyman, S., and Martin, B. (2002). Science studies and activism: Possibilities and problems for reconstructivist agendas. *Social Studies of Science, 32*(2), 297–320.

Woodworth, R. S. (1921). *Psychology.* New York: Holt.

Woodworth, R. S. (1938). *Experimental psychology.* New York: Holt.

Woolgar, S., and Lezaun, J. (2013). A turn to ontology in science and technology studies? [Special issue]. *Social Studies of Science, 43*(3), 321–462.

Wygotski, L. S. (1987). *Ausgewählte Schriften*, bd. 2: *Arbeiten zur psychischen Entwicklung der Persönlichkeit.* Cologne: Pahl-Rugenstein.

Wynne, B. (2002). Risk and environment as legitimatory discourses of technology: Reflexivity inside out? *Current Sociology, 50*(3), 459–477.

Wynne, B. (2003). Seasick on the Third Wave? Subverting the hegemony of propositionalism: Response to Collins and Evans (2002). *Social Studies of Science, 33*(3), 401–417.

Wynne, B. (2005). Reflexing complexity: Post-genomic knowledge and reductionist returns in public science. *Theory, Culture and Society, 22*(5), 67–94.

Yeldan, A. E. (2008). The crisis and the periphery: The view from Turkey. Paper presented at the School of Social Sciences, Cardiff University, December 9, 2008.

York, P. (2002). The ethics of terraforming. *Philosophy Now, 38*, 6–9.

Yusoff, K. (2009). Excess, catastrophe, and climate change. *Environment and Planning D: Society and Space, 27,* 1010–1029.

Yusoff, K. (2013a). The geoengine: Geoengineering and the geopolitics of planetary modification. *Environment and Planning A, 45*(12), 2799–2808. doi:10.1068/a45645.

Yusoff, K. (2013b). Insensible worlds: Postrelational ethics, indeterminacy and the (k)nots of relating. *Environment and Planning D: Society and Space, 31*(2), 208–226. doi:10.1068/d17411.

Zalasiewicz, J., Williams, M., Haywood, L., and Ellis, M. (2011). The Anthropocene: A new epoch of geological time? [Special issue]. *Philosophical Transactions of the Royal Society, 369*(1938), 833–1112.

Zelizer, V. A. R. (1979). *Morals and markets: The development of life insurance in the United States.* New York: Columbia University Press.

Zibechi, R. (2009). Cochabamba: From water war to water management. *America Program, 27.* http://www.cipamericas.org/archives/1723.

Zibechi, R. (2011). *Territorien des Widerstands: Eine politische Kartografie der urbanen Peripherien Lateinamerikas.* Hamburg: Assoziation A.

Zibechi, R., and Ryan, R. (2010). *Dispersing power: Social movements as anti-state forces.* New York: AK Press.

Ziemke, T. (2001). Are robots embodied? Paper presented at the First International Workshop on Epigenetic Robotics: Modeling Cognitive Development in Robotic Systems, Lund University Cognitive Studies.

Žižek, S. (2008). *Violence: Six sideways reflections.* New York: Picador.

Zubrin, R. (1994). The significance of the Martian frontier. *National Space Society.* http://www.nss.org/settlement/mars/zubrin-frontier.html.

INDEX

Anzaldúa, Gloria, 109–110
appropriation, 13, 17, 37, 40, 90, 98, 106–107, 250n96; expropriation, 32–33, 136, 197; of frontier, 14–15
Arduino, 178–179, 180, 191–192
Arduino Certified Windows operating system, 191–192
Arrighi, G., 218n14
articulation, 86
artificial intelligence, 130–131
artisan production, 92, 168–169, 251n109
assemblages, 106–107, 149–150; state as heterogeneous, 60, 81, 96, 98
assetization, 8, 41–43, 191
asymmetry, 151–152
asylum seekers, 67, 76
Austin, George, 124–125
autocreative brain matter, 133–137, *135*
autonomous social movements, 4, 50
autonomy, 3–4, 6, 41, 43, 203, 215n19; as independence from capitalism, 37–38, 223n55, 224n64, 225n73; indigenous, 173–174; material articulations, 47–48; mobility and, 51–52; of transmigration, 59–60
autonomy-of-migration thesis, 59–60, 66

Badiou, Alain, 107
Bakhtin, Mikhail, 107, 112
Barad, K., 20
baroque fieldworking, 4–7
Barrio libre: Criminalizing States and Delinquent Refusals of the New Frontier (Rosas), 53
Bayat, Asaf, 64–65, 194
behaviorism, 123–134, *135*
being-in-common, 200
Benjamin, Walter, 19–20, 111–113, 233–234n58
Berardi, Bifo, 37–38
Bergson, Henri, 225n70
biofinancialization, 7–8, 27–48, 150, 179, 191; assetization and rent, 41–43; code in, 45–46; culture of valuation,

30–31; embodied value production, 33–35, 54; end of refusal of work, 37–39; expropriation of goods and resources, 32–33; future and, 31–32; geocide and geoengineering, 27–28; material articulations, 47–48; performance of, 39–41; postliberalism, 35–37; social science fiction, 43–45, 48; universalizing matrix of, 28–29, 48, 50. *See also* financialization
biohacking, 179
bios, 8, 45. *See also* biofinancialization
biosecuritization, 191
biosubjectification, 134
biovalue production, 135–136
Bishop, H., 73
black communities, 211–212n6
blackmail of precarization, 189–190
Blaser, M., 256n162
body: cerebral, 120, 129; cultural conceptions of, 119; geobody, 45; hereditary, 122–123, 129; immune, 121–122, 129; politics, 85–86, 128; visions of, 120–123. *See also* embodiment
Böhm, S., 41
Bonneuil, C., 192
Borderlands/La Frontera: The New Mestiza (Anzaldúa), 109–110
borders, 8, 53, 57, 65, 143; borderdwelling, 107–110; citizenship as, 50–51, 57; illegalized crossing, 61–63; virtual spaces, 67–68
borderworkers, 72
Boyle, Robert, 92
brain: embodiment, 5, 9, 122, 126–128, *135*
brain (brain matter), 117–137; autocreative, 133–137, *135*, 137; cognitivism, 120, 123–126; connectionism, 119–20, 123–126; ecomorphs, 132–133, 238n79; embodiment, 5, 9, 122, 126–128, *135*; epigenesis and, 5, 9, 12–13, 132–133, 238n79; epochs of, 117, 123, 128, 131, 245n28; experi-

events, 107–108, 110, 182
everyday existence and practices, 4, 8, 47, 80, 198, 212–213n11; AIDS activism, 157; brain matter and, 136; care, 23, 72–74; commoning, 99–103; culture of valuation, 31, 38; financialization of, 35, 249n94; justice and, 65–66, 112–114, 140; mobile commons, 51–52; nonpolitical, 64–65; North African revolutions, 194–195; politics as, 96–97; technoscience, 183–184
exceptionality, 112, 114, 233–234n58
experience, 126–128, 136, 139; continuous, 194; situatedness and indeterminacy of, 151–153; speculative, 19–20
Experimental College of the Twin Cities, 200
experimental politics, 3, 72
experimental practice, 3–10, 13, 92, 182–183, 189–90, 207, 212n10
expertise, 142–145
exploitation, 31–33; of one's own future, 31, 35, 249n94
expropriation, 32–33, 136, 197
extensification, 33–34, 35, 54

fabulation, 45
Facebook, 49, 67, 70, 75
fate, 111
fear, body and, 121–122
Fear of Falling (Ehrenreich), 36
"fear of falling," 36
feminist materialism, 88, 108
feminist movement, 30, 86–87, 106, 128, 135, 151, 224n64
fiction. *See* social science fiction
fieldwork, 4–7
financialization, 7–8; of everyday existence, 35, 249n94; global, 219–220n27; of social justice, 36. *See also* biofinancialization
fold, 10, 192–193, 196–197, 206
Food and Drug Administration (FDA), 139

food and supply chain, 41–42, *42*, *44*
Ford Motor Company's Production Development Center, 170
forest stacking strategy, 175, 184–185
formalist politics, 141, 143–145, 151
foundationalism, 120
Fox, Hannah, 161
framing, 147
free and open software movement, 22–23, 204–205, 255n158
freedom, appropriation of, 98, 105. *See also* labor
Frontex, 67
frontier of matter, 12–15, 20, 28
fundamentalism, 151
future: exploitation of, 31–32, 35, 219n22, 220n31, 249n94; immune body and, 121; "In the Future" campaign (HBSC), 41–42, *42*, *44*; justice and, 113–114; reclaiming past as, 174
F. W. Hampshire chemical factory, 161–162

Geertz, Clifford, 19
gene expression, 12–13
general strike, 112–113, 234n60
genetic providers, 193
genome, 117, 132–133
geocide, 27–28
geocolonialism, 225n71
geoengineering, 27–28, 46
geology, 100
The German Ideology (Marx), 80
Gewalt, 111–113, 233–234n58
Ghelfi, Andrea, 201, 225n73, 247n57
gift economy, 66–68, 73, 185
Gilbert, S. F., 132
Gilroy, Paul, 2, 211–212n6, 212n7
global consciousness, 174
Global North societies: biofinancialization, 28–29; brain imaginaries, 134; freedom and exploitation, mix of, 105; governance, 150; liberal demo-

cratic principles, 36–37; maker culture and, 166–168, 172; material and ecological commons, 33; perpetual crisis, 30
global political activity, 194–195
gnosiological realism, 82, 83, 85
The Good Life Lab: Radical Experiments in Hands-On Living (Tremayne), *186*
Goodnow, Jacqueline, 124–125
Gould, Deborah, 153
governance, 9, 148–151
governmentality, 149–150
Gramsci, Antonio, 97
Grand National Consolidated Trades Union, 197
Grazioli, Margherita, 70
Great Dismal Swamp, 100–101
Greece: Athens, 75–76; transmigrant camps, 61, 66, 67
Guattari, Félix, 88–92, 107, 176, 212n10, 229n24, 254n146

hacking. *See* making
hacking, brain, 128–130
hackspaces, 5, 10, 164–165, 177
Haiti, 104
Haraway, Donna, 47–48, 93
Hartigan, J., 228–229n5, 248n71
Hatch, M., 169
Hayden, Cori, 191
HBSC advertising campaign, 41–42, *42, 44*
Hegel, G. W. F., 97
hegemony, 86
hereditary body, 122–123, 129
Hiss, S., 223–224n60
history, 112–114, 235n69; stacked, 160–162
Hobbes, Thomas, 92
Hoerl, E., 201
Højholt, C., 213n11
Hong Kong, environmental politics, 183

hope, 17, 118
Horstman, J., 118, 134
housekeeping, 94
housing rights movement, 70–71
Human Genome Project, 117, 132–133
humanist universalism, 27–28, 230n6
human rights, 57, 231n21
Hustak, Carla, 183
hybridity, 114, 127, 240n54

idealism, 80–81, 83, 88–90, 97, 176
identity politics, 86, 152
illegalization, 53, 56, 61–63
imaginaries: of body, 120–123; of brain, 119–123; genocentric, 133–134; infrastructural, 200, 203–204; transmigration, 60
immanent critique, 84–87
immanent indeterminacy, 178–179
immune body, 121–122, 129
imperceptible politics, 19, 63–66
impossible citizenship, 56–58
inclusion, 14–18, 151, 199, 215n18; co-optation, 106–107; differential, 54–58; expertise and, 142–143; participatory politics, 145–148
incomputability, 179
indeterminacy, 31–32, 39–40, 91, 178; of biofinancial societies, 35–36; of experience, 151–153
indigenous ecocosmologies, 195–196
indigenous politics, 47, 128
indigenous temporalities, 172–174, *173*
individualism, liberal, 128–130
industry, 32, 166–167, *167*
informal economies, 71–72
infrastructure, 2, 8, 33–34, 255n156; generous, 202–205; infrastructural imagination, 200, 203–204; mobility and, 68–71
innovation, 90, 132, 166, 215n17
Institute for Public Policy Research, 166
institutions, participatory politics and, 145–148

insurgent posthumanism, 8, 9, 94–114; assembling the state, 94–97; borderdwelling, 107–110; eco-commoning, 103–105; ecological commons, 99–103; embodying politics, 105–107; justice/jetztzeit, 110–113; postanthropocentric history, 113–114, 345n69; vagrancy, 97–99. *See also* posthumanism

integrationism, 64

intelligences, 68–71

intensification, 33–34, 39, 54

"In the Future" campaign (HBSC), 41–42, *42*, *44*

invention, 10, 166; distributed invention power, 182–184; experimental labor and, 189–190

investment value, 31–32

invisibilized labor, 12, 16–17, 20, 183; maker culture and, 188–189; situatedness, 151; of transmigrants, 56

involutionary process, 185, 190, 201, 254n146; generous infrastructures, 203–205

irregularity, politicization of, 61–62

Johnson, M. L., 120, 127, 134

justice, 110–113; asymmetry of, 113–114; communities of, 71–72; everyday life and, 65–66, 112–114, 140; generative, 203; making of mobile commons as, 73–74; matter and, 16–18, 114; thick, 18–19, 74

Karfakis, Nikos, 198

Kitson, Woody, 177

knowledge production, 10, 32, 82–83, 185, 201; AIDS activism, 139; alternative knowledge, 204; experimental labor, 189–190; regions of objectivity, 142–143. *See also* production

Kojève, A., 230n10

Kornan, M., 181

Kortright, Chris, 189, 219n23

Kramer, Larry, 157

Kropil, R., 181

Kuster, Brigitta, 65

labor, 5, 217n7, 219–220n27; deregulation of markets, 30, 32; dimensions of exploitation, 33–34; as dressage, 104; end of refusal of work, 37–39; experimental, 189–190; extensification, 33–34, 35, 54; free, 10, 98, 103–105, 188–189, 231n35; informal economies, 71–72; intensification, 33–34, 39, 54; invisibilized, 12, 16–17, 20, 56, 151, 183, 188–189; mobility and, 52–55; postcontractual employment, 35, 219n23, 220n31; precarization of, 5, 10, 66, 92, 189–90, 220n31, 249n94; refusal of, 37–39, 217n7, 250n96; slavery, 55, 98, 100–101, 104, 221n32. *See also* workers

labor process theory, 221–222n41, 221n34

Lakoff, G., 120, 134

Landecker, Hannah, 133

Latin America, 195–196

Latour, Bruno, 107, 114, 148–149, 151

Law, J., 149

The Laws of the Markets (Callon), 39

legitimacy, expertise and, 142–145

Leicester Hackspace, 5, 177, *177*, 201, *202*, 204

Lenin, Vladimir, 82–86, 231n13

Lesbos, Greece, 66

Levidow, L., 192

Lezaun, Javier, 192–193

liberal democratic principles, 36–37, 222n46

liberal predicament, 142–145

liberation, 9; brain matter and, 117–118, 135–136; coloniality and productionism, 13–16

life activity, 80

Linebaugh, Peter, 50, 99, 231n21

literacy, material, 200–202

lived place, 85–86

living labor, 33–34
localizationism, 125
L'Ouverture, Toussaint, 104
Luhmann, Niklas, 106
Lukács, Georg, 230n10

Mackenzie, Adrian, 197
Make magazine, 166
Maker Faire Africa, 166, 172, 173
maker guild, 185
makerspaces and culture, 5, 10,
   163–165, *164*; free labor, 188–189;
   maker faires, *161*, 166, 169, 172–173,
   173, 244n11; nonreciprocal sharing,
   181; tools, 176
making: of actors, 152, 154–155; commensality, 176–182; compositional
   politics, 154–155; constraints,
   201–202; of matter, 176; as movement within technoscience, 165;
   mycomorphs, 187–188, *188*; plasticity
   and, 118; plurality and, 172–173; from
   pluriverse to movement, 175–176;
   provincializing, 170–172; topology
   of, 181. *See also* crafting ontologies;
   remaking
Malabou, Catherine, 118, 135–136
manufacturing, 166–168, *167*, 168
Marazzi, C., 43
*The March of the Modern Makers: An
   Industrial Strategy for the Creative
   Industries* (Straw and Warner), 166
marginalized/neglected experiences,
   10, 95–96, 141, 151–154, 199, 241n69
marronage, 99–101
Martin, Randy, 30
Marx, Karl, 19, 97, 213n2, 214n2, 230n13
Marxism, 79–82
masculinism, 121, 172, 245–246n34
material articulations, 47–48
material conditions of existence, 74,
   80, 159
material engagement, 169, 175, 200,
   244, 245n28, 246n46

materialism, 3, 8, 74, 213–214n2; cultural, 84–87; feminist, 88; idealism
   opposed to, 80–81, 83, 88–90, 97;
   left critiques of, 79–80; literacy and,
   200–202; monist, 82–83, 88–91; not
   possible without activism, 81–82;
   posthumanism and, 109; remaking
   of, 84–85; as theory of knowledge,
   82–83; transformation and, 80–84.
   See also activist materialism
Materialism and Empirio-Criticism
   (Lenin), 82–86
materiality: of code, 178–179; maker
   culture and, 168, 175; of the ordinary,
   112; reclaiming, 47
material spirituality, 6, 185, 207
matter, 214n6, 247n57; commoning,
   100–103; frontier of, 12–15, 20, 28;
   justice and, 16–18, 114; making of,
   176; movements of, 81; multiplicity of, 15–16; as political exit, 91, 93;
   rent-generating, 191–192. *See also*
   brain (brain matter); decolonial
   politics of matter
McFerrin, Grady, *186*
measuring technologies, 34, 39–40, 46
mediation, 113, 124
mestizaje, 109
Metapolitics (Badiou), 107
metropolitan strike, 234n60
Microsoft, 191–192
microworlds, 252n113
Middle East, mobilizations, 64–65
Mignolo, W., 17, 194
migration. *See* transmigration
minor science, 91–93
mobile commons, 8, 49, 50–52, 68–69,
   73–74
mobility, 49, 50–52; autonomous,
   51–52; becoming everyone, 65;
   institutionalization of, 52–55; intelligences and infrastructures of, 68–71;
   types and cases, 59–60. *See also*
   transmigration

Okmûn, Julie, *51*, 52, *54*, *58*, *62*, *69*, *73*, *75*, 226n4
"one-world world," 176
ontogenesis, 118, 123
ontological composition, 46, 140–141
ontological organizing, 8, 22, 47, 49–76, 253n122; imperceptible ontologies, 19, 63–66; informal economies and communities of justice, 71–72; labor and mobility, 52–55; nexus of care, 72–74; politics as, 58; "working for papers," 61–63. *See also* self-organization; transmigration
ontological politics, 11–12, 15, 19, 214n2, 215n16
ontological stacking, 164–165, 175, 205–206
ontology/ies: defined, 74, 162, 243n7; as desire, 17; heterogeneous, 173, 175; limits of, 163; monist, 82–83, 88–91; as movement, 165; multiple, 11, 164–165; plural, 172, 175–176; stacked, 164–165, 175–176. *See also* alterontologies; crafting ontologies
Oriel Canfas gallery, 52
origins, stories of, 169
O'Riordan, Kate, 193
Orr, Jackie, 38, 113
otherness, 86

Pagani camp (Greece), 61, 67, 75
panic, 38
Papadopoulos, Dimitris, 4
"paper market," 62
Paris Commune, 230n13, 251n109
Parisi, Luciana, 134
participatory politics, 145–148, 253n131; conditions of participation, 143, 145–148; partial participants, 151–152
partisan philosophy, 83
Patton, Cindy, 155
peak digitization, 179
permaculture, 175, 177, 181, *182*, 205–206, 246–247n48

permaculture guild, 184–185
Perrier, Maud, 200
Philippines, transgenic rice project, 189
Pickering, A., 127
Pinker, S., 125
plasticity, 9, 117–118, 126; ecological-developmental, 131, 133; proprietarization of, 135–136; recombinant, 119–120, 129–131, 237n58
pluriverse, 175–176, 194, 256n162
political economy, 39, 249–250n96
politics, 214n2; alterontological, 10, 137, 158–159, 195–196; of body, 85–86, 128; of brain, 118–120, 137; embodiment of, 105–109; experimental, 3, 72; formalist, 141, 143–145, 151; maker culture, 190, 195–196; of matter, 17, 176; ontological, 11–12, 214n2, 215n16; as ontological organizing, 58; participatory, 145–148; rehabilitation of, 96–97; types, 141. *See also* activist materialism; compositional politics; decolonial politics of matter
poor, moral economy of, 74
porocracy, 53
postcontractual employment, 35, 219n23, 220n31
posthumanism, 8, 94–95, 109, 211n2, 229–230n3; DIWY (do it without yourself), 181–182. *See also* insurgent posthumanism
postliberalism, 35–37, 58, 134, 222n46, 251n107
poststructuralism, 88, 152
postwelfare metropolis, 70
power: as assembled, 149–150; coloniality of, 14; power over, 17; power with, 17, 108
practice, 19–20. *See also* everyday existence and practices; experimental practice
pragmatism, 124, 126
precarization of labor, 5, 10, 66, 92, 189–90, 220n31, 249n94

prefiguration, 89, 195, 224–225n69, 251n108
private sphere (*res privatae*), 10, 191, 225n72
privatization, 34, 136
problematics, 201
production: artisan, 92, 168–169, 251n109; double architecture, 32–33; externalization of costs, 21, 33; of nonwork life, 34–35, 249n94; relocation outside Global North, 30; third stage, 32. *See also* knowledge production
productionism, 13–16; double sense, 15–16, 20, 28, 32, 150
professionalization, 155
proprietarization, 35; of plasticity, 135–136; of software, 22, 190, 191–192, 255n158
proximity, ecology of, 201
psychology, 123–124
psychopolitics, 38
public, 145–146, 197, 248n71, 250n97
public engagement, 192–193
public sphere (*res publicae*), 10, 191
Puig de la Bellacasa, Maria, 6, 23, 156, 185, 189, 201

queer politics, 86, 89, 128, 135, 224n64

Rabelais (Bakhtin), 108, 112
race and antiracist politics, 13, 30, 55, 65, 75, 85–87, 106, 120, 122, 128, 199, 214n13, 228n5
Rancière, Jacques, 17, 64, 158, 206
Raspberry Pi, 179
rationalities, 12, 28, 92, 97, 106, 239n11
Raven, Lavie, 174
Reagan, Ronald, 138
realism, 80, 82–83, 85
reality, constructed in subject, 87–88
reciprocity, 47, 68, 71, 101, 156, 181

reclaiming, 3, 19, 21–22, 47, 89; future, 174; housing rights movement, 70; relation to material world, 104
recombinant plasticity, 119–20, 129–131, 237n58
recombination, 9; autonomy and, 43; as liberation, 117–118; self-reproducing machines, 130–131
Recovering the Commons (Reid and Taylor), 104
refusal of work, 17, 217n7, 250n96; end of, 37–39
regions of objectivity, 140–141; AIDS activism insertion into, 150, 154–155; alterontological politics, 158–159; composition of, 141; expertise and liberal predicament, 142–145; networks, 148–151; participatory politics, 145–148; time of composition, 153–155. *See also* objectivity
Reid, Herbert, 104
reification, 230n10
relationality, 2, 34, 181, 183, 203; embodiment and, 127–128; situatedness, 151, 153
Re:Make project (Derby Silk Mill), 160–161, 243n4
remaking, 4, 47, 118, 152, 157; of materialism, 84–85. *See also* making
rent, 41–43; rentier technoscience, 190–192
representation, 15, 28, 125–126, 214n15
reprocessing, *177*, 177–178, *178*
res communes, 33, 46, 225n72
resistance, 196, 198, 223n55, 252–253n122
revolution, 83; 1848, 95, 107, 230n13; counterrevolutions, 90–91; state strengthened by, 97–98
rice, transgenic, 189, 219n23
Robbins, B., 229n1
Rome, 70
Rona-Tas, A., 223–224n60
Rose, D. B., 48

Rose, Steven, 131
Ryder, Brett, *167*

Sahlins, Marshall, 83
Sandbothe, M., 122–123
Sapik (interviewee), 49, 50, 68, 75–76
Sayers, Daniel, 100
scale-making, 22–23
Schillmeier, M., 121
Schraube, E., 213n11
science, 252n114; big, state, royal, 22,
    91–93; invention and practice of,
    182–183; minor, 91–93; social con-
    structionist challenges to, 142–143.
    See also technoscience
science wars of the 1990s, 140, 142–143
scientific literacy, 200
securitization, 8, 30, 51, 57, 191, 201
seed bombs, 203–204, *204*
self, exploitation of, 220n31, 249n94
self, pure, 122
self-organization, 6, 12, 37, 96, 127, 166
sexuality, 13, 30, 57, 106, 108, 121, 156, 159
Sharma, Nandita, 57
Shenk, David, 118
sign-mediated-thought, 124
Silk Mill, Derby, England, 160–163, *161*,
    172, 197, *198*; as chemical factory,
    161–162; maker culture and, 163–165,
    164; Re:Make project (Derby Silk
    Mill), 160–161, 243n4; stacked
    histories, 160–162. *See also* Derby,
    England
Sismondo, Sergio, 164
situatedness, 34, 146, 151–153, 173, 241n69
slavery, 55, 98, 100–101, 104, 221n32
Smith, Pamela, 92
Socialist Party, 97
social liberation, 9, 214n2, 250n96
social media, 49, 67–68, 75–76
social movements, 66, 224n64, 253–
    254n132; 1960s and 1970s, 30, 85,
    105–107, 151, 157; 1970s and 1980s, 128;
    after 1990s, 151; between 2008 and

2012, 211n3, 212n7, 222n48; Athens, 75;
autonomous, 4, 50; biofinacialization
of, 29; hackable brain and, 129–130;
imperceptible, 19, 63–66; limitations,
2; material practices of, 7, 29, 80–81;
as movements, 95–96; as organizers,
74; social nonmovements, 64–65;
state targeted by, 95, 105; transmigration as, 66, 68–71; unintended, 61–63;
violence as necessary, 110–113
social movements studies, 107
social reproduction, 34–37
social science, 216n2, 219n25
social science fiction, 6, 43–45, 48,
    224–225n69
social space, 85–86
social studies of finance, 39–41, 224n63
social studies of science, 239n11
societies in movement, 65
Soneryd, Linda, 192–193
sovereignty, 52–53, 56–57, 65–66
space, financialization of, 46–47,
    225n74, 225n77
spatialization: migration and, 52, 54
Speakes, Larry, 138
species-being, 80–81, 228n5
speculation, 6, 19–20, 218nn14–15; about
    migration, 60; autocreative brain
    matter, 133–134; biofinancialization
    and, 43; brain politics, 118; making
    and, 172; social science fiction, 43–45
speed, 53, 56
squats, 70–71
stacking, 10, 190, 193, 246–247n48,
    247n49; histories, 160–162; ontologi-
    cal, 164–165, 175–176, 205–206
standpoint theory, 151, 241n60
Star, Susan Leigh, 12, 188–189, 201
Starhawk, 17, 108
state, 105, 230n10, 230n12; assembling,
    94–97, 98; heterogeneous, 60, 81, 96,
    98; networked neoliberal, 107, 109;
    revolution strengthens, 97–98; tar-
    geted by social movements, 95, 105

state science, 91–92

state socialism, 83–84

Steinfeld, R. J., 104

Stengers, Isabelle, 182

Stephenson, Niamh, 4

Sterling, Bruce, 206

stimulus-response (s-r) model, 123–124

Strathern, M., 149

strike: general, 112–113, 234n60; metropolitan, 234n60

A Study of Thinking (Bruner, Goodnow, and Austin), 124–125

subject, 230n6, 232–233n43; dual, 87–90; historical, 107–108

subjectivity, 87–88, 125, 149

subversion, 106

support, 181

Suvin, D., 224n69

Svampa, M., 196

swampland, 100–101

symmetry, 113–114, 149

synthetic biology, 197

synths, 180–181, *181*

technoscience, 214n2, 248n75; activist materialism and, 88; biosecuritization, 191; coloniality and, 21; community-based, 183–184, 199–200; distributed invention power, 182–184; domination and, 151; embodied, 5; free and open software movement, 22–23, 204–205, 255n158; instituted, 183, 184, 189–190, 192, 197, 201–202; limitations of, 119, 201–202; making as movement within, 165; as more-than-social movement, 3; open, 255–256n160; precarization of labor, 189–190; rentier, 190–192; scale-making, 22–23; social and material transformation, 11. *See also* brain (brain matter); compositional technoscience; plasticity; science

technoscientific guild, 184–185

TechShop (Allen Park, Michigan), *169,* 169–170

temporality: body and, 121–123; of care, 157; indigenous, 172–174, *173*; justice and, 113–114; stacked, 206; technologies of, 40; transmigration and, 52–54

terraformation, 27–48, 225n74; from below, 29, 48, 173; biofinancialization and, 41, 46; geocide and geoengineering, 27–28; Terraforming Earth™, 45–47, 173; universalizing matrix of, 28–29

Terranova, Tiziana, 134

terra nullius, 14, 225n72

territoriality, citizenship and, 57–58

territory, epistemic, 14–18, 20–21, 92

Theses on Feuerbach (Marx), 80

thick justice, 18–19, 74

"thinking," 123–124

third industrial revolution, 166, *167,* 244n15

Thompson, E. P., 74, 101

"thousand ecologies," 201

A Thousand Plateaus (Deleuze and Guattari), 88, 91, 212n10

Tolman, E. C., 124

transformation, 4, 11, 106, 159; everyday existence and, 212–213n11; materialism and, 80–84; regions of objectivity and, 144–145, 147–148

transhumanism, 229–230n3

translation, 185

transmigration, 5, 8, 227n35; asylum seekers, 67, 76; autonomy-of-migration thesis, 59–60, 66; detention and deportation, politics of, 53, 59; differential inclusion, 54–58; gift economy of, 66–68; informal economies and communities of justice, 71–72; political mobilization not intended, 61–63; social media connections, 49, 67–68; sovereignty undermined by, 65–66; specter

of, 63–64; temporal understanding, 52–54; "working for papers," 61–63. *See also* mobility; ontological organizing
Tremayne, Wendy Jehanara, 185, *186*
trust, 185
Tsianos, Vassilis, 4, 5, 8, 69
Tunisian revolution, 2011, 194–195
Turkey, migration from, 66–67

uncommons, 193–197
undecideability, 108, 233n48
universalism, 194, 256n162; biofinancialization and, 28–29, 48, 50; of brain, 126, 129, 134; formalist politics, 145; hereditary body, 122; humanist, 27–28, 230n6; maker culture and, 172–174
un-networking, 173–174, 212n9
urban activism, 85, 251n109
Urbonas, Gediminas, 187–188, *188*
Urbonas, Nomeda, 187–188, *188*
US-Mexico border, 53, 110

vagrancy, 97–99
Vagrancy Act of 1824 (England), 98–99
valuation: performance of, 39–41. *See also* biofinancialization; culture of valuation
value production, 28, 31–35, 54, 90, 221n34; biovalue, 135–136; embodied, 33–34, 35, 54
Vaneigem, Raoul, 168, 169
van Gunsteren, H., 57
Vasta, Ellie, 62
Vercellone, C., 43, 191
verticalization, 34, 37, 222n46, 251n107

violence: *Gewalt*, 111–113, 233–234n58; social justice movements and, 110–113; against state, 110–111
Virilio, P., 225n71
virtualization of economy, 30
virtual spaces, 67–68

Wallerstein, Immanuel, 105
Wallis, Jonathan, 164
Watney, Simon, 157
"we," invisible, 151
West-Eberhard, Mary Jane, 131
Weston, Kath, 191
Wexler, Bruce, 131
What Is to Be Done? (Lenin), 83
Williams, Raymond, 84
Wilson, Rob, 94
Winner, Langdon, 19
Wired magazine, *165*
Withers, Deborah, 200
Wittgenstein, Ludwig, 19, 117
women's movement, 145–146, 199
Woolgar, Steve, 147
workers: citizenship exploited, 34; exit from feudal labor, 98–99; general strike, 112–113, 234n60. *See also* labor
working class: movements, 83, 85, 224n64; transmigration and, 63, 66
World 2, 68, 172, 246n35
World Heritage Site of the Derwent Valley Mills, 160
worlding, 113, 211n4, 229n1
worldly commons, 99–100
Wright, Cynthia, 57

Xenogenesis (Butler), 47

Žižek, Slavoj, 233–234n58

www.ingramcontent.com/pod-product-compliance
Lightning Source LLC
Chambersburg PA
CBHW050333270326
41926CB00016B/3434